THE WORLD

WALDEN

HENRY DAVID THOREAU (1817–62) was born, died, and lived most of his life in Concord, Massachusetts, where the American Revolution against British colonial rule began. Educated at Harvard, Thoreau was an avid reader—in five languages—of everything from classical literature and Hindu and Chinese philosophy through narratives of travel and early American settlement, to works on the flora and fauna of his native region. He was one of the Transcendentalists, that body of quasi-mystical progressive thinkers who in the 1830s began another revolution in Concord, in American social and religious thought. Taking an active part in current political and ethical debates, Thoreau became a courageous, outspoken opponent of federal government policies, such as the expansionist war against Mexico and the refusal of Congress to legislate against southern slavery. His essay, 'Civil Disobedience' (1849), was to become an inspiration to twentieth-century non-violent protesters, from British Fabian Socialists to Mahatma Ghandi and Martin Luther King, Jr. Dismayed at the rapid commercialization of New England life, Thoreau preferred to subsist on the income from odd jobs, helping out in the family pencil factory, and on his work as a surveyor. Though he loved books about discovery and travel, Thoreau wandered neither frequently nor over a great distance. He travelled, as he was fond of saying, 'a good deal in Concord', keeping a journal of his thoughts and observations exceeding two million words by the end of his life, and turning his own narratives of geographical journeys, like *A Week on the Concord and Merrimack Rivers* (1849), *The Maine Woods* and *Cape Cod* (both published posthumously, in 1864 and 1865), into voyages of personal, intellectual discovery. *Walden* (1854), based on his experience at Walden Pond, less than two miles from Concord, where he lived intermittently from 1845 to 1847 in a cabin he built himself, was his local version of the travel books about the American West so popular around the time of the California Gold Rush. The landscape of *Walden* is not the Great Plains, the deserts of Utah, and the steep slopes of the Sierra Nevada, but his economic and ecological environment, and the book is about how to live in it.

STEPHEN FENDER is Professor of American Studies and Director of the Postgraduate Centre in the Humanities at the University of Sussex. His books include *Plotting the Golden West: American Literature and the Rhetoric of the California Trail* and *Sea Changes: British Emigration and American Literature*.

THE WORLD'S CLASSICS

HENRY DAVID THOREAU

Walden

Edited with an Introduction and Notes by
STEPHEN FENDER

Oxford   New York

OXFORD UNIVERSITY PRESS

1997

Oxford University Press, Great Clarendon Street, Oxford OX2 6DP

Oxford New York

Athens Auckland Bangkok Bogota Bombay Buenos Aires
Calcutta Cape Town Dar es Salaam Delhi Florence Hong Kong
Istanbul Karachi Kuala Lumpur Madras Madrid Melbourne
Mexico City Nairobi Paris Singapore Taipei Tokyo Toronto Warsaw

and associated companies in
Berlin Ibadan

Oxford is a trade mark of Oxford University Press

Editorial material © Stephen Fender 1997

First published as a World's Classics paperback 1997

British Library Cataloguing in Publication Data
Data available

Library of Congress Cataloging in Publication Data
Data available
ISBN 0-19-282982-3

1 3 5 7 9 10 8 6 4 2

Typeset by Best-set Typesetter Ltd., Hong Kong
Printed in Great Britain by
Caledonian International Book Manufacturing Ltd.
Glasgow

ACKNOWLEDGEMENTS

The help of Thoreauvians and others has gone into making this edition possible. Above all, I would like to acknowledge the work of someone I have never met, Walter Harding, without whose tireless scholarship—communicated especially through the medium of *The Variorum Walden*—I could not even have begun. Harding's work is signalled again and again in my own notes. John C. Hirsh of Georgetown University sent me tips and hints and xeroxes that guided me through the complexities of the text and elucidated one particularly elusive reference. In Concord I benefited from the friendly, freely given expertise of Thoreau scholars like Tom Blanding, the Curator of the Concord Museum David Wood, Dick O'Connor, Director of the Thoreau Lyceum—above all Sally Chapin and the Research Librarian of the Concord Free Public Library, Marcia Moss, both of whom seem entirely to have dropped their normal work for three weeks in order to root out obscure texts and fugitive articles for the furthering of my education in local history. In London Gina Douglas, Librarian of the Linnean Society, spent hours helping me to track down the modern binomial designations of the flora and fauna mentioned in *Walden*. We *think* we have got them right, but in zoology and botany the nomenclature—especially as to genus—continues to evolve. At Sussex Ian Budden, of the Information Systems division of the University Library, was endlessly patient with my efforts to work the software accessing the electronic text of the Library of America edition, and my colleague Geoff Hemstedt read and re-read the Introduction, advising me most deftly on how to improve it.

Sussex   S.F.
1996

CONTENTS

CONTENTS

INTRODUCTION

Concord and Thoreau

Unlike his (then) more celebrated neighbours, Ralph Waldo Emerson, William Ellery Channing the younger, Bronson Alcott and his daughter Louisa May, Thoreau was a native son of Concord. He was born in 1817 in a house on the old Virginia Road, around two miles east of the town centre. True, the family soon moved away, first to Chelmsford, Massachusetts, then to Boston, but by the time Henry was 7, they had returned to his native town, where his father gave up schoolteaching to go into the business of making lead pencils. So the boy could fairly be said to have grown up in the region, exploring its rivers and surrounding woods even as he received his more formal education. And the man lived there all his life, apart from his four years at Harvard, a year on Staten Island, tutoring the children of William Emerson, brother of Ralph Waldo, and brief trips to Maine, Cape Cod, New Jersey, and the Great Lakes.

There is no town in the United States more densely associated with the political and cultural history of the country. Lying less than twenty miles west-north-west of Boston, the township was settled as early as 1635, only five years after the Great Migration of non-separating Puritans from England. Set in a central position in Massachusetts as settled through the eighteenth century, but far enough from Boston not to be swamped by its economic and political dominance, Concord not only prospered in its modest way as a centre for agriculture and trade, but was chosen as the site of the colony's first provincial congress in 1774–5, the radicals' alternative to the government-dominated General Court of Massachusetts.

By then the American Revolution was all but under way. That provincial congress was voting money to support a militia and pay for military supplies to resist British government forces, should they attempt to break out from Boston. One of the military depots was hidden in Concord; so when in April 1775, the British regulars came looking for them, they found around 400 colonial soldiers, the 'Minute Men' of national legend, facing them across a bridge spanning the Concord River just north of the town centre. In the skirmish that followed, and in their retreat back to Boston, harried by American snipers firing from woods and behind stone walls, the British lost 273 of 700 men.

By the time Thoreau came home from Harvard, Concord had become one of the sacred places of the American Republic. In 1837 the town dedicated its monument to the battle, invoking a tradition of radical freedom-seeking stretching from the Puritans, through the Minute Men down to the progressive thinkers of the day. In 1842 Nathaniel Hawthorne and his young bride moved into the Old Manse, within shouting distance of the Old North Bridge. By then Concord had become the centre of New England transcendentalism. This half-religious, half-philosophical belief in the spiritual unity of the world, accessible by instinct rather than reason, was filtered through German Romanticism and the poetic sensibilities of Coleridge, Wordsworth, and Carlyle, taking root in the New England Puritan belief in direct access to Christ without the mediation of priest or established church hierarchy.

Meetings of the Transcendental Club took place in Emerson's house, attracting participants from Cambridge and Boston, such as Margaret Fuller, Elizabeth Peabody, Orestes Brownson, and Jones Very, as well as the locals, Thoreau, Bronson Alcott, and the younger Channing. Discussion often eventuated in public lectures, or essays in *The Dial*, the transcendentalist quarterly edited by Fuller. Because it was philosophically so diffuse, transcendentalism could be channelled into any variety of practical (or at least real-life) programmes for public improvement: women's rights (Peabody), education reform (Alcott), better conditions for the working man (Brownson), vegetarianism (Alcott again), or—above all and common to all—the abolition of slavery. Thoreau was not alone in his circle, though he was perhaps the most courageous, in opposing, then ignoring the Compromise of 1850 requiring even northern free states to return escaped slaves to their southern owners (he hid them in the family home in Concord, then helped them on their way to Canada), and in championing the case of the abolitionist guerilla John Brown after most others, even of a progressive persuasion, had written him off as a murderous fanatic. With the transcendentalists the movement from theory to practice could also work the other way. The fluidity of their belief admitted any number of bodies of literature as agreeable to their system: German and English Romanticism, of course, but also the teachings of Confucius, the Upanishads and the *Bhagavadgita*, Plato and the neo-Platonists, the metaphysics of Swedenborg. *Walden* shows the influence of many of these enthusiasms for world wisdom.

But Concord was not just the accidental beneficiary of a group of New England intellectuals who liked to live there. Its progressive tradition pre-dated the transcendentalists. The upkeep of the poor, public education, and public libraries were early priorities, dating from colonial times, and much expanded to meet the needs of a responsible republican citizenry after the revolution. In 1848 the town raised its per-capita spending on education from $1.82 to $2.73, the largest increase in any one year of its history.

As with schools, so with less structured adult education. Concord's Lyceum was pre-eminent in a system of public lectures organized throughout New England and wider afield in the nineteenth century. Founded in 1828, the Lyceum soon absorbed the town's debating society. In the winter of 1842/3 alone it featured lectures by such notables as Emerson, Brownson, the historian George Bancroft, Horace Greeley, the Whig editor and founder of the *New York*

Tribune, and seventeen others. That year Thoreau himself, who lectured in the Lyceum nearly twenty times during his life, and was one of its most active organizers, was treasurer of the institution, receiving $109.20 in subscriptions, and leaving a balance of $9.20.[1]

Though progressive, democratic, and part of a commonwealth, Concord still had its class distinctions. The paupers in the almshouse, for instance, were not the only poor to find relief in the town. 'Another class of donations', writes the local historian Lemuel Shattuck, 'has been made to the town for the relief of the *silent* poor,—those individuals who are needy but do not wish to throw themselves on the town for support.' Funding for this more genteel branch of the impoverished came from private donations.[2] Education, too, had its private stream. Those who could afford the modest fees could start their children off at a 'dame school', roughly the equivalent of the modern Massachusetts developmental day-care centre. The next stage was the college-preparatory 'master's school', or Latin grammar school, teaching boys up to 15 reading, spelling, writing, grammar, arithmetic, geography, Latin and Greek. Even the library went through a phase of private funding. The Concord Social Library, established in 1821, which inherited its collection from the Charitable Library and later fed it into the Town Library, was funded by shares, and its use restricted to shareholders. The Lyceum had always been supported by private subscription.

On which side of the private–public line of civic provision did Thoreau see himself? He was privately educated, at the Concord Academy (a college-preparatory 'Latin grammar' school), and at Harvard, then and now one of the best universities in the country. When Henry took over at Concord Academy in 1838, his brother John joining him a year later when

---

[1] Alvah H. Low, 'The Concord Lyceum', *Old-Time New England*, 50 (1959), 29–31.
[2] Lemuel Shattuck, *A History of the Town of Concord* . . . (Boston: Russell, Odiorne and Company; Concord: John Stacy, 1835), 216.

increased enrolment justified the addition to the teaching staff, they introduced Bronson Alcott's progressive principles of education, encouraging discipline without flogging, encouraging free discussion, and getting the students to learn by doing, not memorizing. But it remained a private grammar school.

Money for Thoreau's excellent education had come from his family, though not without sacrifice. His father, a pencil manufacturer, owned his own factory. Along with John, Henry worked in the factory behind the family house on Main St.—part-time, perhaps, though long enough for his lungs to have been fatally damaged by the dust of the lead used in the manufacturing process. He participated willingly and inventively, developing an improved, purified graphite ranging in hardness from draftsmen's to artists' requirements, and a machine for drilling holes lengthwise in the wood holding the lead. These were serious pencils. They sold well. They won prizes for their quality. When his father died in 1859, Henry took over the business.

Thoreau as industrialist; Thoreau as entrepreneur; Thoreau as skilled professional (for he was also an expert surveyor): these are not occupations normally associated with his way of life, normally presented (by him before others) as that of a handyman-loafer. He was not alone among his contemporary American authors in losing money on his books—far from it— but he was unique in making his own living in occupations other than editing, or literary journalism, or government sinecures like consulships abroad and inspectorships in a customs-house.

Walden is famously sceptical about the provisions for education and public services that were expanding so strikingly in mid-century Massachusetts—indeed, about the progressive atmosphere of Concord generally. 'Philanthropy . . . is greatly overrated' (p. 69). A pain in the gut is not sympathy for the underprivileged but the result of eating a green apple; the philanthropist gives to ease his own pain (p. 70). The inmates

of the almshouse were 'wiser than the so called *overseers* of
the poor and the selectmen of the town' (p. 136). 'We have a
comparatively decent system of common schools, schools for
infants only . . . and latterly the puny beginning of a library
[but] . . . no school for ourselves' (p. 99). As for what goes on
in those schools and library, it is mainly 'easy reading':
primers and children's adventure stories and popular ro-
mances, not 'the English classics' or 'the wisdom of mankind'
(pp. 95–7).

So is *Walden* a satire on the town's 'public sector', as seen
from a relatively privileged, middle-class perspective? Not
quite, or not as an end in itself. All the narrator's attacks on
Concord's expanding civic institutions and the progressive
mentality that lay behind them are part of a wider enquiry
into the nature of modernity in general: not just the recent
expansion in libraries and public education and social services, but also the modernization of farming, the new technology of travel, communications, and the production of the
printed word—even the fashions of modern architecture and
interior decoration. 'As with our colleges, so with a hundred
"modern improvements;" there is an illusion about them;
there is not always a positive advance' (p. 48). Now the new
railroad has been built from Boston, through Concord, to
Fitchburg. But suppose he doesn't want to go to Fitchburg?
Or if he does, couldn't he walk the thirty miles in less than the
time it would take him to earn the fare and 'take the cars'?
(pp. 48–9).

Similarly, the newspapers carry no news, the post office no
post. 'I never received more than one or two letters in my
life . . . that were worth the postage', and 'if we read of one
man robbed . . . or one house burned . . . or one steamboat
blown up, or one cow run over on the Western
Railroad . . . we never need read of another. One is enough'
(p. 86). Needless to say, although 'we are in great haste to
construct a magnetic telegraph from Maine to Texas', Maine
and Texas may have nothing to say to each other (p. 48).

In other words, what is the point of the technological shell if it contains no kernel of intellectual or moral purpose? Once thought on, this satire is not as convincing as it is initially compelling. It certainly was not original. Hawthorne had already done it better (and funnier) in *The House of the Seven Gables* (1851), where he has Clifford and Hepzibah Pyncheon 'take the cars', escaping from a crime which they did not commit and which is not a crime, and while en route to nowhere, fall into earnest discussion with a crazed fellow-passenger about all the latest developments: railroads, electricity, the telegraph, even spirit-rapping.

On another level *Walden*'s attack on modernity is only part of a more general critique of commercialization. The strenuous objection to progressive farming and the comic rendition of the Walden ice 'harvest' are premised on a sense of outrage that the natural environment should be so parcelled out, chopped up, ploughed, ditched, walled and manured, and generally so over-utilized—all in the name of profit. But this will not work either. Thoreau himself belonged to the commercial class, and *Walden* certainly underwrites the mechanism of the market in the sadly amusing anecdote (p. 19) of the Indian trying to sell his baskets. Besides, commercialization is not the operative moral distinction that it first seems. Farmers had always farmed for profit. New techniques of intensive farming just helped them to make more of it. That is the difference.

Thoreau's contempt for modernization and commercialization was fed by his anxiety at what was happening to Concord itself. Concord at mid-century was in the process of becoming provincial. As the county seat of Middlesex County, it had been the site of the twice-yearly sittings of the County Court. Leading Massachusetts lawyers and their clerks took up the best rooms in the local inns, spreading their money around, and providing the townspeople with the chance to hear some of the best courtroom oratory in America. The County Sessions Fair was held every May and September. 'Even the children could not be kept in school on court day', according to one local historian. 'Food, rum and cider were sold from booths set up on the common and it was an exciting day although it had its inevitable morning after.'[3]

By the time *Walden* was written, however, the more important legal business had begun to move further east and north. In 1867 Concord finally lost its status as county seat, when the courts moved to Lowell. Although Concord retained its regional importance as a market town, the County Sessions Fairs were contracted into the annual Middlesex Cattle Show, with competitions, exhibits of livestock and other agricultural produce, and even lectures and speeches on topics of local public interest. Thoreau first presented his important scientific study of 'The Succession of Forest Trees' to the farmers convening for the Cattle Show of 1860.

By Thoreau's time also Concord's population had ceased to grow. Between 1845 and 1855 it was static at just over 2,200, and only began to rise (sluggishly) after 1860. Anyone who has driven around the United States will recall signs greeting the traveller at the town limits, proclaiming the population figures as an index of economic health. Thoreau was concerned about this too. Musing on the vanished dwellings in Walden Woods, of which only the cellar dents survive under the scrub, the narrative of *Walden* (p. 236) asks, 'But this small village, germ of something more, why did it fail while Concord kept its ground?' But what he first wrote was, 'why did it fail while Concord grows apace?'[4] By the time *Walden* went to press, its author had taken another look at the town's population figures.

By mid-century it was also obvious that Concord was not going to develop into a manufacturing centre of much significance. Various explanations have been advanced for Concord's indifferent industrial performance—ranging from

---

[3] Ruth R. Wheeler, *Concord: Climate for Freedom* (Concord, Mass.: Concord Antiquarian Society, 1967), 159.

[4] MS of Walden, sheet 217, Huntington Library, HM 924, microfilm.

simple bad management to the financial panic of 1837 and the decade-long depression that followed, hitting the economy of the whole of the American North-East severely. Then there was the railroad. In 1844 the rails reached Concord from the terminal at Charlestown, just across the Charles River from Boston. By the next year the track was complete to Fitchburg, the manufacturing town around thirty miles further west. Before the railroad it had cost 75 cents and taken four hours to get from Concord to Boston. That was on a fast stage-coach; freight took much longer. Now the distance could be covered in one hour for 50 cents: still not cheap, at a time when the day's wage for an unskilled labourer was a dollar, but a great and sudden liberation for the traveller, merchant, and shopper—not to mention the Lyceum lecturer.

The railroad's initial effect on Concord was predictable enough, as 'the stage and wagon traffic declined and business fell off at the taverns, smithies, harness makers, and general stores'.[5] Then it spread to the town's shops, which began to lose business as shoppers could travel to the cheaper, more varied, and more sophisticated offerings in Boston, becoming increasingly exposed to city standards of household furnishings and design. Before the railroad, as a local historian has written, 'many primitive country ways lingered in the town. No sidewalks were constructed, carpets and pictures were rare, and the furniture was plain.'[6]

In some respects, of course, the railroad strengthened Concord's position. It is doubtful whether the town would have been so firmly associated with the transcendentalist movement if the Boston and Cambridge intelligentsia had not been able to travel there so easily for meetings of the 'Hedge Club' and lectures at the Concord Lyceum. But financially speaking

[5] Wheeler, Concord, 178.
[6] John Wilinsky, 'The Impact of the Railroad on Concord, Massachusetts, 1844–1887', unpublished student essay, Concord Free Public Library, C.Pam, 37, item 55, p. 4.

the deficits outweighed the benefits. Railroads opened new market opportunities for the farmer, manufacturer, and trader, but wider markets meant greater specialization. Those towns in which factories were already well established, like Fitchburg and Lowell, could meet the demands of the wider market if they developed their manufactures still further. Farming areas, such as the region for which Concord served as market town, could respond only if they reformed their methods, working the land and keeping their livestock more intensively. As the local speciality grew in strength, other enterprises dwindled or failed. Prices set beyond the town and its surroundings eroded local self-determination. As a local historian puts it, 'much of Concord's independent metropolitan life style vanished' with the coming of the railroad.[7]

Walden reflects many of these themes, from the increasing luxury in household design and furnishing to agricultural reform. Consequently, the cause of all this degeneracy, the railroad, is represented in terms of noise and violence and exploitation: its whistle 'penetrates my woods' and its loads of logs are like 'battering rams . . . against the city's walls'. 'All the Indian huckleberry hills are stripped, all the cranberry meadows are raked into the city. Up comes the cotton, down goes the woven cloth . . . up come the books, but down goes the wit that writes them' (p. 106). Above all, the railroad effaces the local, standardizing time itself, so 'that the farmers set their clocks by [it]' (p. 108). It rides over even what the local people might grow and eat, the produce most suited to their climate and culture:

Every New Englander might easily raise all his own breadstuffs in this land of rye and Indian corn, and not depend on distant and fluctuating markets for them. Yet so far are we from simplicity and independence that, in Concord, fresh and sweet meal is rarely sold in

[7] John Wilinsky, 'The Impact of the Railroad on Concord, Massachusetts, 1844–1887', p. 4.

the shops . . . For the most part the farmer gives to his cattle and hogs the grain of his own producing, and buys flour, which is at least no more wholesome, at a greater cost, at the store. (p. 58)

A satisfactory reading of *Walden*'s unease with the modern world would have to dig deeper than the critique of modernity and commercialism. The narrative discovers the local under threat and finds a way to meet that threat. This theme lies at the heart of *Walden*.

What sort of thing is *Walden*?

It would be impossible to say when Thoreau began to be a writer—or even an author, since his books would never sell well enough to make a living for him. In a sense he had been writing for the whole of his adult life: from the lively essays he produced for his classes at Harvard through his journal, begun in his twentieth year and continued for nearly two million words of natural and philosophical observations to within six months of his death.

But the story of *Walden* really begins in the composition of its predecessor, *A Week on the Concord and Merrimack Rivers* (1849), Thoreau's first published book, prompted by the sudden death of his brother in 1842. Emotionally and physically devastated for over a year, Thoreau finally gathered his spirits and thoughts to write a kind of memorial to his brother. For this he needed time, space, and relatively uninterrupted leisure. He would find them at Walden Pond, an attractive, near-circular lake about a mile and a half south of Concord town centre.

Walden was not Thoreau's first lakeside campsite (in 1836 he had spent six weeks in a hut on the shore of Flint's Pond, just over a mile to the east of Walden, with his Harvard friend Charles Stearns Wheeler), but it offered the unique combination: a beautiful, fairly secure retreat that was also less than half an hour's walk from Concord. It was on the north shore of

Walden Pond that Thoreau's friend and sometime patron
Ralph Waldo Emerson owned a woodlot, on which he was
happy to let Thoreau settle in return for clearing undergrowth
and planting new trees. So in the spring of 1845 Thoreau
began to build his cabin at Walden, moving in on the nation's
birthday, 4 July, of that year. There he would stay for over two
years, though returning frequently to visit family and friends
in town.

Work on *A Week* progressed rapidly. On the surface level a
narrative of a short journey made in a camping dinghy with
John in 1839, the book draws on Thoreau's journal and other
writing of the period for excursions of a more figurative kind:
into religion, philosophy, and personal conduct—and (what is
often overlooked) literary criticism of a high standard. The
book came out to reviews that can most politely be described
as mixed. It had next to no popular appeal. As an untried
author Thoreau had had to forgo his royalties to subsidize a
print run of 1,000 copies, of which 703 remained unsold after
four years. Thoreau himself reclaimed the remainders, com-
menting wryly in his journal (J 5.459 (OS)) that 'I now have a
library of nearly nine hundred volumes, over seven hundred
of which I wrote myself'.

In any case something rather different had begun to com-
pete for Thoreau's attention and energy. While still working
on *A Week* at Walden he conceived the idea of another book
based on his stay there, also drawing on his contemporary
journals but more attentive than the first book to the accidents
of daily life in the natural world around him. A first draft of
Walden he completed while still at the Pond. Later, while still
revising *A Week*, he completed a second draft, then trying out
various parts of the manuscript on the public lecture platform.
By 1854, when the prestigious Boston firm of Ticknor &
Fields agreed reluctantly to publish it (*A Week* having, by this
time, emerged and failed so strikingly), *Walden* had gone into
seven versions, each one larger and more complex than the
one before.

Yet for all its many interbraided discourses, *Walden* follows a fairly simple plan. After the long, mainly discursive first chapter on his 'Economy', the narrative picks up Thoreau's experience at Walden, condensing his twenty-six month's sojourn there to one full year, followed season by season from summer to spring. But like *A Week*, *Walden* is far more than its surface narrative. Even 'Economy' is not quite the material basis of the experiment that it looks to be at first sight, but more of an exemplary metaphor. As for the story itself, it is not likely wholly to satisfy the reader's curiosity as to why Thoreau went to live at Walden and how he managed it. Throughout the book, to quote the two halves of the title of Chapter 2, the 'What I Lived For' is at least as important as the 'Where I Lived'. The book is about Walden, and about Concord and New England and the United States in the middle of the nineteenth century. It is also about a new way to live. Above all, it is about Thoreau in many guises, many moods, many opinions—some of them mutually contradictory.

Perhaps it was this narrative-discursive indirection of its prose, this sometimes bewildering tangle of observations and opinions, that confounded contemporary readers. Though it sold better than *A Week*, *Walden* can hardly be said to have been a popular book in Thoreau's lifetime. Emerson valued its author more as a friend and representative man than a distinguished writer. Contemporary reviews were not unfriendly, but typically characterized the book in slightly condescending terms such as 'shrewed', 'eccentric', 'cynical'. One of the first reviewers to take the book seriously was not an American man but an English woman. Writing in the *Westminster Review* for January 1856, George Eliot praised its 'observations of natural phenomena . . . passing through the medium of a deep poetic sensibility'.[8]

---

[8] Cited in Walter Harding, *The Days of Henry Thoreau: A Biography*, rev. edn. (New York: Dover Publications, 1982), 334–8; see also Lawrence Buell, 'Henry Thoreau Enters the American Canon', in Robert F. Sayre (ed.), *New Essays on Walden* (Cambridge: Cambridge University Press, 1992), 23–52.

That was then. Now it would be hard to exaggerate *Walden*'s fame and influence world-wide. Since its first appearance it has gone through nearly 200 printings in English. In the four decades after World War II alone it came out in over fifty translations into languages as far apart as Malayan and Hungarian. Its spirit of quiet yet stubborn independence is said to have inspired moral and political reformers from Tolstoy and Gandhi to Martin Luther King. Countless conservationists and environmentalists, starting with the author of the American national parks movement, John Muir, have acknowledged *Walden* as their point of departure. In the United States it remains the central canonical text in that ongoing project to define and reinforce the American character known as 'American literature'. A recent survey of American professors found that they considered *Walden* by far the most important nineteenth-century text to teach to their students—well ahead of *The Scarlet Letter* and *Moby Dick*. Yet it remains a popular classic among ordinary American readers too.

Why? On one level, and especially in 'Economy', *Walden* seems to challenge the American ideals of hard work, deferred gratification, and the importance of ownership. At the same time, it seems to re-enact another powerful constituent of national identity, the myth that a man (it is almost always a man) somehow grows to American maturity through an initiation on the isolated frontier. It may be that American readers take the critique of middle-class values rather as people respond to a new diet in the popular press: that is, as a good idea in principle, interesting to read about, but not something they can quite bring themselves to follow in practice. But anyone who really believes that *Walden* is about an isolated 'life in the woods'—let alone on the frontier—is misreading the book.

The central question is, who is 'Thoreau'? There is the man who wrote *Walden* and the voice that talks to us out of its pages. The 'Thoreau' inside *Walden* 'can do without the post

office' and disparages newspapers. But according to his contemporary, and *Walden*'s first editor, Frank Sanborn, 'few
residents in Concord frequented the post office more punctu-
ally, or read the newspapers . . . more eagerly than Thoreau.'[9]
What looks like a solitary life in the woods was nothing of the
kind. The woods themselves were sparser around Concord
and Walden than at any time before or since in the natural
history of the area. (That is why Emerson wanted Thoreau to
plant new trees there.) Walden Pond was not a wilderness,
but a popular resort of the townspeople. Thoreau did not
really 'live' at Walden—more like camp out—and his life was
anything but solitary. He was continually in and out of town,
or receiving visitors at his cabin. Though on p. 56 he slyly
admits to having 'dined out occasionally', in fact he ate regularly at other people's houses in Concord in return for odd
jobs. Walter Harding notes that almost every Sunday dinner
was shared with Emerson and his family, and that on his way
back from Concord to the Pond he would call in on Edmund
Hosmer, often staying for supper. Emerson's house was just
over a mile from Walden down Brister's Hill.[10] With equal
frequency Thoreau would call in at the house of his mother
and sister on Main St., a walk of a mile and a half down the
railroad tracks to Concord, to eat a meal and get his washing
and mending done, sometimes in return for odd jobs.

But what about those carefully entered figures documenting the argument of 'Economy'? Surely if *Walden* tells the
literal truth anywhere, it is here, on this level of the verifiable
facts of material, day-to-day existence. Yet if one checks the
prices given in *Walden* against what is known of contemporary expenditure, one comes up with some odd inconsistencies. The book greatly overestimates the cost of the Harvard
student's rent, and the proportion of tuition costs to overall
expenditure. Though *Walden* makes great play of having

[9] Henry David Thoreau, *Walden: or, Life in the Woods*, ed. F. B. Sanborn
(2 vols.; Boston: Bibliophile Society, 1901), i. 49 n.
[10] Harding, *Days of Henry Thoreau*, 279–80.

secured a bargain in old wooden siding from James Collins, an Irish labourer who had to sell his shanty when the Fitchburg railroad was finished and his work ran out, the actual saving was only about 40 per cent of the price of brand new lumber. The narrator also seems to have paid well over the odds for second-hand bricks and shingles. At $3.90 he bought enough nails to build a full-sized family house.[11]

There are three possible explanations for these discrepancies: that Thoreau was buying prodigious amounts of materials—way more than he needed; that he was a careless keeper of records; or that he was—as the phrase now has it—economical with the truth. The second possibility we can discount immediately. The surveyor whose judgement of the depth and other dimensions of Walden Pond recent electronic measurements have confirmed within a few centimentres, and the author who changed the price of the railroad journey from Concord to Fitchburg three times from first draft to published version (p. 49) to conform to the company's alterations in the fare—not to mention the adjustment between manuscript and book to account for Concord's population levelling off—will not often be caught sleeping in respect of accurate records. There may be some truth to the waste hypothesis. Those nails, for example. According to Harding, the man who discovered and excavated the site of Thoreau's hut in 1946 and 1947 found 'an unusually large number of bent nails, indicating that Thoreau's aim with a hammer was not all it might have been'.[12] So the price for nails may be an accurate reflection of Thoreau's amateur status as a carpenter. His overprovision of—or overpayment for—'refuse' shingles, 'old brick', and weathered shanty-boards might, at a stretch, be attributed to the same cause, since a

[11] Seymour Harris, *The Economics of Harvard* (New York: McGraw Hill, 1970), 40, 72; Account Book of Rufus Hosmer, 1831–39; George L. Prescott, Account Book, 1853–54; Old Store Ledger, 1819, Concord Free Public Library, Concord, Massachusetts.
[12] Harding, *Days of Henry Thoreau*, 277.

builder's success depends as much on accurate measurements and estimates as it does on skill with a hammer and saw.

Anyway, the big mistakes in Thoreau's accounting are the invisible ones. His costs for building, for example, take no account of the value of his labour, nor (more seriously, perhaps) of the cost of land, taxes, and mortgage payments, which usually form the major proportion of any accounts of household or farmstead. His 'title' to (what was, in fact, Emerson's) land is left obscure; on p. 50 he refers to 'not being [its] owner, but merely a squatter'. Later, even more obscurely (p. 119), he says that 'some square miles of unfrequented forest' have been 'abandoned to me by men'.

Then there is the puzzle over *Walden*'s treatment of the Irish. Occasionally this looks sympathetic. The true cost of the new railroad from Boston to Fitchburg must be reckoned in the destroyed or impoverished lives of the labourers who laid the tracks (p. 85): 'Did you ever think what those sleepers are that underlie the railroad? Each one is a man, an Irishman, or a Yankee man. The rails are laid on them, and they are covered with sand, and the cars run smoothly over them. They are sound sleepers, I assure you.' Yet elsewhere the Yankee and the Irishman are set in profound antithesis to each other. Perhaps the most unpleasant passage in *Walden*—and to judge from its rhetorical heightening, one of the most deliberate—is the little story about the narrator's visit to the house of John Field, the Irish 'bogger'. Field's 'several children' include a 'broad-faced boy who assisted his father at his work' and a 'wrinkled, sybil-like cone-headed infant that sat upon its father's knee' (p. 183). 'I had sat there many times of old before the ship was built that floated this family to America', the narrator continues, before zeroing in on the economic theme once again. If only Field and his family could live as he! They had 'floated' over to America as to a promised land offering 'tea, and coffee, and meat every day', but he 'did not use tea, nor coffee, nor butter, nor milk, nor fresh meat, and so did not have to work to get them' (p. 184).

It is no good protesting that these people were refugees from the Great Famine, for whom daily tea, coffee, and fresh meat must have looked like a pretty good deal; that far from 'floating' over the Atlantic as on an easy tide, they had endured hardships and deprivations that to Thoreau were (quite literally) unimaginable; that the economic comparison is not on all fours, given Thoreau's ample infrastructure of food and housing offered him by friends and family; that his preferred alternative motive for emigration to America ('that country where you are at liberty to pursue such a mode of life as may enable you to do without' (p. 185) the commodities that the Fields came in search of) is simply a category mistake, not comparable to the Irish family's reasons for coming over. The mystery is why the narrative works so hard to establish ancillary points not apparently supportive of the overall thematics here. Why is it so important to claim the primacy of his arrival there? Why the strange reversals: the child referred to as 'it' and the chickens 'too humanized to roast well'; the baby as old as the Sibyl?

The larger question raised by the John Field passage is whether the 'economy' on which the text so insists can be extended into a moral dimension. This resistance, or indifference, or whatever it is, first emerges in the little narrative of James Collins's shanty. Here, before his eyes, is one of those very victims of the railroad: a poor Irish navvy who had worked on the Fitchburg line, forced to pull up stakes now that the work had run out and he had lost his income. Here is the same odd reversal of human and animal as in the John Field passage: the chickens that live in the house, and the 'infant in the house where *it* was born'. Far from sympathizing with the Collinses' predicament, the narrative is full of contempt for their pathetic luxuries of 'a silk parasol, gilt-framed looking-glass, and a patent new coffee mill nailed to an oak sapling' (p. 40), and the fact that they were obviously doing a runner to escape ground rent and fuel bills. A full rendering of the economic theme here would take account of the contribution made to Thoreau's (admittedly not very impressive) saving on siding boards by the financial system that first raised huge sums of capital to produce them (the humblest boards required large, steam-powered sawmills to cut them endwise from the log), and then could afford to write off the investment in technology and manpower as so much waste. Capitalism and the exploitation of the Irish labourer combined to make Thoreau's cabin possible.

So Thoreau's economic statistics are selected and modified so as to serve the argument—and a very local argument at that. To understand how local, look again at those misleading figures for the cost of a Harvard education. The argument here is that people can make do with far less living space and fewer modern conveniences than contemporary improvements have led them to expect. If university students could help to build their own housing, it might teach them as much as, or more than, they could learn from more formal instruction. For the time being the cost of food is no part of the argument, since everyone has to eat, whatever the saving on other needs. So when dealing with the example of the Harvard students he can omit the cost of food, even though it was by far the largest proportion of their annual expenses. Very soon he will demonstrate possible savings on food also, but by then he has moved on to another example—himself— and another branch of the argument. By pp. 99–100 he can even reverse his position on practical work as the best education: 'Alas! what with foddering the cattle and tending the store, we are kept from school too long, and our education is sadly neglected.'

This localness of argument may be invoked to explain the text's inconsistency about the Irish. The observation about the Irish serving as ties over which the rails are laid (p. 85) is not about exploitation, or the true cost of the railroad, or the predicament of the Irish. It is part of the book's enquiry into modern improvements in travel and communications: now that we have the railroad and the telegraph, where do we want to go and what news have we got to send? But elsewhere the Irish seem to come in under the general rubric of enterprise. They do not have it. So Collins is condemned to float out with the ebbing economic tide, just as Field came in on the flood. Their values are inverted. They prefer silk parasols, tea, coffee, and fresh beef to wholesome beans, molasses, and contemplation; leaky dilapidated dwellings to the light, clean, dry cabin on the Pond. So with them everything else is turned upside down: animal and human, age and youth. Above all they have come to America perforce, under duress, not as the New England Puritans so often memorialized in *Walden*: as the result of wrestling with their consciences, arguing the alternatives, then planning the journey and subsequent settlement with such deliberation.

The text seems to reverse the old maxim of 'Do what I do, not what I say'. *Walden* is all in the said, or written, not in the done. For all its appearance as a documentary account, it is not an accurate record of Thoreau's experience at Walden Pond. Even its apparent time scheme, the year-long sojourn beginning in summer and ending in the following spring, is a rhetorical shaping of the actuality: the 'real' Thoreau lived on and off at Walden for over two years—from 4 July 1845 to 6 September 1847. *Walden* is not a guidebook on how to live in the woods, though its popular readership has often taken it as such. If the book can be classified as a species of autobiography, that term must be understood in an especially American sense, as in the spiritual autobiographies that the earlier New Englanders were so fond of keeping, or in the way that Benjamin Franklin's autobiography (but, oddly, not *Walden*) has been read by generations of Americans. In other words, *Walden* is a series of exemplary narratives based loosely on the thoughts, feelings, and conscientious meditations, more than the life, of its author. The 'I' in Walden is not Thoreau himself but a narrator, textualized more radically than the 'I' in conventional autobiography, within a network of events arranged so as to present and defend a complex of cultural and moral lessons.

This is not the whole of it, of course. *Walden* engages in ambiguous debate with numerous literary and hortatory conventions—and much more intriguingly, with itself. As for what kind of thing *Walden* is, the best analogy is probably Ezekiel lying on his left side for 390 days, then on his right for 40. This was not a manual of how to survive in the desert, but a physical, symbolic measure of the years of iniquity of the Houses of Israel and Judah, who would not attend to the plain text of God. Or to put it in his own words, Thoreau sojourned on Walden Pond 'as some must work in fields if only for the sake of tropes and expression, to serve a parable-maker one day' (p. 146). *Walden* is those 'tropes and expression'.

Walden's Debates

As the notes to this edition attest, *Walden* is one of the most allusive books in the American literary canon. It is important to distinguish, however, between ideas with which *Walden* engages in lively debate, and an eclectic body of knowledge registered mainly to reinforce an argument already set out in other words. In this latter category is what *Walden*'s narrator calls 'the wisdom of mankind'—the *Bhagavadgita*, Hindu and Persian poetry, classical epic, Egyptian, Greek, Roman, Celtic, and Scandinavian mythology, the bulk of English poetry, as assembled in the twenty-one volumes of Alexander Chalmers's anthology of 1810—not to mention the Bible and Shakespeare.

Other discourses, mainly but not exclusively American, come under closer scrutiny. Above all, *Walden* challenges head-on a central tenet of popular American economic and political thinking about man in the rural environment. Here the text to know for purposes of comparison, although not mentioned by name in the dispute, is St Jean de Crèvecoeur's

Letters from an American Farmer, first published in 1782. According to the Physiocratic economics to which Crèvecœur subscribed, agriculture was the only sure investment, since the dividends were underpinned by natural process, as the seed multiplied itself in its fruit. Aiding this force of nature was the reformed social, economic, and political status of the new settler. Because he owned his own land and could pass it on to his children, he had the incentive to clear it and drain it, to cultivate more and more of it, and generally to improve his capital base and the returns it yielded. Finally, when the work of physical betterment was more or less complete, he and his family could turn to the cultivation of their minds—above all in leisured meditation on the works of nature that surrounded them.

This idea that civic value is grounded on the economics of farming became the basis of Thomas Jefferson's agrarianism and the platform of the Democratic Republican Party, one of the two major ideological orientations of the early Republic. More generally, Crèvecœur's linking of ownership and enterprise is one of the founding assumptions of American identity. Thoreau's challenge to this idea is, therefore, a radical act. The programme of *Walden* could not be more totally opposed to that of *Letters from an American Farmer*. The overriding theme of 'Economy' is the burden of ownership. By the time the farmer has laboured to pay off the 'encumbrances' inherited with his land, or interest on the 'hired money' to buy it (p. 30)—details glossed over in *Letters*—he is even more enslaved by his farm than he was before he took full possession of it. The enterprise and initiative supposedly deriving from pride of ownership merely tighten the noose, for if 'it is hard to have a southern overseer; it is worse to have a northern one; but worst of all when you are the slave-driver of yourself' (p. 9).

Above all, *Walden* challenges the agrarians' assumption that civic virtue, and personal improvement through the earned leisure in which to contemplate the works of nature, somehow grow out of good farming. In *Walden* physical and intellectual cultivation, ownership of land and awed respect for the countryside are opposed, not contiguous. You have to abjure the one in order to gain the other. Again and again the book alludes to Emerson's distinction between farms and landscape, set out in 'Nature' (1836): 'Miller owns this field, Locke that, and Manning the woodland beyond. But none of them owns the landscape. There is a property in the horizon which no man has but he whose eye can integrate all the parts, that is, the poet.'

So much for what might be called the major value system opposed by *Walden*. Other kinds of recent or contemporary discourse are confronted more locally, less generally, and sometimes with a degree of ambivalence. They include the work of other nature writers, accounts of discovery, books and periodicals on progressive farming, the popular house pattern books of the 1840s and 1850s, handbooks on domestic economy, temperance tracts, guidebooks for young men, and other testimonies to, and manuals of, courage, independence, enterprise, and self-help.[13]

Of the other books on nature, Thoreau was probably most influenced by *The Natural History of Selbourne* (1789) by Gilbert White, the English clergyman whose meticulous observation of the fauna and flora around his country village in Hampshire set new standards in natural description. Much as he admired White's patient attention to detail, however, Thoreau was determined to make his own treatment of his immediate environment much more personal: to admit, and

---

[13] The pattern books, as an object of *Walden*'s critique of contemporary architectural styles, are discussed in Richard and Jean Masteller, 'Rural Architecture in Andrew Jackson Downing and Henry David Thoreau: Pattern Book Parody in *Walden*', *New England Quarterly*, 57 (1984), 483–510. Thoreau's use of the young men's guides is discussed extensively in Leonard Neufeldt, *The Economist: Henry Thoreau and Enterprise* (New York: Oxford University Press, 1989), *passim*.

indeed heighten, the collaboration between the facts observed and the subjectivity of the observer.[14]

On the matter of narratives of exploration and newspaper accounts of distant exploration, *Walden* remains as austerely sceptical as it does about travel and newspapers generally. The 'Conclusion' mentions a number of expeditions to the poles and the south seas, the search for the north-west passage, or the source of the Nile, or whatever—only to dismiss them comprehensively. Discover yourself first, it says; only then go further afield, if you still have time, energy, and inclination. Closer to home, the enormous American fascination at mid-century with travel and emigration to California and Oregon, spurred on by the Gold Rush of 1849–52, had produced countless items in the newspapers, and revived the readership for narratives of transcontinental travel and exploration such as John Charles Frémont's *Report of the Exploring Expedition to the Rocky Mountains* (published by the US Government in 1845), Edwin Bryant's *What I Saw in California* (1848), and Francis Parkman's 'The Oregon Trail' (1847, reissued in book form for the gold-rush market as *The California and Oregon Trail* in 1849). Though Thoreau was enormously fond of travel narratives—he read over 200 of them[15]—the 'I' and the 'you' of *Walden* would travel only in the sense of journeying more profoundly into the local: the individual consciousness exploring the topography and ecology of the Pond. 'If I should travel to the prairies, I should much less understand them,' Thoreau wrote in his Journal (10.191 (OS)). 'Many a weed here stands for more of life to me than the big trees of California.' Walden gave him all the 'wildness' he needed; also (less ingenuously) the solitude: 'for

---

[14] The relationship between *Walden* and *The Natural History of Selborne* is outlined in Frank Stewart, *A Natural History of Nature Writing* (Washington DC/Covelo, Calif.: Island Press/ Searwater Books, 1995), 26–33.

[15] John Christie, *Thoreau as World Traveler* (New York: Columbia University Press, 1965), 44–7.

the most part it is as solitary where I live as on the prairies' (p. 119).

Walden's relationship to contemporary manuals on progressive farming is similarly oppositional. Thoreau's hostility to modern developments is unmistakable, as the contempt for the 'model farm' (p. 177) and the chapter on 'The Bean-Field' make clear. Here the statistics documenting the enterprise are unproblematically ridiculous—inaccurate to the extent that they would have been recognized by a knowledgeable and attentive contemporary reader as something approaching a hoax.

Robert Gross has provided the necessary context to show what Thoreau is up to in his story of the bean-field. From the days of first settlement until the late eighteenth century, New England farmers had been faced with cheap land and dear labour, the exact opposite to the conditions in which English farmers worked. As a result, their methods were slovenly, wasteful, and inefficient. They cleared forests and undergrowth by the 'slash and burn' method, counting on the resulting potash to fertilize the soil for a few years. When the land lost its fertility or the population grew too great to be supported by it, people upped stakes and headed West, to the fertile valley of the Ohio River.[16]

It was mainly the desire to stem this depopulation of the New England countryside that turned clergymen like Henry Coleman into agricultural reformers. In a series of four surveys of Massachusetts agriculture from 1838 to 1841, Coleman argued the need for new crops, better fertilizers, more attention to crop rotation, a planting strategy better focused on future demands of the market, and other methods calculated to make more intensive use of the land.

Thoreau's objections to farm reform bear some resemblance to critiques of factory farming today. His disgust at the

---

[16] Robert Gross, 'The Great Bean Field Hoax: Thoreau and the Agricultural Reformers', in Joel Myerson (ed.), *Critical Essays on Henry David Thoreau's* Walden (Boston: G. K. Hall, 1988), 196–7.

'model farm' is a furious reaction to the criterion of intensity. 'Intensive' farming means meadows without flowers (because weedless), animals stabled too close together, men housed as animals, and all of them mixed up in the indistinguishable mire of manure, feed, and sewage; exhausted fields, exhausted lakes, and exhausted men—and always the market driving everything before it, conditioning the choice of crops, their treatment, their timing, and their value in more senses than one.

'The Bean-Field' is another, less direct attack on the agricultural reformers. The joke lies in the commercial improbability of the whole experiment. First of all, the staple crops in the Concord area were Indian corn, oats, and English hay, not beans. The beans were probably more a gesture than a practical necessity, being the chief ingredient of Boston's characteristic dish and also a Native American crop. Secondly, the narrator's figures for his costs and earnings are more or less hopeless. As in building the cabin, he does not cost the rent and other overheads of the land he uses. He makes no allowance for his own labour, and when he buys in the help of others, he gets their labour cheap, despite his complaint, at about half the going rate. Whereas the normal yield on those other farms that did grow beans elsewhere in Massachusetts was twenty bushels per acre, the narrator manages only seven or eight. He needs three times the seed to produce a bushel than other farmers.

In other words, if we are to believe his own figures, the 'Thoreau' in the story fails abjectly to produce a crop which no one in the area wants to buy anyway. He fails because he refuses to employ any of the techniques of 'progressive' intensive farming. He plants too late in the season; he does nothing to improve the soil, using no manure or chemical fertilizer; he sets his seeds too far apart. He even debates at length whether he should allow woodchucks to share his harvest. 'He started with exhausted, barren land,' as Gross puts it, 'did nothing to improve it, obtained little from it, and announced himself quite content.'[17]

Leonard Neufeldt has provided an extensive account of how *Walden* draws on the guidebooks for young men. Thoreau had four of them in his personal library. Typically deploying the exemplary first-person narrator, the young men's guides offered practical as well as moral and religious advice, hints on manners, and the best means of self-instruction. To be avoided were bad company (especially of artful women), bad books, flattery, romantic novels—in short, everything tending towards the 'fantastic, effeminate [and] conceited'. What was to be encouraged could be summarized under the headings of enterprise and economy, two 'terms that imply each other and that touch on every definition and example of success . . . [from] "diligence," "industry," "labour," and "determination" . . . [to] "perseverance," "resolution," and "persevering exertions." '[18]

Here, in *Walden*'s treatment of the young men's guides, however, the relationship between model and parody is not so simple as with farming manuals. As Neufeldt shows, the text both attacks and supports the values advanced in the guides, reinforcing their emphasis on economy and enterprise, but extending these terms beyond their conventional applications. Hence, although *Walden* takes on the responsibility assumed by the guides to advise the young, even co-opting the form's characteristic 'I' and 'you' of teacher and pupil, it does not deal with manners. Its precepts have less to do with advancing a career in commerce or the professions than with personal development generally. So 'enterprise' becomes a habit more radical than merely spotting a gap in the market for goods or services, then working hard to fill it. 'Economy' means more than prudent finances and deferred gratification. Neufeldt shows how thoroughly the chapter on 'Economy' opens out

---

[17] Gross, 196–7, 198–202.

[18] Neufeldt, *Economist*, 103–23.

the Greek sense of the word—household management, husbandry, thrift, rules and precepts, the relationship of parts to
the whole, and stewardship. That is why the chapter devotes
such attention to the buying of the lumber and hardware,
and to the building of the cabin on the Pond: because
economy is grounded on the home. The house is built from
old materials, yet, as Neufeldt explains, it 'was unmistakably
new in location, shape, general outer appearance, inhabitant,
and purpose'.[19]

This emphasis on building the new from old materials is a
clue to those high prices paid for second-hand timber, boards,
shingles, and bricks. Because the events of *Walden* are at once
apparently literal and suggestively symbolic, the documentary
air of balance sheets, if read carefully, will invite an alternative
interpretation. In the more restricted sense of 'economy' the
man who bought second-hand boards for $8.03½ and 1,250
shingles for $4 was careless, or was cheated. By the same
standard, the man who planted beans was a damn fool. But
then it was important to the symbolic action of *Walden* that
the narrator's speculative 'home' be a new structure built of
old materials, and that the experiment with the beans enact
the text's jeremiad against commercialized farming. In the
fuller meaning of 'economy' the narrator's experiments both
succeed.

Walden is most interesting when it argues with itself. This
is because how to write *Walden* is one of *Walden*'s main
subjects. Nowhere is this concern more salient than in the
account of the ice harvest in Chapter 16. Here the events are
narrated three times: first by way of summary (p. 262); then
fancifully (pp. 262–3); finally literally (pp. 263–4). Once the
occasion is established, the narrative moves into the mode of
'let's pretend'. 'A hundred men of Hyperborean extraction
swoop down on to our pond one morning' in winter. They go
to work at once, 'ploughing, harrowing, rolling, furrowing

[19] Neufeldt, *Economist*, 174–5.

. . . as if they were bent on making this a model farm'. Was it to sow winter rye, or some new grain out of Iceland? he feigns to ask. Just as they seem ready to drop seed into furrow, they instead begin 'to hook up the virgin mould itself . . . clean down to the sand, or rather the water . . . and haul it away on sleds, and then I guessed that they must be cutting peat in a bog'. Then the illusion is punctured:

To speak literally, a hundred Irishmen, with Yankee overseers, came from Cambridge every day to get out the ice. They divided it into cakes by methods too well known to require description, and these, being sledded to the shore, were rapidly hauled off on to an ice platform, and raised by grappling irons and block and tackle, worked by horses, on to a stack . . . (pp. 263–4)

The word 'fanciful' is more appropriate to version two than (say) 'imaginative', because it is sandwiched between two other accounts that leave the reader in no doubt as to what is actually happening. So if we know that the narrator is only fooling, what kind of 'truth' can the mock-farming metaphor be said to convey? Or does he just do it to show that he can, as a bravado display of his own craft?

Maybe both. Of all commercial activities mentioned in *Walden*, harvesting ice is the most ephemeral, the least substantial, the most redundant. This ice (at least as he tells the story) will not serve some practical purpose like preserving meat, but 'to cool the summer drink' of 'the prudent landlord'. Here is a commodity that costs nothing in raw materials, adds no value and soon melts back into nothing, yet for the brief moment in which it holds its shape and temperature it is in such demand that it pays the vendor to ship it as far as 'Charleston and New Orleans . . . Madras and Bombay and Calcutta' (p. 265). How does its production contribute to the wealth of nations? Does even the 'gentleman farmer' turned speculator need the revenue, when he already has half a million dollars? Is it really worth disturbing the winter peace of Walden Pond for this?

In other words, the ice trade is itself a fantasy. So 'Thoreau' pays it back in his own currency, confecting a fantasy of his own—a flight of description as redundant to his own narrative as ice is to the economy of the real world—and implicating progressive farming techniques for good measure, as if to say: it is to this absurdity that the commercialization of nature, spurred on by the dream of distant markets, will lead you.

This multiplicity of writing styles generates a sense in the attentive reader that *Walden* is sometimes working so as to test, at times even to subvert, its own discourse. As in the lecture-essays of Ralph Waldo Emerson, the overriding system of that discourse is a secular modification of the sermon and the spiritual autobiography, those twin genres so popular in New England from its first settlement. In both forms an occasion, however trivial or haphazardly noted, or a fact observed in nature, acts as a biblical text promoting a general comment, even a meditation of sorts. So the narrator of *Walden* observes that 'a bucket of water soon becomes putrid, but frozen remains sweet forever' (p. 265). 'Why is this?' he asks, offering no answer, but choosing rather to change the discourse into a preacher's 'application' that might have come from the great, late Puritan, Jonathan Edwards: 'It is commonly said that this is the difference between the affections and the intellect.'

Again, a striped snake runs into a pond hole, remaining under the cold, March water for more than a quarter of an hour, 'perhaps', as the narrative conjectures, 'because he had not yet fairly come out of the torpid state'. Then this lovely text from nature receives its ponderous moralization: 'It appeared to me that for a like reason men remain in their present low and primitive condition; but if they should feel the influence of the spring of springs arousing them, they would of necessity rise to a higher and more ethereal life' (pp. 38–9). This was added to the *Walden* manuscript late in its evolution, in 1852 or early in 1853; so Thoreau was being fairly deliberate in trying to make *Walden* as much like a sermon as possible, fashioning its detailed interpretations so as to strengthen the exemplary nature of its first-person narrative: the 'I' as representative man.

Widen the field of view slightly, though, and the snake becomes less of a parable, more of an ordinary event in a series of closely observed natural phenomena. The narrator has borrowed an axe. Because the head has come off its handle, he decides to feed the shaft back through the hole in the axe, then to drive a wedge in the end to fix it in place by widening it. What first brings him to the water is the need to soak the axe to make the new wood swell. It is only then that he sees the snake; so the snake has not been especially singled out for its analogy to man in his state of spiritual torpor. Nor does the text leave the episode at that little moral lesson:

I had previously seen the snakes in frosty mornings in my path with portions of their bodies still numb and inflexible, waiting for the sun to thaw them. On the 1st of April it rained and melted the ice, and in the early part of the day, which was very foggy, I heard a stray goose groping about over the pond and cackling as if lost, or like the spirit of the fog. (p. 39)

In other words, though Thoreau added the business about men's primitive state and their potential for a 'higher and more ethereal life' at a late stage of revision, as the episode appears in the narrative segment it does not conclude it in some great burst of abstract moral 'application'. This is what a sermon would have done. In *Walden* the description of the snakes' cold-blooded dependence on the heat of their immediate environment floats in and out of focus as the text 'spiritualizes' it, but it returns to the vivid, factual register of their numb, inflexible bodies on the frosty path. As though to consolidate this return to the everyday world of the observable environment, the text then documents the date and weather—the thaw and consequent fog which lead the narrative off into the quite different direction of the sound of the goose overhead. It is hard to explain what makes this passage quite so appealing. It is partly that wonderful sound-and-sense combination of the goose 'groping', but it has even more to do with the argument's willingness to wander, like the goose, without anxiety about just where it will wind up. But it is also that the 'open' impulse which produced this piece of writing, and the observed phenomena which provide a subdued metaphoric echo of its structure—of apprehension, of narration—survive the added homiletic note.

In *The Environmental Imagination* Lawrence Buell argues that Thoreau's development as a writer can be traced along a line moving from 'anthropocentric' to 'ecocentric' concerns, and that *Walden* is where Thoreau's career can be seen to shift from the one to the other. 'Nature was initially more of a pastime for him', Buell writes. 'Increasingly it became the environment in which he felt most comfortable. Then it became an occupation (or rather, occupations, first literary and then botanical) and finally a cause.'[20] This convincing argument is another way of describing what goes on in the striped snake episode. At first the text seems to bend a natural observation to a human lesson; then it relaxes, drawing a renewed human attention to, and strength from, the sights, seasons, and sounds of nature.

The more one looks, the more one finds to support the impression that Thoreau's experience at Walden Pond—or to put it more accurately, his experience of writing about that experience for an audience—challenged and changed the preconceptions he brought with him. This can be demonstrated even in the statement of his purpose in going to the woods in the first place. In his journal he put it like this: 'I wish to meet the facts of life—the vital facts, which were the

[20] Lawrence Buell, *The Environmental Imagination: Thoreau, Nature Writing, and the Formation of American Culture* (Cambridge, Mass.: Harvard University Press, 1995), 138.

phenomena or actuality the Gods meant to show us—face to face. And so I came down here' (J 2.156 (NS)). That has a strongly Platonic feel about it, as though the form of the experience not only preceded the experience itself, but somehow attained a higher—because more general—truth. Yet in the passage in *Walden* that draws on the journal entry this rationale has been made more practical, and expressed in more down-to-earth language: 'I went to the woods because I wished to live deliberately, to front only the essential facts of life, and see if I could not learn what it had to teach, and not, when I came to die, discover that I had not lived. I did not wish to live what was not life, living is so dear' (p. 83).

In an ordinary story of a man going to live in the woods, this paragraph might have been encountered somewhere near the beginning. In fact *Walden* does start with something similar ('When I wrote the following pages, or rather the bulk of them, I lived alone, in the woods, a mile from any neighbor, in a house which I had built myself, on the shore of Walden Pond'); so the latter statement of purpose registers as a beginning again. And it is not the only one to do so. Only twenty or so pages further on, the narrative reverts to what might be seen as introductory, or preliminary information:

My house was on the side of a hill, immediately on the edge of the larger wood, in the midst of a young forest of pitch pines and hickories, and half a dozen rods from the pond, to which a narrow footpath led down the hill. In my front yard grew the strawberry, blackberry, and life-everlasting, johnswort and goldenrod, shrub-oaks and sand-cherry, blueberry and ground-nut. Near the end of May, the sand-cherry (*cerasus pumila*), adorned the sides of the path with its delicate flowers arranged in umbels cylindrically about its short stems, which last, in the fall, weighed down with good sized and handsome cherries, fell over in wreaths like rays on every side. I tasted them out of compliment to Nature, though they were scarcely palatable. (p. 104)

By now we should know where his cabin is. Indeed, he has already told us, in the extensive paragraph on (p. 38), beginning: 'Near the end of March, 1845, I borrowed an axe and went down to the woods by Walden Pond . . .' But this is a beginning renewed, the location now fixed by new co-ordinates: a homely catalogue of second-growth trees, scrub, and other volunteer growth. Common though they are, the plants take on—or are made to yield up—their own grandeur, as the narrative plays their earthy English names off against precise latinate description ('umbels cylindrically') and the actual Latin of their binomial classifications. Who would imagine anyone taking common weeds so seriously? No matter that their delectable promise turns out to be 'scarcely palatable'; he has paid sufficient compliment to Nature by treating her most humdrum productions as rare botanical specimens—which of course they are, if seen afresh. So the disregarded scrub, the scrawny, second-growth trees, are given a new life through close observation and careful writing, just as the second-hand boards and bricks are put to new use in that cabin. As the narrative renews its environment, so it renews itself.

Much of what happens in *Walden* can be traced back to the author's sense of cultural marginality. As Thoreau saw it anyway, Concord was beginning to lose its (albeit minor) metropolitan status. More surprising to those of us used to thinking of mid-nineteenth-century New England as the age of the great American literary renaissance, he was uncertain about the region's cultural presence too. 'I aspire to be acquainted with wiser men than this our Concord soil has produced', the narrator of *Walden* comments (p. 98) in the midst of a critique on the narrowness of his fellow-townspeople's reading.

Thoreau dealt with this sense of marginality in several ways. One was to search out the universals of human thought and feeling in the great books of ancient English and religio-philosophical literature, then to deploy this wisdom as of the ages in his treatment of the local scene. *Walden* provides many examples of this practice, and the earlier *A Week on the Concord and Merrimack Rivers* (1849) even more. But we would not still be reading *Walden* if it were nothing more than an anthology of universal truths. It was his second recourse that confirmed the book in American affections. Like Huckleberry Finn after him, he would, in his own way, 'light out for the Territory'. In other words, instead of fretting at his anxiety about not being at the centre of world culture, he would make the most of it, to the point of turning it into a style. That is why he pretended to live in the woods. He was not lying about his experience so much as ritualizing his marginality.

But what gives *Walden* its special value is Thoreau's third response to his sense of marginality: what he did with his experience at the Pond—above all how he wrote about it. For in the natural world around him he found things and events that were both authentically *local* (as he felt the classics of literature and religion never would be), and also universal, in that they could be plotted against a world-wide system of classification, and a history. This is the 'culture' of *Walden*: natural history and the Linnean taxonomy, brought to life by Thoreau's love of language (itself full of a sense of history), his expansion of conventional meanings—in short, his re-demption of the ordinary through close attention and exalted imagination. This is how Thoreau finally established his own sense of centrality, and how *Walden* itself has become central to our sense of the world.

NOTE ON THE TEXT

The text for this edition is the first edition of *Walden*, published in August 1854, by Ticknor and Fields of Boston. The copy used is British Library, 12356.b.28. This is the only edition of *Walden* to be published during Thoreau's lifetime.

The text does not represent the author's final intention in every detail. The problem is how to distinguish between changes introduced by the author, by the editor, and by the compositor. Six versions of the evolving manuscript survive in the Huntington Library (HM 924). So do the page proofs (HM 925), corrected by both printer and author. The missing evidence is the fair copy that Thoreau sent to the publisher, now lost. Without the fair copy, there is no way of telling which departures from the extant manuscripts in the uncorrected page proofs were introduced by the publisher and which by the author. After the failure of *A Week on the Concord and Merrimack Rivers* five years earlier, Thoreau was not in a strong position to insist on his own version of a disputed reading. Certainly he was not then the canonical writer he has since become. Editors and compositors might well have felt free—indeed seen it as their job—to regularize the text, especially in matters of spelling and punctuation. Besides, all publishers have their house styles, unlikely to be suspended in favour of any but the most prestigious and temperamental author.

Even the corrected proofs may not always represent the author's spelling and punctuation. Of the over 1,000 corrections made to the page proofs, most are to accidentals. Many are in the printer's hand. One or two obvious misprints went uncorrected on the page proofs, surviving into the first edition. Of those substantive corrections made in the author's hand, all but six were incorporated in the first edition. In the absence of a final manuscript with which to compare page proofs and first edition, it is not now possible to distinguish between those corrections that Thoreau would strongly like to have made, those on which he acquiesced in the publishers' judgement, and mistakes that he simply overlooked.

Two further pieces of evidence remain. On 4 March 1862, while negotiating a possible reprint of *Walden* near the end of his life, Thoreau wrote to Ticknor and Fields asking them to drop the half-title of the first edition, '*or, Life in the Woods*'. And to his own copy of the first edition Thoreau made a number of marginal alterations. None of them is mentioned in his letter about the reprint. Some are insertions, others are corrections, others again reminders and other notes to himself that amount almost to belated entries in his Journal. One, a reference to kingfishers following 'white-bellied swallows (*Hirundo bicolor*) skim over it' on p. 167, appears in two forms: once on the page proofs, and again in his own copy, apparently a mis-memory of the insertion he had wished, but failed, to make, in the first edition.

These afterthoughts are of varying value as evidence of authorial intention. Clearly Thoreau wanted to drop the half-title. Clearly too he intended to mention kingfishers, though in what form of words is unclear. Should his own corrections to mistakes made in the first edition be accepted, like 'double spruce' for 'single spruce' on p. 115, and 'black spruce' for 'white-spruce' on p. 181? And what about insertions, such as '*Pomotis obesus*', the binomial Latin designation after the 'breams' mentioned on p. 166, or '(*mus leucopus*)' for the 'mice . . . not the common ones . . . but a wild native kind' on p. 203? And when he inserts 'deer' between 'meadow' and 'mouse' on p. 238, did he mean it as a substitution for 'meadow' or an additional (parenthetical?) adjective?

Of the two modern scholarly texts of *Walden*, the Princeton edition (ed. J. Lyndon Shanley, Princeton: Princeton University Press, 1971) incorporates Thoreau's corrections to the page proofs, most of his marginal additions to his own copy, and modernizes and/or regularizes some of the spelling and punctuation of the first edition. The Library of America edition (ed. Robert F. Sayre, New York: Literary Classics of the United States, Inc.; Cambridge: Cambridge University Press, 1985) is more conservative—and in my opinion, more accurate—omitting the changes made in Thoreau's copy, keeping the spelling and punctuation of the first edition, but correcting the text according to alterations made to page proofs and not incorporated in the first edition, and emending typographical errors set up in proof but not caught by Thoreau.

This text has been edited on a principle even more conservative than that governing the Library of America Edition. I have kept to the first edition without alteration of any kind, except the inevitable errors in transmission between any two texts where the copy is not a photo-offset of the original. All accidentals in the first edition—even typographical errors—have been retained. The Explanatory Notes draw attention to Thoreau's substantive alterations to page proof not incorporated in the first edition, as they do to the marginalia in his own copy. But neither body of data has been accepted as constituting part of the text.

In other words, I consider the textual evidence too sketchy to set about trying to correct the accidental, or even the substantive emendations registered, whether in the page proofs or Thoreau's own copy. As for Thoreau's afterthoughts, I have dropped the half-title, '*or, Life in the Woods*' and admitted kingfishers to the ecological appendix. Beyond that I have accepted none of them—not even his belated insertions of Latin binomial designations for plants and animals or his corrections concerning the nomenclature of spruce trees. 'Double' and 'black' are indeed more accurate than 'single' and 'white', but one of the most interesting features of Thoreau's intellectual and aesthetic development is the extent to which his sensitivity to, and knowledge of ecological detail evolved during the writing of *Walden* and afterwards. *Walden* must survive not only as a universal statement, but also as a historical document of where Thoreau was at when he wrote it.

List of Line-End Hyphenations in the First Edition

Some words broken by hyphens at line endings of the first edition have come through into modern editions variously as single words and hyphenated compounds. The following words inviting this ambiguity, keyed to page number, appeared in the first edition as line-end hyphenations:

40	Door-sill	181	toad-stools
43	in-dweller	181	butter-flies
85	to-morrow	183	pickerel-weed
95	tip-toe	185	huckle-berrying
97	no-body	208	saddle-bow
98	low-lived	216	pitch-pine
103	to-morrow	224	piece-meal
107	drill-barrow	226	sap-wood
114	wood-side	226	wood-chopper's
126	noon-day	230	golden-rod
131	bean-field	235	beggar-ticks
134	home-made	237	fir-trees
141	shrub-oaks	276	johns-wort
145	elm-tree	289	half-witted
169	hill-top		

Further Reading

A general account of the development of the text of *Walden* can be found in J. Lyndon Shanley, *The Making of Walden: With the Text of the First Edition* (Chicago: University of Chicago Press, 1957). Ronald Clapper attempts a reconstruction of the lost fair copy, and traces the text through its printed editions, in his Ph.D. dissertation (University of California at Los Angeles, 1967), 'The Development of *Walden*: A Genetic Text'. Two critiques of modern editions appear in *Proof: The Yearbook of American Bibliographical and Textual Studies*, 4 (1975), both by Joseph R. McElrath, Jr.: 'Practical Editions: Henry D. Thoreau's *Walden*', pp. 175–82, and 'The First Two Volumes of the Writings of Henry D. Thoreau', pp. 215–35.

SELECT BIBLIOGRAPHY

Editions

The Journal of Henry David Thoreau, ed. Bradford Torrey and Francis H. Allen (14 vols.; New York: Houghton Mifflin, 1906).

Journal, ed. Elizabeth Hall Witherell, William L. Howarth, Robert Sattelmeyer, and Thomas Blanding (4 vols. to date; Princeton: Princeton University Press, 1981–).

The Annotated Walden, ed. Philip Van Doren Stern (New York: C. N. Potter, Inc., 1970).

The Variorum Walden, ed. Walter Harding (New York: Twayne, 1962).

Walden, ed. J. Lyndon Shanley (Princeton: Princeton University Press, 1971).

A Week on the Concord and Merrimack Rivers, Walden, The Maine Woods, Cape Cod, ed. Robert F. Sayre (New York: The Library of America, 1985).

Concordances

Marlene Ogden, *Walden: A Concordance* (New York: Garland, 1985).

A Week on the Concord and Merrimack Rivers, Walden, Cape Cod, The Maine Woods [Electronic text based on The Library of America Edition], (Oxford Text Archive, 1647, 1646, 1644, 1645).

Letters

The Correspondence of Henry David Thoreau, ed. Carl Bode and Walter Harding (New York: New York University Press, 1958).

Biography

Walter Harding, *The Days of Henry Thoreau* (New York: Alfred A. Knopf, 1970).

Richard Lebeaux, *Young Man Thoreau* (New York: Harper & Row, 1978).

Robert D. Richardson, Jr., *Henry Thoreau: A Life of the Mind* (Berkeley and Los Angeles: University of California Press, 1986).

Bibliography

Raymond R. Borst, *Henry David Thoreau: A Descriptive Bibliography* (Pittsburg: University of Pittsburg Press, 1982).

Walter Harding and Michael Meyer, *The New Thoreau Handbook* (New York: New York University Press, 1980).

Lewis Leary, 'Henry David Thoreau', in *Eight American Authors* (rev. edn. James Woodress; New York: Norton, 1971), 129–71.

Thoreau Society Bulletin (New York: Geneseo, 1941–).

Collections of Criticism

Walter Harding (ed.), *Thoreau: A Century of Criticism* (Dallas: Southern Methodist University Press, 1954).

Joel Meyerson (ed.), *Critical Essays on Henry David Thoreau's Walden* (Boston: G. K. Hall & Co., 1988).

Sherman Paul (ed.), *Thoreau: A Collection of Critical Essays* (Englewood Cliffs, NJ: Prentice Hall, 1962).

Robert F. Sayre (ed.), *New Essays on Walden* (New York: Cambridge University Press, 1992).

Criticism and Interpretation

Charles Anderson, *The Magic Circle of Walden* (New York: Columbia University Press, 1968).

Richard Bridgman, *Dark Thoreau* (Lincoln: University of Nebraska Press, 1982).

Laurence Buell, *Literary Transcendentalism: Style and Vision in the American Renaissance* (Ithaca, NY: Cornell University Press, 1968).

—— *The Environmental Imagination: Thoreau, Nature Writing, and the Formation of American Culture* (Cambridge, Mass.: Harvard University Press, 1995).

Joan Burbick, *Thoreau's Alternative History: Changing Perspectives on Nature, Culture, and Language* (Philadelphia: University of Pennsylvania Press, 1973).

Stanley Cavell, *The Senses of Walden: An Expanded Edition* (San Francisco: North Point Press, 1981).

James McIntosh, *Thoreau as Romantic Naturalist: His Shifting Stance Toward Nature* (Ithaca, NY: Cornell University Press, 1974).

Leonard Neufeldt, *The Economist: Henry Thoreau and Enterprise* (New York: Oxford University Press, 1989).

Daniel Peck, *Thoreau's Morning Work: Memory and Perception in A Week on the Concord and Merrimack Rivers, the Journal, and Walden* (New Haven: Yale University Press, 1966).

Joel Porte, *Emerson and Thoreau: Transcendentalism in Conflict* (Middletown, Conn.: Wesleyan University Press, 1966).

Robert Sattlemeyer, *Thoreau's Reading: A Study in Intellectual History* (Princeton: Princeton University Press, 1988).

Robert F. Sayre, *Thoreau and the American Indians* (Princeton: Princeton University Press, 1977).

A CHRONOLOGY OF
HENRY DAVID THOREAU

1817   Born in Concord, Massachusetts, third of four children, to John and Cynthia (Dunbar) Thoreau, christened David Henry Thoreau. Father a farmer who also manages a store.

1818   Family moves to Chelmsford, ten miles north of Concord. Father opens a grocery store.

1821   Grocery store closes. Father moves family to Boston, where he works as a schoolteacher.

1823–7   Father returns family to Concord, where he takes over the small pencil manufactory started by his brother-in-law. Thoreau attends private 'dame school', then the town-run primary school.

1828   Thoreau and his brother enter Concord Academy, a private secondary school offering a college-preparatory curriculum.

1833   Thoreau enters Harvard College, his family clubbing together to raise his fees. Exceeds required curriculum by taking courses in sciences and four modern languages.

1835–6   Intermittent leaves of absence from Harvard, to earn money to support his studies and because of ill health. In summer 1836, lives for six weeks on Flint's Pond with Charles Stearns Wheeler. Goes with father to New York to sell pencils.

1837   Graduates from Harvard. Begins teaching at his old primary school, but resigns after being ordered to flog six pupils. Joins informal discussion group of New England transcendentalists meeting at Ralph Waldo Emerson's house. Begins his journal. Works on improving quality of graphite in pencils. Changes his name to Henry David Thoreau.

1838   Makes his first visit to Maine, in search of a teaching job. With his brother John opens a private secondary school in Concord. Later that year takes over the vacant premises and good name of Concord Academy. Lectures for first time at

Concord Lyceum, and begins two-year stint as its secretary and treasurer.

1839   John joins him as fellow-teacher in Concord Academy. John and Henry make two-week excursion up the Concord and Merrimack rivers in their camping dinghy, 'Musketaquid'. Both brothers fall in love with sister of one of their students, Ellen Sewell.

1840   Founding of the transcendentalist journal, *The Dial*, in which Thoreau will publish over thirty essays, poems, and translations during the course of its four-year life. Louisa May Alcott enrols in Concord Academy. Meets Ellery Channing, to become his lifelong friend. Ellen Sewell rejects the suits of both brothers.

1841   John's ill health forces them to close Concord Academy. Henry moves into Emerson's house, earning his keep by working as gardener and handyman. Begins to read widely in Emerson's library, beginning his exploration of Chinese and Indian philosophy, and extending his knowledge of classical and English literature. His essay 'Friendship' printed in *Dial*.

1842   John suffers a razor cut that becomes contaminated; dies of tetanus. Henry is devastated, suffering psychosomatic symptoms of sympathetic paralysis. Gets to know Nathaniel Hawthorne, whose admiration is attracted by his essay 'A Natural History of Massachusetts', published in *Dial*. Sells 'Musketaquid' to Hawthorne for seven dollars.

1843   Lectures on 'Sir Walter Raleigh' in Concord. Takes over as guest editor of *Dial* for April. Leaves Emerson home to become tutor to the children of Emerson's brother William on Staten Island, near New York. Meets the influential New York editor Horace Greeley. 'Walk to Wachusett' published in *Boston Miscellany*. Translations of *Prometheus Bound* and poems of Anacreon, selections from *Laws of Menu* and 'A Winter Walk' published in *Dial*. Critique of technological utopianism, 'Paradise (to be) Regained', published in the literary and political monthly *The United States Magazine and Democratic Review*. Returns to Concord in December, moving back to family home.

1844   'Ethical Scriptures' and 'The Preaching of Buddha' pub-
        lished in *Dial*, as well as an essay defending the abolitionist
        publisher Nathaniel Rogers. Improves technology of pencil
        manufacturing, designing a drilling machine to make an
        accurate hole in the wood to receive the graphite, and
        pioneering grades of graphite hardness to suit uses from
        sketching to draftsman's work. On camping trip accidentally
        sets fire to woods around Concord. Walking trip with Ellery
        Channing through Berkshires and Catskill Mountains.
        Helps construct new family house in the south-western
        quarter of Concord.

1845   Starts to build cabin on Walden Pond. Moves in on 4 July.
        Begins work on manuscript of *A Week on the Concord and
        Merrimack Rivers*, his memoir of the camping-dinghy trip
        with John.

1846   Begins writing *Walden*. Lectures on Thomas Carlyle in
        Concord. Arrested and jailed for one night in Concord for
        refusing to pay his state poll tax, his protest against the role
        of the Commonwealth of Massachusetts in complying with
        slavery. Late summer: excursion to Maine to climb Mount
        Ktaadn with his cousin.

1847   Reads part of his first draft of *Walden* at the Concord
        Lyceum. 'Thomas Carlyle and his Works' published in
        Graham's Magazine, the influential Philadelphia literary
        monthly. Finishes writing *A Week* and leaves the cabin on
        Walden Pond, moving into the Emerson's house to help out
        while Emerson tours Europe.

1848   Delivers lecture at Concord Lyceum on the individual and
        the state, occasioned by his overnight jailing for refusing to
        pay poll tax. 'Ktaadn and the Maine Woods' serialized in the
        New York monthly *Union Magazine* and enthusiastically
        promoted by Horace Greeley in his New York *Tribune*.

1849   Moves back to family home. *A Week* published by James
        Monroe & Co. of Boston, after Thoreau agrees to add his
        royalties to the cost of the book's production. *A Week* sells
        under three hundred copies out of a print run of a thousand.
        Sister Helen dies of tuberculosis. As profits from pencil

business increase, family able to buy a larger house near the centre of Concord. Lyceum lecture on the individual and the state published as 'Resistance to Civil Government' (later, 'Civil Disobedience'). First trip to Cape Cod, with Ellery Channing.

1850   Begins to collect Native American artefacts in family home, and to read widely on American Indians and natural history. Second trip to Cape Cod. Travels to Fire Island, New York, to search for body and manuscripts of Margaret Fuller, shipwrecked while returning from Europe with her husband and child. Thoreau is outraged by the passing of the Fugitive Slave Act, requiring free states to return escaped slaves to their owners in the South. Travels to Montreal and surrounding country with Ellery Channing.

1851   Increasingly active in the 'underground railroad', hiding escaped slaves in the family home and helping them on their way north to Canada.

1852   Excerpts from fourth draft of *Walden* published in *Union Magazine*, which folds without paying Thoreau the agreed fee.

1853   'A Yankee in Canada', based on his trip in 1850, begins to appear in *Putnam's Monthly Magazine*, newly founded in New York to encourage American writing, but after three instalments, Thoreau withdraws the manuscript over a dispute with the magazine's editor. Second trip to Maine.

1854   Arrest of a runaway slave in Boston prompts 'Slavery in Massachusetts', first read at a rally organized by the abolitionist campaigner and editor William Lloyd Garrison, then published in the *Anti-Slavery Standard*, *The Liberator*, and Greeley's New York *Tribune*. *Walden* published by Ticknor and Fields in an edition of 2,000. Reviews are mixed, but sales top 1,700 by the end of the year. English traveller and author Thomas Cholmondeley visits Thoreau in Concord.

1855   'Cape Cod' published in three issues of *Putnam's*. Third trip to Cape Cod, with Ellery Channing. In thanks for his hospitality Cholmondeley sends Thoreau forty-four volumes of Oriental philosophy, religion, and history.

1856   Goes to New Jersey on a large surveying job. On return meets Bronson Alcott in New York. Meets Walt Whitman, who presents him with a copy of the 1856 edition of *Leaves of Grass*.

1857   Meets John Brown, the abolitionist campaigner who would lead the armed raid on Harper's Ferry in 1859, on Brown's visit to Concord. Last trip to Cape Cod. Travels through Maine with a Concord friend and a Native American guide.

1858   'Chesuncook' section of what will appear in 1864 as *The Maine Woods* published in the *Atlantic Monthly*, after some disagreements with editor James Russell Lowell. Cholmondeley visits Concord again. With Concord friend Edward Hoar visits White Mountains in New Hampshire, climbing the highest peak in the range, Mount Washington.

1859   Father's death leaves Thoreau financially responsible for mother and surviving sister. John Brown speaks in Concord. After the raid on Harper's Ferry, Thoreau reads his defence, 'A Plea for John Brown' in Concord, Worcester, and Boston, in the face of growing disapproval of Brown's violent opposition. Visits Mount Monadnock, in southern New Hampshire, with Channing.

1860   'A Plea for John Brown' published in *Echoes of Harper's Ferry*, edited by the abolitionist James Redpath. Climbs Mount Monadnock with Channing. 'The Last Days of John Brown' printed in *The Liberator*, the Boston abolitionist weekly edited by Garrison. 'The Succession of Forest Trees', Thoreau's most significant contribution to botanical science, published in *Transactions of the Middlesex Agricultural Society*. Develops bronchitis.

1861   With Horace Mann, Jr., son of the Massachusetts educator, travels to Minnesota to improve his health. Meets Native Americans of the Sioux Nation, concerned at their treatment by the Federal Government. On return to Concord with health unimproved, begins to make arrangements for posthumous publication of *The Maine Woods* and a revised edition of *A Week*. Makes last visit to Walden Pond.

1862   Dies, 6 May, of tuberculosis.

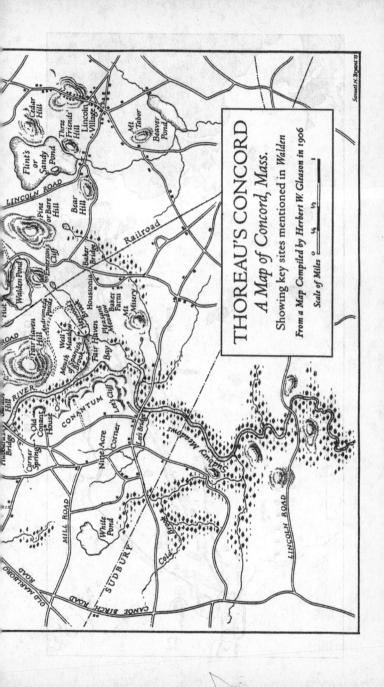

THOREAU'S CONCORD
A Map of Concord, Mass.
Showing key sites mentioned in *Walden*

From a Map Compiled by Herbert W. Gleason in 1906

Scale of Miles

Samuel H. Bryant '51

MAP OF CONCORD AREA

WALDEN;

OR,

LIFE IN THE WOODS.

By HENRY D. THOREAU,

AUTHOR OF "A WEEK ON THE CONCORD AND MERRIMACK RIVERS."

I do not propose to write an ode to dejection, but to brag as lustily as chanticleer in the morning, standing on his roost, if only to wake my neighbors up. — Page 78.

BOSTON:

TICKNOR AND FIELDS.

M DCCC LIV.

CONTENTS

CONTENTS

ECONOMY

I do not propose to write an ode to dejection,* but to brag as lustily as chanticleer* in the morning, standing on his roost, if only to wake my neighbors up.

WHEN I wrote the following pages,* or rather the bulk of them, I lived alone, in the woods, a mile from any neighbor, in a house which I had built myself, on the shore of Walden Pond, in Concord, Massachusetts, and earned my living by the labor of my hands only. I lived there two years and two months. At present I am a sojourner* in civilized life again.

I should not obtrude my affairs so much on the notice of my readers if very particular inquiries had not been made by my townsmen concerning my mode of life, which some would call impertinent, though they do not appear to me at all impertinent, but, considering the circumstances, very natural and pertinent. Some have asked what I got to eat; if I did not feel lonesome;* if I was not afraid; and the like. Others have been curious to learn what portion of my income I devoted to charitable purposes; and some, who have large families, how many poor children I maintained. I will therefore ask those of my readers who feel no particular interest in me to pardon me if I undertake to answer some of these questions in this book. In most books, the *I*, or first person, is omitted; in this it will be retained; that, in respect to egotism, is the main difference. We commonly do not remember that it is, after all, always the first person that is speaking. I should not talk so much about myself if there were any body else whom I knew as well. Unfortunately, I am confined to this theme by the narrowness of my experience. Moreover, I, on my side, require of every writer, first or last, a simple and sincere account of his own life, and not merely what he has heard of other men's lives; some such account as he would send to his kindred from a distant land; for if he has lived sincerely, it must have been in a distant land to me. Perhaps these pages are more particularly addressed to poor students. As for the rest of my readers, they will accept such portions as apply to them. I trust that none will stretch the seams in putting on the coat, for it may do good service to him whom it fits.

I would fain say something, not so much concerning the Chinese and Sandwich Islanders* as you who read these pages, who are said to live in New England; something about your condition, especially your outward condition or circumstances in this world, in this town, what it is, whether it is necessary that it be as bad as it is, whether it cannot be improved as well as not. I have travelled a good deal in Concord; and every where, in shops, and offices, and fields, the inhabitants have appeared to me to be doing penance in a thousand remarkable ways. What I have heard of Bramins* sitting exposed to four fires and looking in the face of the sun; or hanging suspended, with their heads downward, over flames; or looking at the heavens over their shoulders "until it becomes impossible for them to resume their natural position, while from the twist of the neck nothing but liquids can pass into the stomach;" or dwelling, chained for life, at the foot of a tree; or measuring with their bodies, like caterpillars, the breadth of vast empires; or standing on one leg on the tops of pillars,—even these forms of conscious penance are hardly more incredible and astonishing than the scenes which I daily witness. The twelve labors* of Hercules were trifling in comparison with those which my neighbors have undertaken; for they were only twelve, and had an end; but I could never see that these men slew or captured any monster or finished any labor. They have no friend Iolas to burn with a hot iron the root of the hydra's head, but as soon as one head is crushed, two spring up.

I see young men, my townsmen, whose misfortune it is to

have inherited farms, houses, barns, cattle, and farming tools; for these are more easily acquired than got rid of. Better if they had been born in the open pasture and suckled by a wolf,* that they might have seen with clearer eyes what field they were called to labor in. Who made them serfs of the soil? Why should they eat their sixty acres, when man is condemned to eat only his peck of dirt?* Why should they begin digging their graves as soon as they are born? They have got to live a man's life, pushing all these things before them, and get on as well as they can. How many a poor immortal soul have I met well nigh crushed and smothered under its load, creeping down the road of life, pushing before it a barn seventy-five feet by forty, its Augean stables never cleansed, and one hundred acres of land, tillage, mowing, pasture, and wood-lot! The portionless, who struggle with no such unnecessary inherited encumbrances, find it labor enough to subdue and cultivate a few cubic feet of flesh.

But men labor under a mistake. The better part of the man is soon ploughed into the soil for compost. By a seeming fate, commonly called necessity, they are employed, as it says in an old book,* laying up treasures which moth and rust will corrupt and thieves break through and steal. It is a fool's life, as they will find when they get to the end of it, if not before. It is said that Deucalion and Pyrrha* created men by throwing stones over their heads behind them:—

> Inde genus durum sumus, experiensque laborum,
> Et documenta damus quâ simus origine nati.*

Or, as Raleigh rhymes it in his sonorous way,—

> "From thence our kind hard-hearted is, enduring pain and care,
> Approving that our bodies of a stony nature are."

So much for a blind obedience to a blundering oracle, throwing the stones over their heads behind them, and not seeing where they fell.

Most men, even in this comparatively free country, through

mere ignorance and mistake, are so occupied with the factitious cares and superfluously coarse labors of life that its finer fruits cannot be plucked by them. Their fingers, from excessive toil, are too clumsy and tremble too much for that. Actually, the laboring man has not leisure for a true integrity day by day; he cannot afford to sustain the manliest relations to men; his labor would be depreciated in the market. He has no time to be any thing but a machine. How can he remember well his ignorance—which his growth requires—who has so often to use his knowledge? We should feed and clothe him gratuitously sometimes, and recruit him with our cordials, before we judge of him. The finest qualities of our nature, like the bloom on fruits, can be preserved only by the most delicate handling. Yet we do not treat ourselves nor one another thus tenderly.

Some of you, we all know, are poor, find it hard to live, are sometimes, as it were, gasping for breath. I have no doubt that some of you who read this book are unable to pay for all the dinners which you have actually eaten, or for the coats and shoes which are fast wearing or are already worn out, and have come to this page to spend borrowed or stolen time, robbing your creditors of an hour. It is very evident what mean and sneaking lives many of you live, for my sight has been whetted by experience; always on the limits,* trying to get into business and trying to get out of debt, a very ancient slough,* called by the Latins *æs alienum*, another's brass, for some of their coins were made of brass; still living, and dying, and buried by this other's brass; always promising to pay, promising to pay, to-morrow, and dying to-day, insolvent; seeking to curry favor, to get custom, by how many modes, only not state-prison offences; lying, flattering, voting, contracting yourselves into a nutshell* of civility, or dilating into an atmosphere of thin and vaporous generosity, that you may persuade your neighbor to let you make his shoes, or his hat, or his coat, or his carriage, or import his groceries for him; making yourselves sick, that you may lay up something against a sick day, something to be tucked away in an old chest, or in a stocking behind the plastering, or, more safely, in the brick bank; no matter where, no matter how much or how little.

I sometimes wonder that we can be so frivolous, I may almost say, as to attend to the gross but somewhat foreign form of servitude called Negro Slavery, there are so many keen and subtle masters that enslave both north and south. It is hard to have a southern overseer;* it is worse to have a northern one; but worst of all when you are the slave-driver of yourself. Talk of a divinity in man!* Look at the teamster on the highway, wending to market by day or night; does any divinity stir* within him? His highest duty to fodder and water his horses! What is his destiny to him compared with the shipping interests? Does not he drive for Squire Make-a-stir?* How godlike, how immortal, is he? See how he cowers and sneaks, how vaguely all the day he fears, not being immortal nor divine, but the slave and prisoner of his own opinion of himself, a fame won by his own deeds. Public opinion is a weak tyrant compared with our own private opinion. What a man thinks of himself, that it is which determines, or rather indicates, his fate. Self-emancipation even in the West Indian provinces of the fancy and imagination,—what Wilberforce* is there to bring that about? Think, also, of the ladies of the land weaving toilet cushions* against the last day, not to betray too green an interest in their fates! As if you could kill time without injuring eternity.

The mass of men lead lives of quiet desperation. What is called resignation is confirmed desperation. From the desperate city you go into the desperate country, and have to console yourself with the bravery of minks and muskrats. A stereotyped but unconscious despair is concealed even under what are called the games and amusements of mankind. There is no play in them, for this comes after work. But it is a characteristic of wisdom not to do desperate things.

When we consider what, to use the words of the catechism,* is the chief end of man, and what are the true necessaries and means of life, it appears as if men had deliberately chosen the common mode of living because they preferred it to any other. Yet they honestly think there is no choice left. But alert and healthy natures remember that the sun rose clear. It is never too late to give up our prejudices. No way of thinking or doing, however ancient, can be trusted without proof. What every body echoes or in silence passes by as true to-day may turn out to be falsehood to-morrow, mere smoke of opinion, which some had trusted for a cloud that would sprinkle fertilizing rain on their fields. What old people say you cannot do you try and find that you can. Old deeds for old people, and new deeds for new. Old people did not know enough once, perchance, to fetch fresh fuel to keep the fire a-going; new people put a little dry wood under a pot, and are whirled* round the globe with the speed of birds, in a way to kill old people, as the phrase is. Age is no better, hardly so well, qualified for an instructor as youth, for it has not profited so much as it has lost. One may almost doubt if the wisest man has learned any thing of absolute value by living. Practically, the old have no very important advice to give the young, their own experience has been so partial, and their lives have been such miserable failures, for private reasons, as they must believe; and it may be that they have some faith left which belies that experience, and they are only less young than they were. I have lived some thirty years on this planet, and I have yet to hear the first syllable of valuable or even earnest advice from my seniors. They have told me nothing, and probably cannot tell me any thing, to the purpose. Here is life, an experiment to a great extent untried by me; but it does not avail me that they have tried it. If I have any experience which I think valuable, I am sure to reflect that this my Mentors said nothing about.

One farmer says to me, "You cannot live on vegetable food solely, for it furnishes nothing to make bones with;" and so he religiously devotes a part of his day to supplying his system with the raw material of bones; walking all the while he talks behind his oxen, which, with vegetable-made bones, jerk him and his lumbering plough along in spite of every obstacle. Some things are really necessaries of life in some circles, the most helpless and diseased, which in others are luxuries merely, and in others still are entirely unknown.

The whole ground of human life seems to some to have been gone over by their predecessors, both the heights and the valleys, and all things to have been cared for. According to Evelyn, "the wise Solomon prescribed ordinances for the very distances of trees; and the Roman prætors* have decided how often you may go into your neighbor's land to gather the acorns which fall on it without trespass, and what share belongs to that neighbor."* Hippocrates has even left directions how we should cut our nails;* that is, even with the ends of the fingers, neither shorter nor longer. Undoubtedly the very tedium and ennui which presume to have exhausted the variety and the joys of life are as old as Adam. But man's capacities have never been measured; nor are we to judge of what he can do by any precedents, so little has been tried. Whatever have been thy failures hitherto, "be not afflicted, my child, for who shall assign to thee what thou hast left undone?"*

We might try our lives by a thousand simple tests; as, for instance, that the same sun which ripens my beans illumines at once a system of earths like ours. If I had remembered this it would have prevented some mistakes. This was not the light in which I hoed them. The stars are the apexes of what wonderful triangles! What distant and different beings in the various mansions of the universe are contemplating the same one at the same moment! Nature and human life are as various as our several constitutions. Who shall say what prospect life offers to another? Could a greater miracle take place than for us to look through each other's eyes for an instant? We should live in all the ages of the world in an hour; ay, in all the worlds of the ages. History, Poetry, Mythology!—I know of no reading of another's experience so startling and informing as this would be.

The greater part of what my neighbors call good I believe in my soul to be bad, and if I repent of any thing, it is very likely to be my good behavior. What demon possessed me that I behaved so well? You may say the wisest thing you can old man,—you who have lived seventy years, not without honor of a kind,—I hear an irresistible voice which invites me away from all that. One generation abandons the enterprises of another like stranded vessels.

I think that we may safely trust a good deal more than we do. We may waive just so much care of ourselves as we honestly bestow elsewhere. Nature is as well adapted to our weakness as to our strength. The incessant anxiety and strain of some is a well nigh incurable form of disease. We are made to exaggerate the importance of what work we do; and yet how much is not done by us! or, what if we had been taken sick? How vigilant we are! determined not to live by faith if we can avoid it; all the day long on the alert, at night we unwillingly say our prayers and commit ourselves to uncertainties. So thoroughly and sincerely are we compelled to live, reverencing our life, and denying the possibility of change. This is the only way, we say; but there are as many ways as there can be drawn radii from one centre. All change is a miracle to contemplate; but it is a miracle which is taking place every instant. Confucius said, "To know that we know what we know, and that we do not know what we do not know, that is true knowledge."* When one man has reduced a fact of the imagination to be a fact to his understanding, I foresee that all men will at length establish their lives on that basis.

Let us consider for a moment what most of the trouble and anxiety which I have referred to is about, and how much it is necessary that we be troubled, or, at least, careful. It would be some advantage to live a primitive and frontier life, though in the midst of an outward civilization, if only to learn what are the gross necessaries of life and what methods have been taken to obtain them; or even to look over the old day-books of the merchants, to see what it was that men most commonly bought at the stores, what they stored, that is, what are the grossest groceries. For the improvements of ages have had but little influence on the essential laws of man's existence; as our skeletons, probably, are not to be distinguished from those of our ancestors.

By the words, *necessary of life*, I mean whatever, of all that man obtains by his own exertions, has been from the first, or from long use has become, so important to human life that few, if any, whether from savageness, or poverty, or philosophy, ever attempt to do without it. To many creatures there is in this sense but one necessary of life, Food. To the bison of the prairie it is a few inches of palatable grass, with water to drink; unless he seeks the Shelter of the forest or the mountain's shadow. None of the brute creation requires more than Food and Shelter. The necessaries of life for man in this climate may, accurately enough, be distributed under the several heads of Food, Shelter, Clothing, and Fuel; for not till we have secured these are we prepared to entertain the true problems of life with freedom and a prospect of success. Man has invented, not only houses, but clothes and cooked food; and possibly from the accidental discovery of the warmth of fire, and the consequent use of it, at first a luxury, arose the present necessity to sit by it. We observe cats and dogs acquiring the same second nature. By proper Shelter and Clothing we legitimately retain our own internal heat; but with an excess of these, or of Fuel, that is, with an external heat greater than our own internal, may not cookery properly be said to begin? Darwin, the naturalist, says of the inhabitants of Tierra del Fuego, that while his own party, who were well clothed and sitting close to a fire, were far from too warm, these naked savages, who were farther off, were observed, to his great surprise, "to be streaming with perspiration at undergoing such a roasting."* So, we are told, the New Hollander* goes naked with impunity, while the European shivers in his clothes. Is it impossible to combine the hardiness of these savages with the intellectualness of the civilized man? According to Liebig,* man's body is a stove, and food the fuel which keeps up the internal combustion in the lungs. In cold weather we eat more, in warm less. The animal heat is the result of a slow combustion, and disease and death take place when this is too rapid; or for want of fuel, or from some defect in the draught, the fire goes out. Of course the vital heat is not to be confounded with fire; but so much for analogy. It appears, therefore, from the above list, that the expression, *animal life*, is nearly synonymous with the expression, *animal heat*; for while Food may be regarded as the Fuel which keeps up the fire within us,—and Fuel serves only to prepare that Food or to increase the warmth of our bodies by addition from without,—Shelter and Clothing also serve only to retain the *heat* thus generated and absorbed.

The grand necessity, then, for our bodies, is to keep warm, to keep the vital heat in us. What pains we accordingly take, not only with our Food, and Clothing, and Shelter, but with our beds, which are our nightclothes, robbing the nests and breasts of birds to prepare this shelter within a shelter, as the mole has its bed of grass and leaves at the end of its burrow! The poor man is wont to complain that this is a cold world; and to cold, no less physical than social, we refer directly a great part of our ails. The summer, in some climates, makes possible to man a sort of Elysian life.* Fuel, except to cook his Food, is then unnecessary; the sun is his fire, and many of the fruits are sufficiently cooked by its rays; while Food generally is more various, and more easily obtained, and Clothing and Shelter are wholly or half unnecessary. At the present day, and in this country, as I find by my own experience, a few implements, a knife, an axe, a spade, a wheelbarrow, &c., and for the studious, lamplight, stationery, and access to a few books, rank next to necessaries, and can all be obtained at a trifling cost. Yet some,* not wise, go to the other side of the globe, to barbarous and unhealthy regions, and devote themselves to trade for ten or twenty years, in order that they may live,—that is, keep comfortably warm,—and die in New England at last. The luxuriously rich are not simply kept comfortably warm, but unnaturally hot; as I implied before, they are cooked, of course *à la mode.**

Most of the luxuries, and many of the so called comforts of life, are not only not indispensable, but positive hinderances to the elevation of mankind. With respect to luxuries and comforts, the wisest have ever lived a more simple and meagre life than the poor. The ancient philosophers, Chinese, Hindoo, Persian, and Greek, were a class than which none has been poorer in outward riches, none so rich in inward. We know not much about them. It is remarkable that *we* know so much of them as we do. The same is true of the more modern reformers and benefactors of their race. None can be an impartial or wise observer of human life but from the vantage ground of what *we* should call voluntary poverty. Of a life of luxury the fruit is luxury, whether in agriculture, or commerce, or literature, or art. There are nowadays professors of philosophy, but not philosophers. Yet it is admirable to profess because it was once admirable to live. To be a philosopher is not merely to have subtle thoughts, nor even to found a school, but so to love wisdom as to live according to its dictates, a life of simplicity, independence, magnanimity, and trust. It is to solve some of the problems of life, not only theoretically, but practically. The success of great scholars and thinkers is commonly a courtier-like success, not kingly, not manly. They make shift to live merely by conformity, practically as their fathers did, and are in no sense the progenitors of a nobler race of men. But why do men degenerate ever?* What makes families run out? What is the nature of the luxury which enervates and destroys nations? Are we sure that there is none of it in our own lives? The philosopher is in advance of his age even in the outward form of his life. He is not fed, sheltered, clothed, warmed, like his contemporaries. How can a man be a philosopher and not maintain his vital heat by better methods than other men?

When a man is warmed by the several modes which I have described, what does he want next? Surely not more warmth of the same kind, as more and richer food, larger and more splendid houses, finer and more abundant clothing, more numerous incessant and hotter fires, and the like. When he has obtained those things which are necessary to life, there is another alternative than to obtain the superfluities; and that is, to adventure on life now, his vacation from humbler toil having commenced. The soil, it appears, is suited to the seed, for it has sent its radicle downward, and it may now send its shoot upward also with confidence. Why has man rooted himself thus firmly in the earth, but that he may rise in the same proportion into the heavens above?—for the nobler plants* are valued for the fruit they bear at last in the air and light, far from the ground, and are not treated like the humbler esculents,* which, though they may be biennials, are cultivated only till they have perfected their root, and often cut down at top for this purpose, so that most would not know them in their flowering season.

I do not mean to prescribe rules to strong and valiant natures, who will mind their own affairs whether in heaven or hell, and perchance build more magnificently and spend more lavishly than the richest, without ever impoverishing themselves, not knowing how they live,—if, indeed, there are any such, as has been dreamed; nor to those who find their encouragement and inspiration in precisely the present condition of things, and cherish it with the fondness and enthusiasm of lovers,—and, to some extent, I reckon myself in this number; I do not speak to those who are well employed, in whatever circumstances, and they know whether they are well employed or not;—but mainly to the mass of men who are discontented, and idly complaining of the hardness of their lot or of the times, when they might improve them. There are some who complain most energetically and inconsolably of any, because they are, as they say, doing their duty. I also have in my mind that seemingly wealthy, but most terribly impoverished class of all, who have accumulated dross, but know not how to use it, or get rid of it, and thus have forged their own golden or silver fetters.

If I should attempt to tell how I have desired to spend my life in years past, it would probably surprise those of my readers who are somewhat acquainted with its actual history; it would certainly astonish those who know nothing about it. I will only hint at some of the enterprises which I have cherished.

In any weather, at any hour of the day or night, I have been anxious to improve the nick of time, and notch* it on my stick too; to stand on the meeting of two eternities,* the past and future, which is precisely the present moment; to toe that line. You will pardon some obscurities, for there are more secrets in my trade than in most men's, and yet not voluntarily kept, but inseparable from its very nature. I would gladly tell all that I know about it, and never paint "No Admittance" on my gate.

I long ago lost a hound, a bay horse, and a turtle-dove, and am still on their trail. Many are the travellers I have spoken concerning them, describing their tracks and what calls they answered to. I have met one or two who had heard the hound, and the tramp of the horse, and even seen the dove disappear behind a cloud, and they seemed as anxious to recover them as if they had lost them themselves.*

To anticipate, not the sunrise and the dawn merely, but, if possible, Nature herself! How many mornings, summer and winter, before yet any neighbor was stirring about his business, have I been about mine! No doubt, many of my townsmen have met me returning from this enterprise, farmers starting for Boston in the twilight, or woodchoppers going to their work. It is true, I never assisted the sun materially in his rising, but, doubt not, it was of the last importance only to be present at it.

So many autumn, ay, and winter days, spent outside the town, trying to hear what was in the wind, to hear and carry it express! I well-nigh sunk all my capital* in it, and lost my own breath into the bargain, running in the face of it. If it had concerned either of the political parties, depend upon it, it would have appeared in the Gazette with the earliest intelligence. At other times watching from the observatory of some cliff or tree, to telegraph any new arrival; or waiting at evening on the hill-tops for the sky to fall, that I might catch something, though I never caught much, and that, mannawise,* would dissolve again in the sun.

For a long time I was reporter to a journal,* of no very wide circulation, whose editor has never yet seen fit to print the bulk of my contributions, and, as is too common with writers, I got only my labor for my pains. However, in this case my pains were their own reward.

For many years I was self-appointed inspector of snow storms and rain storms, and did my duty faithfully; surveyor, if not of highways, then of forest paths and all across-lot routes, keeping them open, and ravines bridged and passable at all seasons, where the public heel had testified to their utility.*

I have looked after the wild stock of the town, which give a faithful herdsman a good deal of trouble by leaping fences; and I have had an eye to the unfrequented nooks and corners of the farm; though I did not always know whether Jonas or Solomon worked in a particular field to-day; that was none of my business. I have watered the red huckleberry, the sand cherry and the nettle tree, the red pine and the black ash, the white grape and the yellow violet,* which might have withered else in dry seasons.

In short, I went on thus for a long time, I may say it without boasting, faithfully minding my business, till it became more and more evident that my townsmen would not after all admit me into the list of town officers,* nor make my place a sinecure with a moderate allowance. My accounts, which I can swear to have kept faithfully, I have, indeed, never got audited, still less accepted, still less paid and settled. However, I have not set my heart on that.

Not long since, a strolling Indian went to sell baskets at the house of a well-known lawyer in my neighborhood. "Do you wish to buy any baskets?" he asked. "No, we do not want any," was the reply. "What!" exclaimed the Indian as he went out the gate, "do you mean to starve us?" Having seen his industrious white neighbors so well off,—that the lawyer had only to weave arguments, and by some magic wealth and standing followed, he had said to himself; I will go into business; I will weave baskets; it is a thing which I can do. Thinking that when he had made the baskets he would have done his part, and then it would be the white man's to buy them. He had not discovered that it was necessary for him to make it worth the other's while to buy them, or at least make him think that it was so, or to make something else which it would be worth his while to buy. I too had woven a kind of basket of a delicate texture, but I had not made it worth any one's while to buy them. Yet not the less, in my case, did I think it worth my while to weave them, and instead of studying how to make it worth men's while to buy my baskets, I studied rather how to avoid the necessity of selling them. The life which men praise and regard as successful is but one kind. Why should we exaggerate any one kind at the expense of the others?

Finding that my fellow-citizens were not likely to offer me any room in the court house,* or any curacy or living any where else, but I must shift for myself, I turned my face more exclusively than ever to the woods, where I was better known. I determined to go into business at once, and not wait to acquire the usual capital, using such slender means as I had already got. My purpose in going to Walden Pond was not to live cheaply nor to live dearly there, but to transact some private business* with the fewest obstacles; to be hindered from accomplishing which for want of a little common sense, a little enterprise and business talent, appeared not so sad as foolish.

I have always endeavored to acquire strict business habits;* they are indispensable to every man. If your trade is with the Celestial Empire,* then some small counting house on the coast, in some Salem harbor,* will be fixture enough. You will export such articles as the country affords, purely native products, much ice and pine timber and a little granite, always in native bottoms.* These will be good ventures. To oversee all the details yourself in person; to be at once pilot and captain, and owner and underwriter; to buy and sell and keep the accounts; to read every letter received, and write or read every letter sent; to superintend the discharge of imports night and day; to be upon many parts of the coast almost at the same time;—often the richest freight will be discharged upon a Jersey shore;*—to be your own telegraph,* unweariedly sweeping the horizon, speaking all passing vessels* bound coastwise; to keep up a steady despatch of commodities, for the supply of such a distant and exorbitant market; to keep yourself informed of the state of the markets, prospects of war and peace every where, and anticipate the tendencies of trade and civilization,—taking advantage of the results of all exploring expeditions, using new passages and all improvements in navigation;—charts to be studied, the position of reefs and new lights and buoys to be ascertained, and ever, and ever, the logarithmic tables to be corrected,* for by the error of some calculator the vessel often splits upon a rock that should have reached a friendly pier,—there is the untold fate of La Perouse;*—universal science to be kept pace with, studying the lives of all great discoverers and navigators, great adventurers and merchants, from Hanno* and the Phœnicians down to our day; in fine, account of stock to be taken from time to time, to know how you stand. It is a labor to task the faculties of a man,—such problems of profit and loss, of interest, of tare and tret,* and gauging of all kinds in it, as demand a universal knowledge.

I have thought that Walden Pond would be a good place for business, not solely on account of the railroad and the ice trade; it offers advantages which it may not be good policy to divulge; it is a good post* and a good foundation. No Neva marshes* to be filled; though you must every where build on piles of your own driving. It is said that a flood-tide, with a westerly wind, and ice in the Neva, would sweep St. Petersburg from the face of the earth.

As this business was to be entered into without the usual capital, it may not be easy to conjecture where those means, that will still be indispensable to every such undertaking, were to be obtained. As for Clothing,* to come at once to the practical part of the question, perhaps we are led oftener by the love of novelty, and a regard for the opinions of men, in procuring it, than by a true utility. Let him who has work to do recollect that the object of clothing is, first, to retain the vital heat, and secondly, in this state of society, to cover nakedness, and he may judge how much of any necessary or important work may be accomplished without adding to his wardrobe. Kings and queens who wear a suit but once, though made by some tailor or dressmaker to their majesties, cannot know the comfort of wearing a suit that fits. They are no better than wooden horses to hang the clean clothes on. Every day our garments become more assimilated to ourselves, receiving the impress of the wearer's character, until we hesitate to lay them aside, without such delay and medical appliances and some such solemnity even as our bodies. No man ever stood the lower in my estimation for having a patch in his clothes; yet I am sure that there is greater anxiety, commonly, to have fashionable, or at least clean and unpatched clothes, than to have a sound conscience. But even if the rent is not mended, perhaps the worst vice betrayed is improvidence. I sometimes try my acquaintance by such tests as this;—who could wear a patch, or two extra seams only, over the knee? Most behave as if they believed that their prospects for life would be ruined if they should do it. It would be easier for them to hobble to town with a broken leg than with a broken pantaloon. Often if an accident happens to a gentleman's legs, they can be mended; but if a similar accident happens to the legs of his pantaloons, there is no help for it; for he considers, not what is truly respectable, but what is respected. We know but few men, a great many coats and breeches. Dress a scarecrow in your last shift, you standing shiftless by, who would not soonest salute the scarecrow? Passing a cornfield the other day, close by a hat and coat on a stake, I recognized the owner of the farm. He was only a little more weather-beaten than when I saw him last. I have heard of a dog that barked at every stranger who approached his master's premises with clothes on, but was easily quieted by a naked thief. It is an interesting question how far men would retain their relative rank if they were divested of their clothes. Could you, in such a case, tell surely of any company of civilized men, which belonged to the most respected class? When Madam Pfeiffer,* in her adventurous travels round the world, from east to west, had got so near home as Asiatic Russia, she says that she felt the necessity of wearing other than a travelling dress, when she went to meet the authorities, for she "was now in a civilized country, where——people are judged of by their clothes." Even in our democratic New England towns the accidental possession of wealth, and its manifestation in dress and equipage alone, obtain for the possessor almost universal respect. But they who yield such respect, numerous as they are, are so far heathen, and need to have a missionary sent to them. Beside, clothes introduced sewing, a kind of work which you may call endless; a woman's dress, at least, is never done.*

A man who has at length found something to do will not need to get a new suit to do it in; for him the old will do, that has lain dusty in the garret for an indeterminate period. Old shoes will serve a hero* longer than they have served his valet,—if a hero ever has a valet,—bare feet are older than shoes, and he can make them do. Only they who go to soirées and legislative halls must have new coats, coats to change as often as the man changes in them. But if my jacket and trousers, my hat and shoes, are fit to worship God in, they will do; will they not? Who ever saw his old clothes,—his old coat, actually worn out, resolved into its primitive elements, so that it was not a deed of charity to bestow it on some poor boy, by him perchance to be bestowed on some poorer still, or shall we say richer, who could do with less? I say, beware of all enterprises that require new clothes, and not rather a new wearer of clothes. If there is not a new man, how can the new clothes be made to fit? If you have any enterprise before you, try it in your old clothes. All men want, not something to *do with*, but something to *do*, or rather something to *be*. Perhaps we should never procure a new suit, however ragged or dirty the old, until we have so conducted, so enterprised or sailed in some way, that we feel like new men in the old, and that to retain it would be like keeping new wine in old bottles.* Our moulting season, like that of the fowls, must be a crisis in our lives. The loon retires to solitary ponds to spend it. Thus also the snake casts its slough, and the caterpillar its wormy coat, by an internal industry and expansion; for clothes are but our outmost cuticle and mortal coil.* Otherwise we shall be found sailing under false colors, and be inevitably cashiered at last by our own opinion, as well as that of mankind.

We don garment after garment, as if we grew like exogenous plants* by addition without. Our outside and often thin and fanciful clothes are our epidermis or false skin, which partakes not of our life, and may be stripped off here and there without fatal injury; our thicker garments, constantly worn, are our cellular integument,* or cortex;* but our shirts are our liber* or true bark, which cannot be removed without girdling and so destroying the man. I believe that all races at some seasons wear something equivalent to the shirt. It is desirable that a man be clad so simply that he can lay his hands on himself in the dark, and that he live in all respects so compactly and preparedly, that, if an enemy take the town, he can, like the old philosopher,* walk out the gate empty-handed without anxiety. While one thick garment is, for most purposes, as good as three thin ones, and cheap clothing can be obtained at prices really to suit customers; while a thick coat can be bought for five dollars, which will last as many years, thick pantaloons for two dollars, cowhide boots for a dollar and a half a pair, a summer hat for a quarter of a dollar, and a winter cap* for sixty-two and a half cents, or a better be made at home at a nominal cost, where is he so poor that, clad in such a suit, *of his own earning*, there will not be found wise men to do him reverence?

When I ask for a garment of a particular form, my tailoress tells me gravely, "They do not make them so now," not emphasizing the "They" at all, as if she quoted an authority as impersonal as the Fates,* and I find it difficult to get made what I want, simply because she cannot believe that I mean what I say, that I am so rash. When I hear this oracular sentence, I am for a moment absorbed in thought, emphasizing to myself each word separately that I may come at the meaning of it, that I may find out by what degree of consanguinity *They* are related to *me*, and what authority they may have in an affair which affects me so nearly; and, finally, I am inclined to answer her with equal mystery, and without any more emphasis of the "they,"—"It is true, they did not make them so recently, but they do now." Of what use this measuring of me if she does not measure my character, but only the breadth of my shoulders, as it were a peg to hang the coat on? We worship not the Graces,* nor the Parcæ, but Fashion. She spins and weaves and cuts with full authority. The head monkey* at Paris puts on a traveller's cap, and all the monkeys in America do the same. I sometimes despair of getting any thing quite simple and honest done in this world by the help of men. They would have to be passed through a powerful press first, to squeeze their old notions out of them, so that they would not soon get upon their legs again, and then there

would be some one in the company with a maggot in his head,
hatched from an egg deposited there nobody knows when, for
not even fire kills these things, and you would have lost your
labor. Nevertheless, we will not forget that some Egyptian
wheat* was* handed down to us by a mummy.

On the whole, I think that it cannot be maintained that
dressing has in this or any country risen to the dignity of an
art. At present men make shift to wear what they can get. Like
shipwrecked sailors, they put on what they can find on the
beach, and at a little distance, whether of space or time, laugh
at each other's masquerade. Every generation laughs at the
old fashions, but follows religiously the new. We are amused
at beholding the costume of Henry VIII., or Queen Elizabeth,
as much as if it was that of the King and Queen of the
Cannibal Islands. All costume off a man is pitiful or grotesque.
It is only the serious eye peering from and the sincere life
passed within it, which restrain laughter and consecrate the
costume of any people. Let Harlequin* be taken with a fit of
the colic and his trappings will have to serve that mood too.
When the soldier is hit by a cannon ball rags are as becoming
as purple.

The childish and savage taste of men and women for new
patterns keeps how many shaking and squinting through
kaleidoscopes that they may discover the particular figure
which this generation requires to-day. The manufacturers
have learned that this taste is merely whimsical. Of two patterns which differ only by a few threads more or less of a
particular color, the one will be sold readily, the other lie on
the shelf, though it frequently happens that after the lapse of
a season the latter becomes the most fashionable. Comparatively, tattooing is not the hideous custom which it is called. It
is not barbarous merely because the printing is skin-deep and
unalterable.

I cannot believe that our factory system is the best mode by
which men may get clothing. The condition of the operatives
is becoming every day more like that of the English; and it

cannot be wondered at, since, as far as I have heard or observed, the principal object is, not that mankind may be well and honestly clad, but, unquestionably, that the corporations may be enriched. In the long run men hit only what they aim at. Therefore, though they should fail immediately, they had better aim at something high.

As for a Shelter, I will not deny that this is now a necessary of life, though there are instances of men having done without it for long periods in colder countries than this. Samuel Laing* says that "The Laplander in his skin dress, and in a skin bag which he puts over his head and shoulders, will sleep night after night on the snow—in a degree of cold which would extinguish the life of one exposed to it in any woollen clothing." He had seen them asleep thus. Yet he adds, "They are not hardier than other people." But, probably, man did not live long on the earth without discovering the convenience which there is in a house, the domestic comforts, which phrase may have originally signified the satisfactions of the house more than of the family; though these must be extremely partial and occasional in those climates where the house is associated in our thoughts with winter or the rainy season chiefly, and two thirds of the year, except for a parasol, is unnecessary. In our climate, in the summer, it was formerly almost solely a covering at night. In the Indian gazettes a wigwam was the symbol of a day's march, and a row of them cut or painted on the bark of a tree signified that so many times they had camped. Man was not made so large limbed and robust but that he must seek to narrow his world, and wall in a space such as fitted him. He was at first bare and out of doors; but though this was pleasant enough in serene and warm weather, by daylight, the rainy season and the winter, to say nothing of the torrid sun, would perhaps have nipped his race in the bud if he had not made haste to clothe himself with the shelter of a house. Adam and Eve, according to the fable, wore the bower before other clothes. Man wanted a home, a place of warmth, or comfort, first of physical warmth, then the warmth of the affections.

We may imagine a time when, in the infancy of the human race, some enterprising mortal crept into a hollow in a rock for shelter. Every child begins the world again, to some extent, and loves to stay out doors, even in wet and cold. It plays house, as well as horse, having an instinct for it. Who does not remember the interest with which when young he looked at shelving rocks, or any approach to a cave? It was the natural yearning of that portion of our most primitive ancestor which still survived in us. From the cave we have advanced to roofs of palm leaves, of bark and boughs, of linen woven and stretched, of grass and straw, of boards and shingles, of stones and tiles. At last, we know not what it is to live in the open air, and our lives are domestic in more senses than we think. From the hearth to the field is a great distance. It would be well perhaps if we were to spend more of our days and nights without any obstruction between us and the celestial bodies, if the poet did not speak so much from under a roof, or the saint dwell there so long. Birds do not sing in caves, nor do doves cherish their innocence in dovecots.

However, if one designs to construct a dwelling house, it behooves him to exercise a little Yankee shrewdness, lest after all he find himself in a workhouse, a labyrinth* without a clew, a museum, an almshouse, a prison,* or a splendid mausoleum instead. Consider first how slight a shelter is absolutely necessary. I have seen Penobscot Indians,* in this town, living in tents of thin cotton cloth, while the snow was nearly a foot deep around them, and I thought that they would be glad to have it deeper to keep out the wind. Formerly, when how to get my living honestly, with freedom left for my proper pursuits, was a question which vexed me even more than it does now, for unfortunately I am become somewhat callous, I used to see a large box by the railroad, six feet long by three wide, in which the laborers locked up their tools at night, and it suggested to me that every man who was hard pushed might get such a one for a dollar, and, having bored a few auger holes in it, to admit the air at least, get into it when it rained and at night, and hook down the lid, and so have freedom in his love,* and in his soul be free. This did not appear the worst, nor by any means a despicable alternative. You could sit up as late as you pleased, and, whenever you got up, go abroad without any landlord or house-lord dogging you for rent. Many a man is harassed to death to pay the rent of a larger and more luxurious box who would not have frozen to death in such a box as this. I am far from jesting. Economy is a subject which admits of being treated with levity, but it cannot so be disposed of. A comfortable house for a rude and hardy race, that lived mostly out of doors, was once made here almost entirely of such materials as Nature furnished ready to their hands. Gookin, who was superintendent of the Indians subject to the Massachusetts Colony, writing in 1674, says, "The best of their houses are covered very neatly, tight and warm, with barks of trees, slipped from their bodies at those seasons when the sap is up, and made into great flakes, with pressure of weighty timber, when they are green. . . . The meaner sort are covered with mats which they make of a kind of bulrush, and are also indifferently tight and warm, but not so good as the former. . . . Some I have seen, sixty or a hundred feet long and thirty feet broad. . . . I have often lodged in their wigwams, and found them as warm as the best English houses."* He adds, that they were commonly carpeted and lined within with well-wrought embroidered mats, and were furnished with various utensils. The Indians had advanced so far as to regulate the effect of the wind by a mat suspended over the hole in the roof and moved by a string. Such a lodge was in the first instance constructed in a day or two at most, and taken down and put up in a few hours; and every family owned one, or its apartment in one.

In the savage state every family owns a shelter as good as the best, and sufficient for its coarser and simpler wants; but I think that I speak within bounds when I say that, though the

birds of the air have their nests, and the foxes their holes,* and
the savages their wigwams, in modern civilized society not
more than one half the families own a shelter. In the large
towns and cities, where civilization especially prevails, the
number of those who own a shelter is a very small fraction of
the whole. The rest pay an annual tax for this outside garment
of all, become indispensable summer and winter, which
would buy a village of Indian wigwams, but now helps to keep
them poor as long as they live. I do not mean to insist here on
the disadvantage of hiring compared with owning, but it is
evident that the savage owns his shelter because it costs so
little, while the civilized man hires his commonly because he
cannot afford to own it; nor can he, in the long run, any better
afford to hire. But, answers one, by merely paying this tax the
poor civilized man secures an abode which is a palace compared with the savage's. An annual rent of from twenty-five to
a hundred dollars, these are the country rates, entitles him to
the benefit of the improvements of centuries, spacious apartments, clean paint and paper, Rumford fireplace,* back plastering,* Venetian blinds, copper pump, spring lock, a
commodious cellar, and many other things. But how happens
it that he who is said to enjoy these things is so commonly a
poor civilized man, while the savage, who has them not, is rich
as a savage? If it is asserted that civilization is a real advance
in the condition of man,—and I think that it is, though only
the wise improve their advantages,—it must be shown that it
has produced better dwellings without making them more
costly; and the cost of a thing is the amount of what I will call
life which is required to be exchanged for it, immediately or in
the long run. An average house in this neighborhood costs
perhaps eight hundred dollars,* and to lay up this sum will
take from ten to fifteen years of the laborer's life, even if he is
not encumbered* with a family;—estimating the pecuniary
value of every man's labor at one dollar a day, for if some
receive more, others receive less;*—so that he must have
spent more than half his life commonly before *his* wigwam

will be earned. If we suppose him to pay a rent instead, this is but a doubtful choice of evils. Would the savage have been wise to exchange his wigwam for a palace on these terms?

It may be guessed that I reduce almost the whole advantage of holding this superfluous property as a fund in store against the future, so far as the individual is concerned, mainly to the defraying of funeral expenses. But perhaps a man is not required to bury himself. Nevertheless this points to an important distinction between the civilized man and the savage; and, no doubt, they have designs on us for our benefit, in making the life of a civilized people an *institution*, in which the life of the individual is to a great extent absorbed, in order to preserve and perfect that of the race. But I wish to show at what a sacrifice this advantage is at present obtained, and to suggest that we may possibly so live as to secure all the advantage without suffering any of the disadvantage. What mean ye by saying that the poor ye have always with you,* or that the fathers have eaten sour grapes, and the children's teeth are set on edge?

"As I live, saith the Lord God, ye shall not have occasion any more to use this proverb in Israel."

"Behold all souls are mine; as the soul of the father, so also the soul of the son is mine: the soul that sinneth it shall die."*

When I consider my neighbors, the farmers of Concord, who are at least as well off as the other classes, I find that for the most part they have been toiling twenty, thirty, or forty years, that they may become the real owners of their farms, which commonly they have inherited with encumbrances,* or else bought with hired money,—and we may regard one third of that toil as the cost of their houses,—but commonly they have not paid for them yet. It is true, the encumbrances sometimes outweigh the value of the farm, so that the farm itself becomes one great encumbrance, and still a man is found to inherit it, being well acquainted with it, as he says. On applying to the assessors,* I am surprised to learn that they cannot at once name a dozen in the town who own their farms free and clear. If you would know the history of these homesteads, inquire at the bank where they are mortgaged. The man who has actually paid for his farm with labor on it is so rare that every neighbor can point to him. I doubt if there are three such men in Concord. What has been said of the merchants, that a very large majority, even ninety-seven in a hundred, are sure to fail, is equally true of the farmers. With regard to the merchants, however, one of them says pertinently that a great part of their failures are not genuine pecuniary failures, but merely failures to fulfil their engagements, because it is inconvenient; that is, it is the moral character that breaks down. But this puts an infinitely worse face on the matter, and suggests, beside, that probably not even the other three succeed in saving their souls, but are perchance bankrupt in a worse sense than they who fail honestly. Bankruptcy and repudiation* are the spring-boards from which much of our civilization vaults and turns its somersets, but the savage stands on the unelastic plank of famine. Yet the Middlesex Cattle Show* goes off here with *éclat* annually, as if all the joints of the agricultural machine were suent.*

The farmer is endeavoring to solve the problem of a livelihood by a formula more complicated than the problem itself. To get his shoestrings he speculates in herds of cattle. With consummate skill he has set his trap with a hair springe* to catch comfort and independence, and then, as he turned away, got his own leg into it. This is the reason he is poor; and for a similar reason we are all poor in respect to a thousand savage comforts, though surrounded by luxuries. As Chapman sings,—

> "The false society of men—
> —for earthly greatness
> All heavenly comforts rarefies to air.*

And when the farmer has got his house, he may not be the richer but the poorer for it, and it be the house that has got

him. As I understand it, that was a valid objection urged by
Momus* against the house which Minerva* made, that she
"had not made it movable, by which means a bad
neighborhood might be avoided;" and it may still be urged, for
our houses are such unwieldy property that we are often
imprisoned rather than housed in them; and the bad
neighborhood to be avoided is our own scurvy selves. I know
one or two families, at least, in this town, who, for nearly a
generation, have been wishing to sell their houses in the
outskirts and move into the village, but have not been able to
accomplish it, and only death will set them free.

Granted that the *majority* are able at last either to own or
hire the modern house with all its improvements. While civilization has been improving our houses, it has not equally
improved the men who are to inhabit them. It has created
palaces, but it was not so easy to create noblemen and kings.
And *if the civilized man's pursuits are no worthier than the*
savage's, if he is employed the greater part of his life in obtain-
ing gross necessaries and comforts merely, why should he
have a better dwelling than the former?

But how do the poor *minority* fare? Perhaps it will be
found, that just in proportion as some have been placed in
outward circumstances above the savage, others have been
degraded below him. The luxury of one class is counterbalanced by the indigence of another. On the one side is the
palace, on the other are the almshouse and "silent poor."* The
myriads who built the pyramids to be the tombs of the Phar-
aohs were fed on garlic, and it may be were not decently
buried themselves. The mason who finishes the cornice of the
palace returns at night perchance to a hut not so good as a
wigwam. It is a mistake to suppose that, in a country where
the usual evidences of civilization exist, the condition of a very
large body of the inhabitants may not be as degraded as that
of savages. I refer to the degraded poor, not now to the
degraded rich. To know this I should not need to look farther
than to the shanties* which every where border our railroads,

that last improvement in civilization; where I see in my daily walks human beings living in sties, and all winter with an open door, for the sake of light, without any visible, often imaginable, wood pile, and the forms of both old and young are permanently contracted by the long habit of shrinking from cold and misery, and the development of all their limbs and faculties is checked. It certainly is fair to look at that class by whose labor the works which distinguish this generation are accomplished. Such too, to a greater or less extent, is the condition of the operatives of every denomination in England, which is the great workhouse of the world.* Or I could refer you to Ireland, which is marked as one of the white or enlightened spots on the map. Contrast the physical condition of the Irish* with that of the North American Indian, or the South Sea Islander, or any other savage race before it was degraded by contact with the civilized man. Yet I have no doubt that people's rulers are as wise as the average of civilized rulers. Their condition only proves what squalidness may consist with civilization. I hardly need refer now to the laborers in our Southern States who produce the staple exports of this country, and are themselves a staple production of the South. But to confine myself to those who are said to be in *moderate* circumstances.

Most men appear never to have considered what a house is, and are actually though needlessly poor all their lives because they think that they must have such a one as their neighbors have. As if one were to wear any sort of coat which the tailor might cut out for him, or, gradually leaving off palmleaf hat or cap of woodchuck skin, complain of hard times because he could not afford to buy him a crown! It is possible to invent a house still more convenient and luxurious than we have, which yet all would admit that man could not afford to pay for. Shall we always study to obtain more of these things, and not sometimes to be content with less? Shall the respectable citizen thus gravely teach, by precept and example, the necessity of the young man's providing a certain number of superfluous glowshoes,* and umbrellas, and empty guest chambers for empty guests, before he dies? Why should not our furniture be as simple as the Arab's or the Indian's? When I think of the benefactors of the race, whom we have apotheosized as messengers from heaven, bearers of divine gifts to man, I do not see in my mind any retinue at their heels, any car-load of fashionable furniture. Or what if I were to allow—would it not be a singular allowance?—that our furniture should be more complex than the Arab's, in proportion as we are morally and intellectually his superiors! At present our houses are cluttered and defiled with it, and a good housewife would sweep out the greater part into the dust hole, and not leave her morning's work undone. Morning work! By the blushes of Aurora* and the music of Memnon,* what should be man's *morning work* in this world? I had three pieces of limestone on my desk, but I was terrified to find that they required to be dusted daily, when the furniture of my mind was all undusted still, and I threw them out the window in disgust. How, then, could I have a furnished house? I would rather sit in the open air, for no dust gathers on the grass, unless where man has broken ground.

It is the luxurious and dissipated who set the fashions which the herd so diligently follow. The traveller who stops at the best houses, so called, soon discovers this, for the publicans presume him to be a Sardanapalus,* and if he resigned himself to their tender mercies he would soon be completely emasculated. I think that in the railroad car we are inclined to spend more on luxury than on safety and convenience, and it threatens without attaining these to become no better than a modern drawing room, with its divans, and ottomans, and sunshades, and a hundred other oriental* things, which we are taking west with us, invented for the ladies of the harem and the effeminate natives of the Celestial Empire, which Jonathan* should be ashamed to know the names of. I would rather sit on a pumpkin and have it all to myself, than be crowded on a velvet cushion. I would rather ride on earth in

an ox cart with a free circulation, than go to heaven in the
fancy car of an excursion train and breathe a *malaria** all the
way.

The very simplicity and nakedness of man's life in the
primitive ages imply this advantage at least, that they left him
still but a sojourner in nature. When he was refreshed with
food and sleep he contemplated his journey again. He dwelt,
as it were, in a tent in this world, and was either threading the
valleys, or crossing the plains, or climbing the mountain tops.
But lo! men have become the tools of their tools. The man
who independently plucked the fruits when he was hungry is
become a farmer; and he who stood under a tree for shelter,
a housekeeper. We now no longer camp as for a night, but
have settled down on earth and forgotten heaven. We have
adopted Christianity merely as an improved method of *agriculture.** We have built for this world a family mansion,* and
for the next a family tomb. The best works of art are the
expression of man's struggle to free himself from this condition, but the effect of our art is merely to make this low state
comfortable and that higher state to be forgotten. There is
actually no place in this village for a work of *fine* art, if any had
come down to us, to stand, for our lives, our houses and
streets, furnish no proper pedestal for it. There is not a nail to
hang a picture on, nor a shelf to receive the bust of a hero or
a saint. When I consider how our houses are built and paid
for, or not paid for, and their internal economy managed and
sustained, I wonder that the floor does not give way under the
visitor while he is admiring the gewgaws upon the mantelpiece, and let him through into the cellar, to some solid and
honest though earthy foundation. I cannot but perceive that
this so called rich and refined life is a thing jumped at, and I
do not get on in the enjoyment of the *fine* arts which adorn it,
my attention being wholly occupied with the jump; for I
remember that the greatest genuine leap, due to human
muscles alone, on record, is that of certain wandering Arabs,
who are said to have cleared twenty-five feet on level ground.

Without factitious support, man is sure to come to earth again beyond that distance. The first question which I am tempted to put to the proprietor of such great impropriety is, Who bolsters you? Are you one of the ninety-seven who fail, or the three* who succeed? Answer me these questions, and then perhaps I may look at your bawbles and find them ornamental. The cart before the horse is neither beautiful nor useful. Before we can adorn our houses with beautiful objects the walls must be stripped, and our lives must be stripped, and beautiful housekeeping and beautiful living be laid for a foundation: now, a taste for the beautiful is most cultivated out of doors, where there is no house and no housekeeper.

Old Johnson, in his "Wonder-Working Providence," speaking of the first settlers of this town, with whom he was contemporary, tells us that "they burrow themselves in the earth for their first shelter under some hillside, and, casting the soil aloft upon timber, they make a smoky fire against the earth, at the highest side." They did not "provide them houses," says he, "till the earth, by the Lord's blessing, brought forth bread to feed them," and the first year's crop was so light that "they were forced to cut their bread very thin for a long season."* The secretary of the Province of New Netherland, writing in Dutch, in 1650, for the information of those who wished to take up land there, states more particularly, that "those in New Netherland, and especially in New England, who have no means to build farm houses at first according to their wishes, dig a square pit in the ground, cellar fashion, six or seven feet deep, as long and as broad as they think proper, case the earth inside with wood all round the wall, and line the wood with the bark of trees or something else to prevent the caving in of the earth; floor this cellar with plank, and wainscot it overhead for a ceiling, raise a roof of spars clear up, and cover the spars with bark or green sods, so that they can live dry and warm in these houses with their entire families for two, three, and four years, it being understood that partitions are run through those cellars which are adapted to the size of the family. The wealthy and principal men in New England, in the beginning of the colonies, commenced their first dwelling houses in this fashion for two reasons; firstly, in order not to waste time in building, and not to want food the next season; secondly, in order not to discourage poor laboring people whom they brought over in numbers from Fatherland. In the course of three or four years, when the country became adapted to agriculture, they built themselves handsome houses, spending on them several thousands."*

In this course which our ancestors took there was a show of prudence at least, as if their principle were to satisfy the more pressing wants first. But are the more pressing wants satisfied now? When I think of acquiring for myself one of our luxurious dwellings, I am deterred, for, so to speak, the country is not yet adapted to *human* culture, and we are still forced to cut our *spiritual* bread far thinner than our forefathers did their wheaten. Not that all architectural ornament is to be neglected even in the rudest periods; but let our houses first be lined with beauty, where they come in contact with our lives, like the tenement of the shellfish, and not overlaid with it. But, alas! I have been inside one or two of them, and know what they are lined with.

Though we are not so degenerate* but that we might possibly live in a cave or a wigwam or wear skins to-day, it certainly is better to accept the advantages, though so dearly bought, which the invention and industry of mankind offer. In such a neighborhood as this, boards and shingles, lime and bricks, are cheaper and more easily obtained than suitable caves, or whole logs, or bark in sufficient quantities, or even well-tempered clay or flat stones. I speak understandingly on this subject, for I have made myself acquainted with it both theoretically and practically. With a little more wit we might use these materials so as to become richer than the richest now are, and make our civilization a blessing. The civilized man is a more experienced and wiser savage. But to make haste to my own experiment.

Near the end of March, 1845, I borrowed an axe* and went down to the woods by Walden Pond, nearest to where I intended to build my house, and began to cut down some tall arrowy white pines, still in their youth, for timber. It is difficult to begin without borrowing, but perhaps it is the most generous course thus to permit your fellow-men to have an interest in your enterprise. The owner of the axe, as he released his hold on it, said that it was the apple of his eye; but I returned it sharper than I received it. It was a pleasant hillside where I worked, covered with pine woods, through which I looked out on the pond, and a small open field in the woods where pines and hickories were springing up. The ice in the pond was not yet dissolved, though there were some open spaces, and it was all dark colored and saturated with water. There were some slight flurries of snow during the days that I worked there; but for the most part when I came out on to the railroad, on my way home, its yellow sand heap stretched away gleaming in the hazy atmosphere, and the rails shone in the spring sun, and I heard the lark and pewee and other birds already come to commence another year with us. They were pleasant spring days, in which the winter of man's discontent* was thawing as well as the earth, and the life that had lain torpid began to stretch itself. One day, when my axe had come off and I had cut a green hickory for a wedge, driving it with a stone, and had placed the whole to soak in a pond hole in order to swell the wood, I saw a striped snake run into the water, and he lay on the bottom, apparently without inconvenience, as long as I staid there, or more than a quarter of an hour; perhaps because he had not yet fairly come out of the torpid state. It appeared to me that for a like reason men remain in their present low and primitive condition; but if they should feel the influence of the spring of springs arousing them, they would of necessity rise to a higher and more ethereal life. I had previously seen the snakes in frosty mornings in my path with portions of their bodies still numb and inflexible, waiting for the sun to thaw them. On the 1st of April it rained and melted the ice, and in the early part of the day, which was very foggy, I heard a stray goose groping about over the pond and cackling as if lost, or like the spirit of the fog.

So I went on for some days cutting and hewing timber, and also studs and rafters,* all with my narrow axe, not having many communicable or scholar-like thoughts, singing to myself,—

> Men say they know many things;
> But lo! they have taken wings,—
> The arts and sciences,
> And a thousand appliances;
> The wind that blows
> Is all that any body knows.*

I hewed the main timbers six inches square, most of the studs on two sides only, and the rafters and floor timbers on one side, leaving the rest of the bark on,* so that they were just as straight and much stronger than sawed ones. Each stick was carefully mortised or tenoned* by its stump, for I had borrowed other tools by this time. My days in the woods were not very long ones; yet I usually carried my dinner of bread and butter, and read the newspaper in which it was wrapped, at noon, sitting amid the green pine boughs which I had cut off, and to my bread was imparted some of their fragance, for my hands were covered with a thick coat of pitch. Before I had done I was more the friend than the foe of the pine tree, though I had cut down some of them, having become better acquainted with it. Sometimes a rambler in the wood was attracted by the sound of my axe, and we chatted pleasantly over the chips which I had made.

By the middle of April, for I made no haste in my work, but

rather made the most of it, my house was framed and ready
for the raising. I had already bought the shanty of James
Collins, an Irishman who worked on the Fitchburg Railroad,*
for boards. James Collins' shanty was considered an uncom-
monly fine one. When I called to see it he was not at home. I
walked about the outside, at first unobserved from within, the
window was so deep and high. It was of small dimensions,
with a peaked cottage roof, and not much else to be seen, the
dirt being raised five feet all around as if it were a compost
heap. The roof was the soundest part, though a good deal
warped and made brittle by the sun. Door-sill there was none,
but a perennial passage for the hens under the door board.
Mrs. C. came to the door and asked me to view it from the
inside. The hens were driven in by my approach. It was dark,
and had a dirt floor for the most part, dank, clammy, and
aguish, only here a board and there a board which would not
bear removal. She lighted a lamp to show me the inside of the
roof and the walls, and also that the board floor extended
under the bed, warning me not to step into the cellar, a sort of
dust hole two feet deep. In her own words, they were "good
boards overhead, good boards all around, and a good window,"—of two whole squares originally, only the cat had
passed out that way lately. There was a stove, a bed, and a
place to sit, an infant in the house where it was born, a silk
parasol, gilt-framed looking-glass, and a patent new coffee
mill nailed to an oak sapling, all told. The bargain was soon
concluded, for James had in the mean while returned. I to pay
four dollars and twenty-five cents* to-night, he to vacate at five
to-morrow morning, selling to nobody else meanwhile: I to
take possession at six. It were well, he said, to be there early,
and anticipate certain indistinct but wholly unjust claims on
the score of ground rent and fuel. This he assured me was the
only encumbrance. At six I passed him and his family on the
road. One large bundle held their all,—bed, coffee-mill, looking-glass, hens,—all but the cat, she took to the woods and

became a wild cat, and, as I learned afterward, trod in a trap set for woodchucks, and so became a dead cat at last.

I took down this dwelling the same morning, drawing the nails, and removed it to the pond side by small cartloads, spreading the boards on the grass there to bleach and warp back again in the sun. One early thrush gave me a note or two as I drove along the woodland path. I was informed treacherously by a young Patrick* that neighbor Seeley, an Irishman, in the intervals of the carting, transferred the still tolerable, straight, and drivable nails, staples, and spikes to his pocket, and then stood when I came back to pass the time of day, and look freshly up, unconcerned, with spring thoughts, at the devastation; there being a dearth of work, as he said. He was there to represent spectatordom, and help make this seemingly insignificant event one with the removal of the gods of Troy.*

I dug my cellar in the side of a hill sloping to the south, where a woodchuck had formerly dug his burrow, down through sumach and blackberry roots, and the lowest stain of vegetation, six feet square by seven deep, to a fine sand where potatoes would not freeze in any winter. The sides were left shelving, and not stoned; but the sun having never shone on them, the sand still keeps its place. It was but two hours' work. I took particular pleasure in this breaking of ground, for in almost all latitudes men dig into the earth for an equable temperature. Under the most splendid house in the city is still to be found the cellar where they store their roots as of old, and long after the superstructure has disappeared posterity remark its dent in the earth. The house is still but a sort of porch at the entrance of a burrow.

At length, in the beginning of May, with the help of some of my acquaintances,* rather to improve so good an occasion for neighborliness than from any necessity, I set up the frame of my house. No man was ever more honored in the character of his raisers than I. They are destined, I trust, to assist at the raising of loftier structures one day. I began to occupy my house on the 4th of July,* as soon as it was boarded and roofed, for the boards were carefully feather-edged and lapped,* so that it was perfectly impervious to rain; but before boarding I laid the foundation of a chimney at one end, bringing two cartloads of stones up the hill from the pond in my arms. I built the chimney after my hoeing in the fall, before a fire became necessary for warmth, doing my cooking in the mean while out of doors on the ground, early in the morning: which mode I still think is in some respects more convenient and agreeable than the usual one. When it stormed before my bread was baked, I fixed a few boards over the fire, and sat under them to watch my loaf, and passed some pleasant hours in that way. In those days, when my hands were much employed, I read but little, but the least scraps of paper which lay on the ground, my holder, or table-cloth, afforded me as much entertainment, in fact answered the same purpose as the Iliad.

It would be worth the while to build still more deliberately than I did, considering, for instance, what foundation a door, a window, a cellar, a garret, have in the nature of man, and perchance never raising any superstructure until we found a better reason for it than our temporal necessities even. There is some of the same fitness in a man's building his own house that there is in a bird's building its own nest. Who knows but if men constructed their dwellings with their own hands, and provided food for themselves and families simply and honestly enough, the poetic faculty would be universally developed, as birds universally sing when they are so engaged? But alas! we do like cowbirds and cuckoos, which lay their eggs in nests which other birds have built, and cheer no traveller with their chattering and unmusical notes. Shall we forever resign the pleasure of construction to the carpenter? What does architecture amount to in the experience of the mass of men? I never in all my walks came across a man engaged in so simple and natural an occupation as building his house. We belong to the community. It is not the tailor alone who is the ninth part of a man;* it is as much the preacher, and the merchant, and the farmer. Where is this division of labor to end? and what object does it finally serve? No doubt another *may* also think for me; but it is not therefore desirable that he should do so to the exclusion of my thinking for myself.

True, there are architects so called in this country, and I have heard of one at least possessed with the idea of making architectural ornaments have a core of truth, a necessity, and hence a beauty, as if it were a revelation to him. All very well perhaps from his point of view, but only a little better than the common dilettantism. A sentimental reformer in architecture, he began at the cornice, not at the foundation. It was only how to put a core of truth within the ornaments, that every sugar plum in fact might have an almond or caraway seed in it,— though I hold that almonds are most wholesome without the sugar,—and not how the inhabitant, the indweller, might build truly within and without, and let the ornaments take care of themselves. What reasonable man ever supposed that ornaments were something outward and in the skin merely,— that the tortoise got his spotted shell, or the shellfish its mother-o'-pearl tints, by such a contract as the inhabitants of Broadway their Trinity Church?* But a man has no more to do with the style of architecture* of his house than a tortoise with that of its shell: nor need the soldier be so idle as to try to paint the precise *color* of his virtue on his standard. The enemy will find it out. He may turn pale when the trial comes. This man seemed to me to lean over the cornice, and timidly whisper his half truth to the rude occupants who really knew it better than he. What of architectural beauty I now see, I know has gradually grown from within outward, out of the necessities and character of the indweller, who is the only builder,—out of some unconscious truthfulness, and nobleness, without ever a thought for the appearance; and whatever additional beauty of this kind is destined to be produced will be preceded by a like unconscious beauty of life. The most interesting dwellings in this country, as the painter knows, are the most unpretending, humble log huts and cottages of the poor commonly; it is the life of the inhabitants whose shells they are, and not any peculiarity in their surfaces merely, which makes them *picturesque*; and equally interesting will be the citizen's suburban box, when his life shall be as simple and as agreeable to the imagination, and there is as little straining after effect in the style of his dwelling. A great proportion of architectural ornaments are literally hollow, and a September gale would strip them off, like borrowed plumes, without injury to the substantials. They can do without *architecture* who have no olives nor wines in the cellar. What if an equal ado were made about the ornaments of style in literature, and the architects of our bibles spent as much time about their cornices as the architects of our churches do? So are made the *belles-lettres* and the *beaux-arts** and their professors. Much it concerns a man, forsooth, how a few sticks are slanted over him or under him, and what colors are daubed upon his box. It would signify somewhat, if, in any earnest sense, *he* slanted them and daubed it; but the spirit having departed out of the tenant, it is of a piece with constructing his own coffin,—the architecture of the grave, and "carpenter," is but another name for "coffin-maker."* One man says, in his despair or indifference to life, take up a handful of the earth at your feet, and paint your house that color. Is he thinking of his last and narrow house? Toss up a copper for it as well. What an abundance of leisure he must have! Why do you take up a handful of dirt? Better paint your house your own complexion; let it turn pale or blush for you. An enterprise to improve the style of cottage architecture! When you have got my ornaments ready I will wear them.

Before winter I built a chimney, and shingled the sides of my house, which were already impervious to rain, with imperfect and sappy shingles made of the first slice of the log, whose edges I was obliged to straighten with a plane.

I have thus a tight shingled and plastered house, ten feet wide by fifteen long, and eight-feet posts, with a garret and a closet, a large window on each side, two trap doors, one door at the end, and a brick fireplace opposite. The exact cost of my house, paying the usual price for such materials as I used, but not counting the work, all of which was done by myself, was as follows; and I give the details because very few are able to tell exactly what their houses cost, and fewer still, if any, the separate cost of the various materials which compose them:—

Boards, ...	$8 03½,	mostly shanty boards.
Refuse shingles for roof and sides, ...	4 00	
Laths, ...	1 25	
Two second-hand windows with glass, ...	2 43	
One thousand old brick, . .	4 00	
Two casks of lime, ...	2 40	That was high.
Hair, ...	0 31	More than I needed.
Mantle-tree iron, ...	0 15	
Nails, ...	3 90	
Hinges and screws, . . .	0 14	
Latch, ...	0 10	
Chalk, ...	0 01	
Transportation, ...	1 40	I carried a good part on my back.
In all,	$28 12½*	

These are all the materials excepting the timber, stones and sand, which I claimed by squatter's rights. I have also a small wood-shed adjoining, made chiefly of the stuff which was left after building the house.

I intend to build me a house which will surpass any on the main street in Concord in grandeur and luxury, as soon as it pleases me as much and will cost me no more than my present one.

I thus found that the student who wishes for a shelter can obtain one for a lifetime at an expense not greater than the rent which he now pays annually. If I seem to boast more than is becoming, my excuse is that I brag for humanity rather than for myself; and my shortcomings and inconsistencies do not affect the truth of my statement. Notwithstanding much cant and hypocrisy,—chaff which I find it difficult to separate from my wheat, but for which I am as sorry as any man,—I will breathe freely and stretch myself in this respect, it is such a relief to both the moral and physical system; and I am resolved that I will not through humility become the devil's attorney. I will endeavor to speak a good word for the truth. At Cambridge College* the mere rent of a student's room, which is only a little larger than my own, is thirty dollars each year, though the corporation had the advantage of building thirty-two side by side and under one roof, and the occupant suffers the inconvenience of many and noisy neighbors, and perhaps a residence in the fourth story. I cannot but think that if we had more true wisdom in these respects, not only less education would be needed, because, forsooth, more would already have been acquired, but the pecuniary expense of getting an education would in a great measure vanish. Those conveniences which the student requires at Cambridge or elsewhere cost him or somebody else ten times as great a sacrifice of life as they would with proper management on both sides. Those things for which the most money is demanded are never the things which the student most wants. Tuition, for instance, is an important item in the term bill, while for the far more valuable education which he gets by associating with the most cultivated of his contemporaries no charge is made. The mode of founding a college is, commonly, to get up a subscription of dollars and cents, and then following blindly the principles of a division of labor to its extreme, a principle which should never be followed but with circumspection,—to call in a contractor who makes this a subject of speculation, and he employs Irishmen or other operatives actually to lay the foundations, while the students that are to be are said to be fitting themselves for it; and for these oversights successive generations have to pay. I think that it would be *better than this*, for the students, or those who desire to be benefited by it, even to lay the foundation themselves. The student who secures his coveted leisure and retirement by systematically shirking any labor necessary to man obtains but an ignoble and unprofitable leisure, defrauding himself of the experience which alone can make leisure fruitful. "But," says one, "you do not mean that the students should go to work with their hands instead of their heads?" I do not mean that exactly, but I mean something which he might think a good deal like that; I mean that they should not *play* life, or *study* it merely, while the community supports them at this expensive game, but earnestly *live* it from beginning to end. How could youths better learn to live than by at once trying the experiment of living? Methinks this would exercise their minds as much as mathematics. If I wished a boy to know something about the arts and sciences, for instance, I would not pursue the common course, which is merely to send him into the neighborhood of some professor, where any thing is professed and practised but the art of life;—to survey the world through a telescope or a microscope, and never with his natural eye; to study chemistry, and not learn how his bread is made, or mechanics, and not learn how it is earned; to discover new satellites to Neptune, and not detect the motes* in his eyes, or to what vagabond he is a satellite himself; or to be devoured by the monsters that swarm all around him, while contemplating the monsters in a drop of vinegar.* Which would have advanced the most at the end of a month,—the boy who had made his own jackknife from the ore which he had dug and smelted, reading as much as would be necessary for this,—or the boy who had attended the lectures on metallurgy at the Institute in the mean while, and had received a Rogers' penknife* from

his father? Which would be most likely to cut his
fingers? . . . To my astonishment I was informed on leaving
college that I had studied navigation!*—why, if I had taken
one turn down the harbor I should have known more about it.
Even the *poor* student studies and is taught only *political*
economy,* while that economy of living which is synonymous
with philosophy is not even sincerely professed in our colleges. The consequence is, that while he is reading Adam
Smith, Ricardo, and Say,* he runs his father in debt
irretrievably.

As with our colleges, so with a hundred "modern improvements;" there is an illusion about them; there is not always a
positive advance. The devil goes on exacting compound interest to the last for his early share and numerous succeeding
investments in them. Our inventions are wont to be pretty
toys, which distract our attention from serious things. They
are but improved means to an unimproved end, an end which
it was already but too easy to arrive at; as railroads lead to
Boston or New York. We are in great haste to construct a
magnetic telegraph from Maine to Texas; but Maine and
Texas, it may be, have nothing important to communicate.
Either is in such a predicament as the man who was earnest to
be introduced to a distinguished deaf woman, but when he
was presented, and one end of her ear trumpet was put into
his hand, had nothing to say. As if the main object were to talk
fast and not to talk sensibly. We are eager to tunnel under the
Atlantic and bring the old world some weeks nearer to the
new; but perchance the first news that will leak through into
the broad, flapping American ear will be that the Princess
Adelaide* has the whooping cough. After all, the man whose
horse trots a mile in a minute does not carry the most important messages; he is not an evangelist, nor does he come
round eating locusts and wild honey.* I doubt if Flying
Childers* ever carried a peck of corn to mill.

One says to me, "I wonder that you do not lay up money;
you love to travel; you might take the cars* and go to

Fitchburg* to-day and see the country." But I am wiser than that. I have learned that the swiftest traveller is he that goes afoot. I say to my friend, Suppose we try who will get there first. The distance is thirty miles; the fare ninety cents. That is almost a day's wages. I remember when wages were sixty cents a day for laborers on this very road. Well, I start now on foot, and get there before night; I have travelled at that rate by the week together. You will in the mean while have earned your fare, and arrive there some time to-morrow, or possibly this evening, if you are lucky enough to get a job in season. Instead of going to Fitchburg, you will be working here the greater part of the day. And so, if the railroad reached round the world, I think that I should keep ahead of you; and as for seeing the country and getting experience of that kind, I should have to cut your acquaintance altogether.

Such is the universal law, which no man can ever outwit, and with regard to the railroad even we may say it is as broad as it is long. To make a railroad round the world available to all mankind is equivalent to grading the whole surface of the planet. Men have an indistinct notion that if they keep up this activity of joint stocks and spades long enough all will at length ride somewhere, in next to no time, and for nothing; but though a crowd rushes to the depot, and the conductor shouts "All aboard!" when the smoke is blown away and the vapor condensed, it will be perceived that a few are riding, but the rest are run over,—and it will be called, and will be, "A melancholy accident."* No doubt they can ride at last who shall have earned their fare, that is, if they survive so long, but they will probably have lost their elasticity and desire to travel by that time. This spending of the best part of one's life earning money in order to enjoy a questionable liberty during the least valuable part of it, reminds me of the Englishman* who went to India to make a fortune first, in order that he might return to England and live the life of a poet. He should have gone up garret at once. "What!" exclaim a million Irish-men starting up from all the shanties in the land, "is not this railroad which we have built a good thing?" Yes, I answer, *comparatively* good, that is, you might have done worse; but I wish, as you are brothers of mine, that you could have spent your time better than digging in this dirt.

Before I finished my house, wishing to earn ten or twelve dollars by some honest and agreeable method, in order to meet my unusual expenses, I planted about two acres and a half of light and sandy soil near it chiefly with beans, but also a small part with potatoes, corn, peas, and turnips. The whole lot contains eleven acres, mostly growing up to pines and hickories, and was sold the preceding season for eight dollars and eight cents an acre. One farmer said that it was "good for nothing but to raise cheeping squirrels on." I put no manure whatever on* this land, not being the owner, but merely a squatter,* and not expecting to cultivate so much again, and I did not quite hoe it all once. I got out several cords of stumps in ploughing, which supplied me with fuel for a long time, and left small circles of virgin mould, easily distinguishable through the summer by the greater luxuriance of the beans there. The dead and for the most part unmerchantable wood behind my house, and the driftwood from the pond, have supplied the remainder of my fuel. I was obliged to hire a team and a man for the ploughing, though I held the plough myself. My farm outgoes for the first season were, for implements, seed, work, &c., $14 72½. The seed corn was given me. This never costs any thing to speak of, unless you plant more than enough. I got twelve bushels of beans, and eighteen bushels of potatoes, beside some peas and sweet corn. The yellow corn and turnips were too late to come to any thing. My whole income from the farm was

$23 44.
Deducting the outgoes, . .   14 72½
There are left, ... $ 8 71½,*

beside produce consumed and on hand at the time this estimate was made of the value of $4 50,—the amount on hand
much more than balancing a little grass which I did not raise.
All things considered, that is, considering the importance of a
man's soul and of to-day, notwithstanding the short time occupied by my experiment, nay, partly even because of its transient character, I believe that that was doing better than any
farmer in Concord did that year.

The next year I did better still, for I spaded up all the land
which I required, about a third of an acre, and I learned from
the experience of both years, not being in the least awed by
many celebrated works on husbandry, Arthur Young* among
the rest, that if one would live simply and eat only the crop
which he raised, and raise no more than he ate, and not
exchange it for an insufficient quantity of more luxurious and
expensive things, he would need to cultivate only a few rods of
ground, and that it would be cheaper to spade up that than to
use oxen to plough it, and to select a fresh spot from time to
time than to manure the old, and he could do all his necessary
farm work as it were with his left hand at odd hours in the
summer; and thus he would not be tied to an ox, or horse, or
cow, or pig, as at present. I desire to speak impartially on this
point, and as one not interested in the success or failure of the
present economical and social arrangements. I was more independent than any farmer in Concord, for I was not anchored to a house or farm, but could follow the bent of my
genius, which is a very crooked one, every moment. Beside
being better off than they already, if my house had been
burned or my crops had failed, I should have been nearly as
well off as before.

I am wont to think that men are not so much the keepers of
herds as herds are the keepers of men, the former are so
much the freer. Men and oxen exchange work; but if we
consider necessary work only, the oxen will be seen to have
greatly the advantage, their farm is so much the larger. Man

does some of his part of the exchange work in his six weeks of haying, and it is no boy's play. Certainly no nation that lived simply in all respects, that is, no nation of philosophers, would commit so great a blunder as to use the labor of animals. True, there never was and is not likely soon to be a nation of philosophers, nor am I certain it is desirable that there should be. However, *I* should never have broken a horse or bull and taken him to board for any work he might do for me, for fear I should become a horse-man or a herds-man merely; and if society seems to be the gainer by so doing, are we certain that what is one man's gain is not another's loss, and that the stable-boy has equal cause with his master to be satisfied? Granted that some public works would not have been constructed without this aid, and let man share the glory of such with the ox and horse; does it follow that he could not have accomplished works yet more worthy of himself in that case? When men begin to do, not merely unnecessary or artistic, but luxurious and idle work, with their assistance, it is inevitable that a few do all the exchange work with the oxen, or, in other words, become the slaves of the strongest. Man thus not only works for the animal within him, but, for a symbol of this, he works for the animal without him. Though we have many substantial houses of brick or stone, the prosperity of the farmer is still measured by the degree to which the barn overshadows the house. This town is said to have the largest houses for oxen, cows, and horses hereabouts, and it is not behindhand in its public buildings; but there are very few halls for free worship or free speech* in this county. It should not be by their architecture, but why not even by their power of abstract thought, that nations should seek to commemorate themselves? How much more admirable the Bhagvat-Geeta* than all the ruins of the East! Towers and temples are the luxury of princes. A simple and independent mind does not toil at the bidding of any prince. Genius is not a retainer to any emperor, nor is its material silver, or gold, or marble, except to a trifling extent. To what end, pray, is so much stone hammered? In Arcadia,* when I was there, I did not see any hammering stone. Nations are possessed with an insane ambition to perpetuate the memory of themselves by the amount of hammered stone they leave. What if equal pains were taken to smooth and polish their manners? One piece of good sense would be more memorable than a monument as high as the moon. I love better to see stones in place. The grandeur of Thebes* was a vulgar grandeur. More sensible is a rod* of stone wall that bounds an honest man's field than a hundred-gated Thebes that has wandered farther from the true end of life. The religion and civilization which are barbaric and heathenish build splendid temples; but what you might call Christianity does not. Most of the stone a nation hammers goes toward its tomb only. It buries itself alive. As for the Pyramids, there is nothing to wonder at in them so much as the fact that so many men could be found degraded enough to spend their lives constructing a tomb for some ambitious booby, whom it would have been wiser and manlier to have drowned in the Nile, and then given his body to the dogs. I might possibly invent some excuse for them and him, but I have no time for it. As for the religion and love of art of the builders, it is much the same all the world over, whether the building be an Egyptian temple or the United States Bank.* It costs more than it comes to. The mainspring is vanity, assisted by the love of garlic and bread and butter.* Mr. Balcom, a promising young architect, designs it on the back of his Vitruvius,* with hard pencil and ruler, and the job is let out to Dobson & Sons,* stonecutters. When the thirty centuries begin to look down on it, mankind begin to look up at it. As for your high towers and monuments, there was a crazy fellow once in this town who undertook to dig through to China, and he got so far that, as he said, he heard the Chinese pots and kettles rattle; but I think that I shall not go out of my way to admire the hole which he made. Many are concerned about the monuments of the West and the East,—to know who built them. For my part, I should like to know who in those days did

not build them,—who were above such trifling. But to pro-
ceed with my statistics.

By surveying, carpentry, and day-labor of various other
kinds in the village in the mean while, for I have as many
trades as fingers, I had earned $13 34. The expense of food for
eight months, namely, from July 4th to March 1st, the time
when these estimates were made, though I lived there more
than two years,—not counting potatoes, a little green corn,
and some peas, which I had raised, nor considering the value
of what was on hand at the last date, was

Rice,	$1 73½	
Molasses,	1 73	Cheapest form of the saccharine.
Rye meal,	1 04¾	
Indian meal,	0 99¾	Cheaper than rye.
Pork,	0 22	
Flour,	0 88	Costs more than Indian meal, both money and trouble.
Sugar,	0 80	
Lard,	0 65	
Apples,	0 25	
Dried apple,	0 22	
Sweet potatoes,	0 10	
One pumpkin,	0 6	
One watermelon,	0 2	
Salt,	0 3*	

All experiments which failed.

Yes, I did eat $8 74, all told; but I should not thus
unblushingly publish my guilt, if I did not know that most of
my readers were equally guilty with myself, and that their
deeds would look no better in print. The next year I sometimes caught a mess of fish for my dinner, and once I went so
far as to slaughter a woodchuck which ravaged my bean-
field,—effect his transmigration, as a Tartar* would say,—and
devour him, partly for experiment's sake; but though it afforded me a momentary enjoyment, notwithstanding a musky
flavor, I saw that the longest use would not make that a good

practice, however it might seem to have your woodchucks ready dressed by the village butcher.

Clothing and some incidental expenses within the same dates, though little can be inferred from this item, amounted to

| | $8 40¾ |
| Oil* and some household utensils, . . . | 2 00 |

So that all the pecuniary outgoes, excepting for washing and mending,* which for the most part were done out of the house, and their bills have not yet been received,—and these are all and more than all the ways by which money necessarily goes out in this part of the world,—were

House, ...	$28 12½
Farm one year, ...	14 72½
Food eight months, ...	8 74
Clothing, &c., eight months, . . .	8 40¾
Oil, &c., eight months, ...	2 00
In all,	$61 99¾

I address myself now to those of my readers who have a living to get. And to meet this I have for farm produce sold

	$23 44
Earned by day-labor, ...	13 34
In all, ...	$36 78,

which subtracted from the sum of the outgoes leaves a balance of $25 21¾ on the one side,—this being very nearly the means with which I started, and the measure of expenses to be incurred,—and on the other, beside the leisure and independence and health thus secured, a comfortable house for me as long as I choose to occupy it.

These statistics, however accidental and therefore

uninstructive they may appear, as they have a certain completeness,* have a certain value also. Nothing was given me of which I have not rendered some account. It appears from the above estimate, that my food alone cost me in money about twenty-seven cents a week. It was, for nearly two years after this, rye and Indian meal without yeast, potatoes, rice, a very little salt pork, molasses, and salt, and my drink water. It was fit that I should live on rice, mainly, who loved so well the philosophy of India. To meet the objections of some inveterate cavillers, I may as well state, that if I dined out occasionally, as I always had done, and I trust shall have opportunities to do again, it was frequently to the detriment of my domestic arrangements. But the dining out, being, as I have stated, a constant element, does not in the least affect a comparative statement like this.

I learned from my two years' experience that it would cost incredibly little trouble to obtain one's necessary food, even in this latitude; that a man may use as simple a diet as the animals, and yet retain health and strength. I have made a satisfactory dinner, satisfactory on several accounts, simply off a dish of purslane (*Portulaca oleracea*) which I gathered in my cornfield, boiled and salted. I give the Latin on account of the savoriness of the trivial name. And pray what more can a reasonable man desire, in peaceful times, in ordinary noons, than a sufficient number of ears of green sweet-corn boiled, with the addition of salt? Even the little variety which I used was a yielding to the demands of appetite, and not of health. Yet men have come to such a pass that they frequently starve, not for want of necessaries, but for want of luxuries; and I know a good woman who thinks that her son lost his life because he took to drinking water* only.

The reader will perceive that I am treating the subject rather from an economic than a dietetic point of view, and he will not venture to put my abstemiousness to the test unless he has a well-stocked larder.

Bread I at first made of pure Indian meal and salt, genuine

hoe-cakes,* which I baked before my fire out of doors on a shingle or the end of a stick of timber sawed off in building my house; but it was wont to get smoked and to have a piny flavor. I tried flour also; but have at last found a mixture of rye and Indian meal most convenient and agreeable. In cold weather it was no little amusement to bake several small loaves of this in succession, tending and turning them as carefully as an Egyptian his hatching eggs.* They were a real cereal fruit which I ripened, and they had to my senses a fragrance like that of other noble fruits, which I kept in as long as possible by wrapping them in cloths. I made a study of the ancient and indispensable art of bread-making, consulting such authorities as offered, going back to the primitive days and first invention of the unleavened kind, when from the wildness of nuts and meats men first reached the mildness and refinement of this diet, and travelling gradually down in my studies through that accidental souring of the dough* which, it is supposed, taught the leavening process, and through the various fermentations thereafter, till I came to "good, sweet, wholesome bread," the staff of life. Leaven, which some deem the soul of bread, the *spiritus** which fills its cellular tissue, which is religiously preserved like the vestal fire,*—some precious bottle-full, I suppose, first brought over in the Mayflower, did the business for America, and its influence is still rising, swelling, spreading, in cerealian* billows over the land,—this seed I regularly and faithfully procured from the village, till at length one morning I forgot the rules, and scalded my yeast; by which accident I discovered that even this was not indispensable,—for my discoveries were not by the synthetic but analytic process,—and I have gladly omitted it since, though most housewives earnestly assured me that safe and wholesome bread without yeast might not be, and elderly people prophesied a speedy decay of the vital forces. Yet I find it not to be an essential ingredient, and after going without it for a year am still in the land of the living; and I am glad to escape the trivialness of carrying a bottle-full in my pocket, which would sometimes pop and discharge its contents to my discomfiture. It is simpler and more respectable to omit it. Man is an animal who more than any other can adapt himself to all climates and circumstances. Neither did I put any sal soda, or other acid or alkali,* into my bread. It would seem that I made it according to the recipe which Marcus Porcius Cato gave about two centuries before Christ. "Panem depsticium sic facito. Manus mortariumque bene lavato. Farinam in mortarium indito, aquæ paulatim addito, subigitoque pulchre. Ubi bene subegeris, defingito, coquitoque sub testu."* Which I take to mean—"Make kneaded bread thus. Wash your hands and trough well. Put the meal into the trough, add water gradually, and knead it thoroughly. When you have kneaded it well, mould it, and bake it under a cover," that is, in a baking-kettle. Not a word about leaven. But I did not always use this staff of life. At one time, owing to the emptiness of my purse, I saw none of it for more than a month.

Every New Englander might easily raise all his own breadstuffs in this land of rye and Indian corn, and not depend on distant and fluctuating markets for them. Yet so far are we from simplicity and independence that, in Concord, fresh and sweet meal is rarely sold in the shops, and hominy* and corn in a still coarser form are hardly used by any. For the most part the farmer gives to his cattle and hogs the grain of his own producing, and buys flour, which is at least no more wholesome, at a greater cost, at the store. I saw that I could easily raise my bushel or two of rye and Indian corn, for the former will grow on the poorest land, and the latter does not require the best, and grind them in a hand-mill, and so do without rice and pork; and if I must have some concentrated sweet, I found by experiment that I could make a very good molasses either of pumpkins or beets,* and I knew that I needed only to set out a few maples to obtain it more easily still, and while these were growing I could use various substitutes beside those which I have named. "For," as the Fore-fathers sang,—

> "we can make liquor to sweeten our lips
> Of pumpkins and parsnips and walnut-tree chips."*

Finally, as for salt, that grossest of groceries, to obtain this might be a fit occasion for a visit to the seashore, or, if I did without it altogether, I should probably drink the less water. I do not learn that the Indians ever troubled themselves to go after it.

Thus I could avoid all trade and barter, so far as my food was concerned, and having a shelter already, it would only remain to get clothing and fuel. The pantaloons which I now wear were woven in a farmer's family,—thank Heaven there is so much virtue still in man; for I think the fall from the farmer to the operative* as great and memorable as that from the man to the farmer;—and in a new country fuel is an encumbrance.* As for a habitat, if I were not permitted still to squat, I might purchase one acre at the same price for which the land I cultivated was sold—namely, eight dollars and eight cents. But as it was, I considered that I enhanced the value of the land by squatting on it.

There is a certain class of unbelievers who sometimes ask me such questions as, if I think that I can live on vegetable food alone; and to strike at the root of the matter at once,—for the root is faith,—I am accustomed to answer such, that I can live on board nails. If they cannot understand that, they cannot understand much that I have to say. For my part, I am glad to hear of experiments of this kind being tried; as that a young man tried for a fortnight to live on hard, raw corn on the ear, using his teeth for all mortar. The squirrel tribe tried the same and succeeded. The human race is interested in these experiments, though a few old women who are incapacitated for them, or who own their thirds in mills,* may be alarmed.

My furniture,* part of which I made myself, and the rest cost me nothing of which I have not rendered an account, consisted of a bed, a table, a desk, three chairs, a looking-glass three inches in diameter, a pair of tongs and andirons, a kettle, a skillet, and a frying-pan, a dipper, a wash-bowl, two knives and forks, three plates, one cup, one spoon, a jug for oil, a jug for molasses, and a japanned lamp. None is so poor that he need sit on a pumpkin. That is shiftlessness. There is a plenty of such chairs as I like best in the village garrets to be had for taking them away. Furniture! Thank God, I can sit and I can stand without the aid of a furniture warehouse. What man but a philosopher would not be ashamed to see his furniture packed in a cart and going up country exposed to the light of heaven and the eyes of men, a beggarly account of empty boxes? That is Spaulding's furniture. I could never tell from inspecting such a load whether it belonged to a so called rich man or a poor one; the owner always seemed poverty-stricken. Indeed, the more you have of such things the poorer you are. Each load looks as if it contained the contents of a dozen shanties; and if one shanty is poor, this is a dozen times as poor. Pray, for what do we *move* ever but to get rid of our furniture, our *exuviæ;*￼* at last to go from this world to another newly furnished, and leave this to be burned? It is the same as if all these traps were buckled to a man's belt, and he could not move over the rough country where our lines are cast without dragging them,—dragging his trap. He was a lucky fox that left his tail in the trap. The muskrat will gnaw his third leg off to be free.* No wonder man has lost his elasticity. How often he is at a dead set! "Sir, if I may be so bold, what do you mean by a dead set?"* If you are a seer, whenever you meet a man you will see all that he owns, ay, and much that he pretends to disown, behind him, even to his kitchen furniture and all the trumpery which he saves and will not burn, and he will appear to be harnessed to it and making what headway he can. I think that the man is at a dead set who has got through a knot hole or gateway where his sledge load of furniture cannot follow him. I cannot but feel compassion when I hear some trig,* compact-looking man, seemingly free, all girded and ready, speak of his "furniture," as whether it is insured or not. "But what shall I do with my furniture?" My gay butterfly is entangled in a spider's web then. Even those who seem for a long while not to have any, if you inquire more narrowly you will find have some stored in somebody's barn. I look upon England today as an old gentleman who is travelling with a great deal of baggage, trumpery which has accumulated from long housekeeping, which he has not the courage to burn; great trunk, little trunk, bandbox and bundle. Throw away the first three at least. It would surpass the powers of a well man nowadays to take up his bed and walk,* and I should certainly advise a sick one to lay down his bed and run. When I have met an immigrant tottering under a bundle which contained his all—looking like an enormous wen which had grown out of the nape of his neck—I have pitied him, not because that was his all, but because he had all *that* to carry. If I have got to drag my trap, I will take care that it be a light one and do not nip me in a vital part. But perchance it would be wisest never to put one's paw into it.

I would observe, by the way, that it costs me nothing for curtains, for I have no gazers to shut out but the sun and moon, and I am willing that they should look in. The moon will not sour milk nor taint meat of mine,* nor will the sun injure my furniture or fade my carpet,* and if he is sometimes too warm a friend, I find it still better economy to retreat behind some curtain which nature has provided, than to add a single item to the details of housekeeping. A lady once offered me a mat, but as I had no room to spare within the house, nor time to spare within or without to shake it, I declined it, preferring to wipe my feet on the sod before my door. It is best to avoid the beginnings of evil.

Not long since I was present at the auction of a deacon's effects, for his life had not been ineffectual:—

"The evil that men do lives after them."*

As usual, a great proportion was trumpery which had begun to accumulate in his father's day. Among the rest was a dried tapeworm. And now, after lying half a century in his garret and other dust holes, these things were not burned; instead of a *bonfire*,* or purifying destruction of them, there was an *auction*,* or increasing of them. The neighbors eagerly collected to view them, bought them all, and carefully transported them to their garrets and dust holes,* to lie there till their estates are settled, when they will start again. When a man dies he kicks the dust.*

The customs of some savage nations might, perchance, be profitably imitated by us, for they at least go through the semblance of casting their slough annually; they have the idea of the thing, whether they have the reality or not. Would it not be well if we were to celebrate such a "busk," or "feast of first fruits," as Bartram* describes to have been the custom of the Mucclasse Indians? "When a town celebrates the busk," says he, "having previously provided themselves with new clothes, new pots, pans, and other household utensils and furniture, they collect all their worn out clothes and other despicable things, sweep and cleanse their houses, squares, and the whole town, of their filth, which with all the remaining grain and other old provisions they cast together into one common heap, and consume it with fire. After having taken medicine, and fasted for three days, all the fire in the town is extinguished. During this fast they abstain from the gratification of every appetite and passion whatever. A general amnesty is proclaimed; all malefactors may return to their town.—"

"On the fourth morning, the high priest, by rubbing dry wood together, produces new fire in the public square, from whence every habitation in the town is supplied with the new and pure flame."

They then feast on the new corn and fruits and dance and sing for three days, "and the four following days they receive visits and rejoice with their friends from neighboring towns who have in like manner purified and prepared themselves."

The Mexicans also practised a similar purification at the end of every fifty-two years, in the belief that it was time for the world to come to an end.

I have scarcely heard of a truer sacrament, that is, as the dictionary defines it,* "outward and visible sign of an inward and spiritual grace," than this, and I have no doubt that they were originally inspired directly from Heaven to do thus, though they have no biblical record of the revelation.

For more than five years I maintained myself thus solely by the labor of my hands, and I found, that by working* about six weeks in a year, I could meet all the expenses of living. The whole of my winters, as well as most of my summers, I had free and clear for study. I have thoroughly tried school-keeping, and found that my expenses were in proportion, or rather out of proportion, to my income, for I was obliged to dress and train, not to say think and believe, accordingly, and I lost my time into the bargain. As I did not teach for the good of my fellow-men, but simply for a livelihood, this was a failure. I have tried trade; but I found that it would take ten years to get under way in that, and that then I should probably be on my way to the devil. I was actually afraid that I might by that time be doing what is called a good business. When formerly I was looking about to see what I could do for a living, some sad experience in conforming to the wishes of friends being fresh in my mind to tax my ingenuity, I thought often and seriously of picking huckleberries;* that surely I could do, and its small profits might suffice,—for my greatest skill has been to want but little,—so little capital it required, so little distraction from my wonted moods, I foolishly thought. While my acquaintances went unhesitatingly into trade or the professions, I contemplated this occupation as most like theirs; ranging the hills all summer to pick the berries which came in my way, and thereafter carelessly dispose of them; so, to keep the flocks of Admetus.* I also dreamed that I might gather the wild herbs, or carry evergreens to such villagers as loved to be reminded of the woods, even to the city, by hay-cart loads. But I have since learned that trade curses every thing it handles; and though you trade in messages from heaven, the whole curse of trade attaches to the business.

As I preferred some things to others, and especially valued my freedom, as I could fare hard and yet succeed well, I did not wish to spend my time in earning rich carpets or other fine furniture, or delicate cookery, or a house in the Grecian or the Gothic* style just yet. If there are any to whom it is no interruption to acquire these things, and who know how to use them when acquired, I relinquish to them the pursuit. Some are "industrious," and appear to love labor for its own sake, or perhaps because it keeps them out of worse mischief; to such I have at present nothing to say. Those who would not know what to do with more leisure than they now enjoy, I might advise to work twice as hard as they do,—work till they pay for themselves, and get their free papers.* For myself I found that the occupation of a day-laborer was the most independent of any, especially as it required only thirty or forty days in a year to support one. The laborer's day ends with the going down of the sun, and he is then free to devote himself to his chosen pursuit, independent of his labor; but his employer, who speculates from month to month, has no respite from one end of the year to the other.

In short, I am convinced, both by faith and experience, that to maintain one's self on this earth is not a hardship but a pastime, if we will live simply and wisely; as the pursuits of the simpler nations are still the sports of the more artificial. It is not necessary that a man should earn his living by the sweat of his brow, unless he sweats easier than I do.*

One young man of my acquaintance, who has inherited some acres, told me that he thought he should live as I did, *if he had the means*. I would not have any one adopt *my* mode of living on any account; for, beside that before he has fairly learned it I may have found out another for myself, I desire that there may be as many different persons in the world as possible; but I would have each one be very careful to find out and pursue *his own* way, and not his father's or his mother's or his neighbor's instead. The youth may build or plant or sail, only let him not be hindered from doing that which he tells me he would like to do. It is by a mathematical point only that we are wise, as the sailor or the fugitive slave keeps the pole-star in his eye;* but that is sufficient guidance for all our life. We may not arrive at our port within a calculable period, but we would preserve the true course.

Undoubtedly, in this case, what is true for one is truer still for a thousand, as a large house is not proportionally more expensive than a small one,* since one roof may cover, one cellar underlie, and one wall separate several apartments. But for my part, I preferred the solitary dwelling. Moreover, it will commonly be cheaper to build the whole yourself than to convince another of the advantage of the common wall; and when you have done this, the common partition, to be much cheaper, must be a thin one, and that other may prove a bad neighbor, and also not keep his side in repair. The only coöperation which is commonly possible is exceedingly partial and superficial; and what little true coöperation there is, is as if it were not, being a harmony inaudible to men. If a man has faith he will coöperate with equal faith every where; if he has not faith, he will continue to live like the rest of the world, whatever company he is joined to. To coöperate, in the highest as well as the lowest sense, means *to get our living together*. I heard it proposed lately that two young men should travel together over the world, the one without money, earning his means as he went, before the mast and behind the plough, the other carrying a bill of exchange in his pocket. It was easy to see that they could not long be companions or coöperate, since one would not *operate* at all.* They would part at the first interesting crisis in their adventures. Above all, as I have implied, the man who goes alone can start to-day; but he who travels with another must wait till that other is ready, and it may be a long time before they get off.

But all this is very selfish, I have heard some of my townsmen say. I confess that I have hitherto indulged very little in philanthropic enterprises. I have made some sacrifices to a sense of duty, and among others have sacrificed this pleasure also. There are those who have used all their arts to persuade me to undertake the support of some poor family in the town; and if I had nothing to do,—for the devil finds employment for the idle,*—I might try my hand at some such pastime as that. However, when I have thought to indulge myself in this respect, and lay their Heaven under an obligation by maintaining certain poor persons in all respects as comfortably as I maintain myself, and have even ventured so far as to make them the offer, they have one and all unhesitatingly preferred to remain poor. While my townsmen and women are devoted in so many ways to the good of their fellows, I trust that one at least may be spared to other and less humane pursuits. You must have a genius for charity as well as for any thing else. As for Doing-good, that is one of the professions which are full. Moreover, I have tried it fairly, and, strange as it may seem, am satisfied that it does not agree with my constitution. Probably I should not consciously and deliberately forsake my particular calling to do the good which society demands of me, to save the universe from annihilation; and I believe that a like but infinitely greater steadfastness elsewhere is all that now preserves it. But I would not stand between any man and his genius; and to him who does this work, which I decline, with his whole heart and soul and life, I would say, Persevere, even if the world call it doing evil, as it is most likely they will.

I am far from supposing that my case is a peculiar one; no doubt many of my readers would make a similar defence. At doing something,—I will not engage that my neighbors shall pronounce it good,—I do not hesitate to say that I should be a capital fellow to hire; but what that is, it is for my employer to find out. What *good* I do, in the common sense of that word, must be aside from my main path, and for the most part wholly unintended. Men say, practically, Begin where you are and such as you are, without aiming mainly to become of more worth, and with kindness aforethought go about doing good. If I were to preach at all in this strain, I should say rather, Set about being good. As if the sun should stop when he had kindled his fires up to the splendor of a moon or a star of the sixth magnitude, and go about like a Robin Goodfellow,* peeping in at every cottage window, inspiring lunatics, and tainting meats, and making darkness visible,* instead of steadily increasing his genial heat and beneficence till he is of such brightness that no mortal can look him in the face, and then, and in the mean while too, going about the world in his own orbit, doing it good, or rather, as a truer philosophy* has discovered, the world going about him getting good. When Phaeton,* wishing to prove his heavenly birth by his beneficence, had the sun's chariot but one day, and drove out of the beaten track, he burned several blocks of houses in the lower streets of heaven, and scorched the surface of the earth, and dried up every spring, and made the great desert of Sahara, till at length Jupiter hurled him headlong to the earth with a thunderbolt, and the sun, through grief at his death, did not shine for a year.

There is no odor so bad as that which arises from goodness tainted. It is human, it is divine, carrion. If I knew for a certainty that a man was coming to my house with the conscious design of doing me good, I should run for my life, as from that dry and parching wind of the African deserts called the simoom, which fills the mouth and nose and ears and eyes with dust till you are suffocated, for fear that I should get some of his good done to me,—some of its virus mingled with my blood. No,—in this case I would rather suffer evil the natural way. A man is not a good *man* to me because he will feed me if I should be starving, or warm me if I should be freezing, or pull me out of a ditch if I should ever fall into one. I can find you a Newfoundland dog that will do as much. Philanthropy is not love for one's fellow-man in the broadest sense. Howard* was no doubt an exceedingly kind and worthy man in his way, and has his reward; but, comparatively speaking, what are a hundred Howards to *us*, if their philanthropy do not help *us* in our best estate, when we are most worthy to be helped? I never heard of a philanthropic meeting in which it was sincerely proposed to do any good to me, or the like of me.

The Jesuits were quite balked by those Indians who, being burned at the stake, suggested new modes of torture to their tormentors. Being superior to physical suffering, it sometimes chanced that they were superior to any consolation which the missionaries could offer; and the law to do as you would be done by fell with less persuasiveness on the ears of those, who, for their part, did not care how they were done by, who loved their enemies after a new fashion, and came very near freely forgiving them all they did.

Be sure that you give the poor the aid they most need, though it be your example which leaves them far behind. If you give money, spend yourself with it, and do not merely abandon it to them. We make curious mistakes sometimes. Often the poor man is not so cold and hungry as he is dirty and ragged and gross. It is partly his taste, and not merely his misfortune. If you give him money, he will perhaps buy more rags with it. I was wont to pity the clumsy Irish laborers who cut ice* on the pond, in such mean and ragged clothes, while I shivered in my more tidy and somewhat more fashionable garments, till, one bitter cold day, one who had slipped into the water came to my house to warm him, and I saw him strip off three pairs of pants and two pairs of stockings ere he got down to the skin, though they were dirty and ragged enough, it is true, and that he could afford to refuse the *extra* garments which I offered him, he had so many *intra** ones. This ducking was the very thing he needed. Then I began to pity myself, and I saw that it would be a greater charity to bestow on me a flannel shirt than a whole slop-shop on him. There are a thousand hacking at the branches of evil to one who is striking at the root, and it may be that he who bestows the largest amount of time and money on the needy is doing the most by his mode of life to produce that misery which he strives in vain to relieve. It is the pious slave-breeder devoting the proceeds of every tenth slave to buy a Sunday's liberty for the rest. Some show their kindness to the poor by employing them in their kitchens. Would they not be kinder if they employed themselves there? You boast of spending a tenth part* of your income in charity; may be you should spend the nine tenths so, and done with it. Society recovers only a tenth part of the property then. Is this owing to the generosity of him in whose possession it is found, or to the remissness of the officers of justice?

Philanthropy is almost the only virtue which is sufficiently appreciated by mankind. Nay, it is greatly overrated; and it is our selfishness which overrates it. A robust poor man, one sunny day here in Concord, praised a fellow-townsman to me, because, as he said, he was kind to the poor; meaning himself. The kind uncles and aunts of the race are more esteemed than its true spiritual fathers and mothers. I once heard a reverend lecturer on England, a man of learning and intelligence, after enumerating her scientific, literary, and political worthies, Shakspeare, Bacon, Cromwell, Milton, Newton, and others, speak next of her Christian heroes, whom, as if his profession required it of him, he elevated to a place far above all the rest, as the greatest of the great. They were Penn, Howard, and Mrs. Fry.* Every one must feel the falsehood and cant of this. The last were not England's best men and women; only, perhaps, her best philanthropists.

I would not subtract any thing from the praise that is due to philanthropy, but merely demand justice for all who by their lives and works are a blessing to mankind. I do not value chiefly a man's uprightness and benevolence, which are, as it were, his stem and leaves. Those plants of whose greenness withered we make herb tea for the sick, serve but a humble use, and are most employed by quacks. I want the flower and fruit of a man; that some fragrance be wafted over from him to me, and some ripeness flavor our intercourse. His goodness must not be a partial and transitory act, but a constant superfluity, which costs him nothing and of which he is unconscious. This is a charity that hides a multitude of sins.* The philanthropist too often surrounds mankind with the remembrance of his own cast-off griefs as an atmosphere, and calls it sympathy. We should impart our courage, and not our despair, our health and ease, and not our disease, and take care that this does not spread by contagion. From what southern plains comes up the voice of wailing? Under what latitudes reside the heathen to whom we would send light? Who is that intemperate and brutal man whom we would redeem? If any thing ail a man, so that he does not perform his functions, if he have a pain in his bowels even,—for that is the seat of sympathy,—he forthwith sets about reforming—the world. Being a microcosm himself, he discovers, and it is a true discovery, and he is the man to make it,—that the world has been eating green apples; to his eyes, in fact, the globe itself is a great green apple, which there is danger awful to think of that the children of men will nibble before it is ripe; and straightway his drastic philanthropy seeks out the Esquimaux and the Patagonian, and embraces the populous Indian and Chinese villages; and thus, by a few years of philanthropic activity, the powers in the mean while using him for their own ends, no doubt, he cures himself of his dyspepsia, the globe acquires a faint blush on one or both of its cheeks, as if it were beginning to be ripe, and life loses its crudity and is once more sweet and wholesome to live. I never dreamed of any enormity greater than I have committed. I never knew, and never shall know, a worse man than myself.

I believe that what so saddens the reformer is not his

sympathy with his fellows in distress, but, though he be the holiest son of God, is his private ail. Let this be righted, let the spring come to him, the morning rise over his couch, and he will forsake his generous companions without apology. My excuse for not lecturing against the use of tobacco is, that I never chewed it; that is a penalty which reformed tobacco-chewers have to pay; though there are things enough I have chewed, which I could lecture against. If you should ever be betrayed into any of these philanthropies, do not let your left hand* know what your right hand does, for it is not worth knowing. Rescue the drowning and tie your shoe-strings. Take your time, and set about some free labor.

Our manners have been corrupted by communication with the saints.* Our hymn-books resound with a melodious curs-ing of God and enduring him forever. One would say that even the prophets and redeemers had rather consoled the fears than confirmed the hopes of man. There is nowhere recorded a simple and irrepressible satisfaction with the gift of life, any memorable praise of God. All health and success does me good, however far off and withdrawn it may appear; all disease and failure helps to make me sad and does me evil, however much sympathy it may have with me or I with it. If, then, we would indeed restore mankind by truly Indian, bo-tanic, magnetic, or natural means, let us first be as simple and well as Nature ourselves, dispel the clouds which hang over our own brows, and take up a little life into our pores. Do not stay to be an overseer of the poor, but endeavor to become one of the worthies of the world.

I read in the Gulistan, or Flower Garden, of Sheik Sadi of Shiraz, that "They asked a wise man, saying; Of the many celebrated trees which the Most High God has created lofty and umbrageous, they call none azad, or free, excepting the cypress, which bears no fruit; what mystery is there in this? He replied; Each has its appropriate produce, and appointed season, during the continuance of which it is fresh and bloom-ing, and during their absence dry and withered; to neither of which states is the cypress exposed, being always flourishing; and of this nature are the azads, or religious independents.— Fix not thy heart on that which is transitory; for the Dijlah, or Tigris, will continue to flow through Bagdad after the race of caliphs is extinct; if thy hand has plenty, be liberal as the date tree; but if it affords nothing to give away, be an azad, or free man, like the cypress."*

COMPLEMENTAL VERSES

THE PRETENSIONS OF POVERTY

"Thou dost presume too much, poor needy wretch,
   To claim a station in the firmament,
   Because thy humble cottage, or thy tub,
Nurses some lazy or pedantic virtue
   In the cheap sunshine or by shady springs,
With roots and pot-herbs; where thy right hand,
Tearing those humane passions from the mind,
Upon whose stocks fair blooming virtues flourish,
Degradeth nature, and benumbeth sense,
And, Gorgon-like, turns active men to stone,
   We not require the dull society
   Of your necessitated temperance,
   Or that unnatural stupidity
That knows nor joy nor sorrow; nor your fore'd
Falsely exalted passive fortitude
Above the active. This low abject brood,
That fix their seats in mediocrity,
Become your servile minds; but we advance
Such virtues only as admit excess,
Brave, bounteous acts, regal magnificence,
All-seeing prudence, magnanimity
That knows no bound, and that heroic virtue
For which antiquity hath left no name,
But patterns only, such as Hercules,
Achilles, Theseus. Back to thy loath'd cell;
And when thou seest the new enlightened sphere,
Study to know but what those worthies were."

T. CAREW.

COMPLEMENTAL VERSES

THE PRETENSIONS OF POVERTY

"Thou dost presume too much, poor needy wretch,
To claim a station in the firmament,
Because thy humble cottage, or thy tub,
Nurses some lazy or pedantic virtue
In the cheap sunshine or by shady springs,
With roots and pot-herbs; where thy right hand,
Tearing those humane passions from the mind,
Upon whose stocks fair blooming virtues flourish,
Degradeth nature, and benumbeth sense,
And, Gorgon-like, turns active men to stone.
We not require the dull society
Of your necessitated temperance,
Or that unnatural stupidity
That knows nor joy nor sorrow; nor your forc'd
Falsely exalted passive fortitude
Above the active. This low abject brood,
That fix their seats in mediocrity,
Become your servile minds; but we advance
Such virtues only as admit excess,
Brave, bounteous acts, regal magnificence,
All-seeing prudence, magnanimity
That knows no bound, and that heroic virtue
For which antiquity hath left no name,
But patterns only, such as Hercules,
Achilles, Theseus. Back to thy loath'd cell;
And when thou seest the new enlightened sphere,
Study to know but what those worthies were."

T. CAREW.*

WHERE I LIVED, AND
WHAT I LIVED FOR

AT a certain season of our life we are accustomed to consider
every spot as the possible site of a house. I have thus surveyed
the country on every side within a dozen miles of where I
live. In imagination I have bought all the farms in succession,
for all were to be bought, and I knew their price. I walked
over each farmer's premises, tasted his wild apples, dis-
coursed on husbandry with him, took his farm at his price,
at any price, mortgaging it to him in my mind; even put a
higher price on it,—took every thing but a deed of it,—took
his word for his deed, for I dearly love to talk,—cultivated
it, and him too to some extent, I trust, and withdrew when
I had enjoyed it long enough, leaving him to carry it on.
This experience entitled me to be regarded as a sort of
real-estate broker by my friends. Wherever I sat, there I
might live, and the landscape radiated from me accordingly.
What is a house but a *sedes*, a seat?—better if a country seat.
I discovered many a site for a house not likely to be soon
improved, which some might have thought too far from
the village, but to my eyes the village was too far from it.
Well, there I might live, I said; and there I did live, for an
hour, a summer and a winter life; saw how I could let the
years run off, buffet the winter through, and see the spring
come in. The future inhabitants of this region, wherever
they may place their houses, may be sure that they have
been anticipated. An afternoon sufficed to lay out the land
into orchard, woodlot, and pasture, and to decide what fine
oaks or pines should be left to stand before the door, and
whence each blasted tree could be seen to the best advantage;
and then I let it lie, fallow perchance, for a man is rich in

proportion to the number of things which he can afford to let alone.

My imagination carried me so far that I even had the refusal of several farms,—the refusal was all I wanted,—but I never got my fingers burned by actual possession. The nearest that I came to actual possession was when I bought the Hollowell place,* and had begun to sort my seeds, and collected materials with which to make a wheelbarrow to carry it on or off with; but before the owner gave me a deed of it, his wife— every man has such a wife—changed her mind and wished to keep it, and he offered me ten dollars to release him. Now, to speak the truth, I had but ten cents in the world, and it surpassed my arithmetic to tell, if I was that man who had ten cents, or who had a farm, or ten dollars, or all together. However, I let him keep the ten dollars and the farm too, for I had carried it far enough; or rather, to be generous, I sold him the farm for just what I gave for it, and, as he was not a rich man, made him a present of ten dollars, and still had my ten cents, and seeds, and materials for a wheelbarrow left. I found thus that I had been a rich man without any damage to my poverty. But I retained the landscape, and I have since annually carried off what it yielded without a wheelbarrow. With respect to landscapes,—

> "I am monarch of all I *survey*,
> My right there is none to dispute."*

I have frequently seen a poet withdraw, having enjoyed the most valuable part of a farm, while the crusty farmer supposed that he had got a few wild apples only. Why, the owner does not know it for many years when a poet has put his farm in rhyme, the most admirable kind of invisible fence, has fairly impounded it, milked it, skimmed it, and got all the cream, and left the farmer only the skimmed milk.

The real attractions of the Hollowell farm, to me, were; its complete retirement, being about two miles from the village, half a mile from the nearest neighbor, and separated from the highway by a broad field; its bounding on the river, which the owner said protected it by its fogs from frosts in the spring, though that was nothing to me; the gray color and ruinous state of the house and barn, and the dilapidated fences, which put such an interval between me and the last occupant; the hollow and lichen-covered apple trees, gnawed by rabbits, showing what kind of neighbors I should have; but above all, the recollection I had of it from my earliest voyages up the river, when the house was concealed behind a dense grove of red maples, through which I heard the house-dog bark. I was in haste to buy it, before the proprietor finished getting out some rocks, cutting down the hollow apple trees, and grubbing up some young birches which had sprung up in the pasture, or, in short, had made any more of his improvements. To enjoy these advantages I was ready to carry it on; like Atlas,* to take the world on my shoulders,—I never heard what compensation he received for that,—and do all those things which had no other motive or excuse but that I might pay for it and be unmolested in my possession of it; for I knew all the while that it would yield the most abundant crop of the kind I wanted if I could only afford to let it alone. But it turned out as I have said.

All that I could say, then, with respect to farming on a large scale, (I have always cultivated a garden,) was, that I had had my seeds ready. Many think that seeds improve with age. I have no doubt that time discriminates between the good and the bad; and when at last I shall plant, I shall be less likely to be disappointed. But I would say to my fellows, once for all, As long as possible live free and uncommitted. It makes but little difference whether you are committed to a farm or the county jail.

Old Cato, whose "De Re Rusticâ" is my "Cultivator,"* says, and the only translation I have seen makes sheer nonsense of the passage, "When you think of getting a farm, turn it thus in your mind, not to buy greedily; nor spare your pains to look at it, and do not think it enough to go round it once.

The oftener you go there the more it will please you, if it is good." I think I shall not buy greedily, but go round and round it as long as I live, and be buried in it first, that it may please me the more at last.

The present was my next experiment of this kind, which I purpose to describe more at length; for convenience, putting the experience of two years into one.* As I have said, I do not propose to write an ode to dejection,* but to brag as lustily as chanticleer in the morning, standing on his roost, if only to wake my neighbors up.

When first I took up my abode in the woods, that is, began to spend my nights as well as days there, which, by accident, was on Independence day,* or the fourth of July, 1845, my house was not finished for winter, but was merely a defence against the rain, without plastering or chimney, the walls being of rough weather-stained boards, with wide chinks, which made it cool at night. The upright white hewn studs and freshly planed door and window casings gave it a clean and airy look, especially in the morning, when its timbers were saturated with dew, so that I fancied that by noon some sweet gum would exude from them. To my imagination it retained throughout the day more or less of this auroral character, reminding me of a certain house* on a mountain which I had visited the year before. This was an airy and unplastered cabin, fit to entertain a travelling god, and where a goddess might trail her garments. The winds which passed over my dwelling were such as sweep over the ridges of mountains, bearing the broken strains, or celestial parts only, of terrestrial music. The morning wind forever blows, the poem of creation is uninterrupted; but few are the ears that hear it. Olympus* is but the outside of the earth every where.

The only house I had been the owner of before, if I except a boat,* was a tent, which I used occasionally when making excursions in the summer, and this is still rolled up in my garret; but the boat, after passing from hand to hand, has gone down the stream of time. With this more substantial shelter about me, I had made some progress toward settling in the world. This frame, so slightly clad, was a sort of crystallization around me, and reacted on the builder. It was suggestive somewhat as a picture in outlines. I did not need to go out doors to take the air, for the atmosphere within had lost none of its freshness. It was not so much within doors as behind a door where I sat, even in the rainiest weather. The Harivansa* says, "An abode without birds is like a meat without seasoning." Such was not my abode, for I found myself suddenly neighbor to the birds; not by having imprisoned one, but having caged myself near them. I was not only nearer to some of those which commonly frequent the garden and the orchard, but to those wilder and more thrilling songsters of the forest which never, or rarely, serenade a villager,—the wood-thrush, the veery, the scarlet tanager, the field-sparrow, the whippoorwill, and many others.

I was seated by the shore of a small pond, about a mile and a half* south of the village of Concord and somewhat higher than it, in the midst of an extensive wood between that town and Lincoln, and about two miles south of that our only field known to fame, Concord Battle Ground;* but I was so low in the woods that the opposite shore, half a mile off, like the rest, covered with wood, was my most distant horizon. For the first week, whenever I looked out on the pond it impressed me like a tarn* high up on the side of a mountain, its bottom far above the surface of other lakes, and, as the sun arose, I saw it throwing off its nightly clothing of mist, and here and there, by degrees, its soft ripples or its smooth reflecting surface was revealed, while the mists, like ghosts, were stealthily withdrawing in every direction into the woods, as at the breaking up of some nocturnal conventicle.* The very dew seemed to hang upon the trees later into the day than usual, as on the sides of mountains.

This small lake was of most value as a neighbor in the intervals of a gentle rain storm in August, when, both air and water being perfectly still, but the sky overcast, mid-afternoon had all the serenity of evening, and the wood-thrush sang around, and was heard from shore to shore. A lake like this is never smoother than at such a time; and the clear portion of the air above it being shallow and darkened by clouds, the water, full of light and reflections, becomes a lower heaven itself so much the more important. From a hill top near by, where the wood had been recently cut off, there was a pleasing vista southward across the pond, through a wide indentation in the hills which form the shore there, where their opposite sides sloping toward each other suggested a stream flowing out in that direction through a wooded valley, but stream there was none. That way I looked between and over the near green hills to some distant and higher ones in the horizon, tinged with blue. Indeed, by standing on tiptoe I could catch a glimpse of some of the peaks of the still bluer and more distant mountain ranges* in the north-west, those true-blue coins from heaven's own mint, and also of some portion of the village. But in other directions, even from this point, I could not see over or beyond the woods which surrounded me. It is well to have some water in your neighborhood, to give buoyancy to and float the earth. One value even of the smallest well is, that when you look into it you see that earth is not continent but insular. This is as important as that it keeps butter cool.* When I looked across the pond from this peak toward the Sudbury meadows,* which in time of flood I distinguished elevated perhaps by a mirage in their seething valley, like a coin in a basin, all the earth beyond the pond appeared like a thin crust insulated and floated even by this small sheet of intervening water, and I was reminded that this on which I dwelt was but *dry land*.

Though the view from my door was still more contracted, I did not feel crowded or confined in the least. There was pasture enough for my imagination. The low shrub-oak plateau to which the opposite shore arose, stretched away toward the prairies of the West and the steppes of Tartary, affording ample room for all the roving families of men. "There are none happy in the world but beings who enjoy freely a vast horizon,"—said Damodara,* when his herds required new and larger pastures.

Both place and time were changed, and I dwelt nearer to those parts of the universe and to those eras in history which had most attracted me. Where I lived was as far off as many a region viewed nightly by astronomers. We are wont to imagine rare and delectable places in some remote and more celestial corner of the system, behind the constellation of Cassiopeia's Chair, far from noise and disturbance. I discovered that my house actually had its site in such a withdrawn, but forever new and unprofaned, part of the universe. If it were worth the while to settle in those parts near to the Pleiades or the Hyades, to Aldebaran or Altair,* then I was really there, or at an equal remoteness from the life which I had left behind, dwindled and twinkling with as fine a ray to my nearest neighbor, and to be seen only in moonless nights by him. Such was that part of creation where I had squatted;—

> "There was a shepherd that did live,
>   And held his thoughts as high
> As were the mounts whereon his flocks
>   Did hourly feed him by."*

What should we think of the shepherd's life if his flocks always wandered to higher pastures than his thoughts?

Every morning was a cheerful invitation to make my life of equal simplicity, and I may say innocence, with Nature herself. I have been as sincere a worshipper of Aurora* as the Greeks. I got up early and bathed in the pond; that was a religious exercise, and one of the best things which I did. They say that characters were engraven on the bathing tub of king Tching-thang to this effect: "Renew thyself* completely each day; do it again, and again, and forever again." I can understand that. Morning brings back the heroic ages.* I was as much affected by the faint hum of a mosquito making its invisible and unimaginable tour through my apartment at earliest dawn, when I was sitting with door and windows open, as I could be by any trumpet that ever sang of fame.* It was Homer's requiem; itself an Iliad and Odyssey in the air, singing its own wrath and wanderings.* There was something cosmical about it; a standing advertisement, till forbidden,* of the everlasting vigor and fertility of the world. The morning, which is the most memorable season of the day, is the awakening hour. Then there is least somnolence in us; and for an hour, at least, some part of us awakes which slumbers all the rest of the day and night. Little is to be expected of that day, if it can be called a day, to which we are not awakened by our Genius, but by the mechanical nudgings of some servitor, are not awakened by our own newly-acquired force and aspirations from within, accompanied by the undulations of celestial music, instead of factory bells, and a fragrance filling the air— to a higher life than we fell asleep from; and thus the darkness bear its fruit, and prove itself to be good, no less than the light. That man who does not believe that each day contains an earlier, more sacred, and auroral hour than he has yet profaned, has despaired of life, and is pursuing a descending and darkening way. After a partial cessation of his sensuous life, the soul of man, or its organs rather, are reinvigorated each day, and his Genius tries again what noble life it can make. All memorable events, I should say, transpire in morning time and in a morning atmosphere. The Vedas* say, "All intelligences awake with the morning." Poetry and art, and the fairest and most memorable of the actions of men, date from such an hour. All poets and heroes, like Memnon,* are the children of Aurora, and emit their music at sunrise. To him whose elastic and vigorous thought keeps pace with the sun, the day is a perpetual morning. It matters not what the clocks say or the attitudes and labors of men. Morning is when I am awake and there is a dawn in me. Moral reform is the effort to throw off sleep. Why is it that men give so poor an account of their day if they have not been slumbering? They are not such poor calculators. If they had not been overcome with drowsiness they would have performed something. The millions are awake enough for physical labor; but only one in a million is awake enough for effective intellectual exertion, only one in a hundred millions to a poetic or divine life. To be awake is to be alive. I have never yet met a man who was quite awake. How could I have looked him in the face?

We must learn to reawaken and keep ourselves awake, not by mechanical aids, but by an infinite expectation of the dawn, which does not forsake us in our soundest sleep. I know of no more encouraging fact than the unquestionable ability of man to elevate his life by a conscious endeavor. It is something to be able to paint a particular picture, or to carve a statue, and so to make a few objects beautiful; but it is far more glorious to carve and paint the very atmosphere and medium through which we look, which morally we can do. To affect the quality of the day, that is the highest of arts. Every man is tasked to make his life, even in its details, worthy of the contemplation of his most elevated and critical hour. If we refused, or rather used up, such paltry information as we get, the oracles would distinctly inform us how this might be done.

I went to the woods because I wished to live deliberately, to front only the essential facts of life, and see if I could not learn what it had to teach, and not, when I came to die, discover that I had not lived. I did not wish to live what was not life, living is so dear; nor did I wish to practise resignation, unless it was quite necessary. I wanted to live deep and suck out all the marrow of life, to live so sturdily and Spartan-like* as to put to rout all that was not life, to cut a broad swath and shave close, to drive life into a corner, and reduce it to its lowest terms, and, if it proved to be mean, why then to get the whole and genuine meanness of it, and publish its meanness to the world; or if it were sublime, to know it by experience, and be able to give a true account of it in my next excursion. For most men, it appears to me, are in a strange uncertainty about it, whether it is of the devil or of God, and have *somewhat hastily* concluded that it is the chief end of man here to "glorify God and enjoy him forever."*

Still we live meanly, like ants; though the fable tells us that we were long ago changed into men;* like pygmies we fight with cranes;* it is error upon error, and clout upon clout,* and our best virtue has for its occasion a superfluous and evitable wretchedness. Our life is frittered away by detail. An honest man has hardly need to count more than his ten fingers, or in extreme cases he may add his ten toes, and lump the rest. Simplicity, simplicity, simplicity! I say, let your affairs be as two or three, and not a hundred or a thousand; instead of a million count half a dozen, and keep your accounts on your thumb nail. In the midst of this chopping sea of civilized life, such are the clouds and storms and quicksands and thousand-and-one items to be allowed for, that a man has to live, if he would not founder and go to the bottom and not make his port at all, by dead reckoning,* and he must be a great calculator indeed who succeeds. Simplify, simplify. Instead of three meals a day, if it be necessary eat but one; instead of a hundred dishes, five; and reduce other things in proportion. Our life is like a German Confederacy, made up of petty states, with its boundary forever fluctuating, so that even a German cannot tell you how it is bounded at any moment. The nation itself, with all its so called internal improvements,* which, by the way, are all external and superficial, is just such an unwieldy and overgrown establishment, cluttered with furniture and tripped up by its own traps, ruined by luxury and heedless expense, by want of calculation and a worthy aim, as the million households in the land; and the only cure for it as for them is a rigid economy, a stern and more than Spartan simplicity of life and elevation of purpose. It lives too fast. Men think that it is essential that the *Nation* have commerce, and export ice, and talk through a telegraph, and ride thirty miles an hour, without a doubt, whether *they* do or not; but whether we should live like baboons or like men, is a little uncertain. If we do not get out sleepers,* and forge rails, and devote days and nights to the work, but go to tinkering upon our *lives* to improve *them*, who will build railroads? And if railroads are not built, how shall we get to heaven in season?* But if we stay at home and mind our business, who will want railroads? We do not ride on the railroad; it rides upon us. Did you ever think what those sleepers are that underlie the railroad? Each one is a man, an Irishman, or a Yankee man. The rails are laid on them, and they are covered with sand, and the cars run smoothly over them. They are sound sleepers, I assure you. And every few years a new lot is laid down and run over; so that, if some have the pleasure of riding on a rail,* others have the misfortune to be ridden upon. And when they run over a man that is walking in his sleep, a supernumerary sleeper in the wrong position, and wake him up, they suddenly stop the cars, and make a hue and cry about it, as if this were an exception. I am glad to know that it takes a gang of men for every five miles to keep the sleepers down and level in their beds as it is, for this is a sign that they may sometime get up again.

Why should we live with such hurry and waste of life? We are determined to be starved before we are hungry. Men say that a stitch in time saves nine, and so they take a thousand stitches to-day to save nine to-morrow. As for *work*, we haven't any of any consequence. We have the Saint Vitus' dance,* and cannot possibly keep our heads still. If I should only give a few pulls at the parish bell-rope, as for a fire, that is, without setting the bell,* there is hardly a man on his farm in the outskirts of Concord, notwithstanding that press of engagements which was his excuse so many times this morning, nor a boy, nor a woman, I might almost say, but would forsake all and follow that sound, not mainly to save property from the flames, but, if we will confess the truth, much more to see it burn, since burn it must, and we, be it known, did not set it on fire,*—or to see it put out, and have a hand in it, if that is done as handsomely; yes, even if it were the parish church itself. Hardly a man takes a half hour's nap after dinner, but when he wakes he holds up his head and asks, "What's the news?" as if the rest of mankind had stood his sentinels. Some give directions to be waked every half hour, doubtless for no other purpose; and then, to pay for it, they tell what they have dreamed. After a night's sleep the news is as indispensable as the breakfast. "Pray tell me any thing new that has happened to a man any where on this globe,"—and he reads it over his coffee and rolls, that a man has had his eyes gouged out this morning on the Wachito River;* never dreaming the while that he lives in the dark unfathomed mammoth cave* of this world, and has but the rudiment of an eye himself.

For my part, I could easily do without the post-office. I think there are very few important communications made through it. To speak critically, I never received more than one or two letters in my life—I wrote this some years ago—that were worth the postage. The penny-post* is, commonly, an institution through which you seriously offer a man that penny for his thoughts which is so often safely offered in jest. And I am sure that I never read any memorable news in a newspaper. If we read of one man robbed, or murdered, or killed by accident, or one house burned, or one vessel wrecked, or one steamboat blown up, or one cow run over on the Western Railroad, or one mad dog killed, or one lot of grasshoppers in the winter,—we never need read of another. One is enough. If you are acquainted with the principle, what do you care for a myriad instances and applications? To a philosopher all *news*, as it is called, is gossip, and they who edit and read it are old women over their tea. Yet not a few are greedy after this gossip. There was such a rush, as I hear, the other day at one of the offices to learn the foreign news by the last arrival, that several large squares of plate glass belonging to the establishment were broken by the pressure,—news which I seriously think a ready wit might write a twelvemonth or twelve years beforehand with sufficient accuracy. As for Spain, for instance, if you know how to throw in Don Carlos and the Infanta, and Don Pedro and Seville and Granada,* from time to time in the right proportions,—they may have changed the names a little since I saw the papers,—and serve up a bull-fight when other entertainments fail, it will be true to the letter, and give us as good an idea of the exact state or ruin of things in Spain as the most succinct and lucid reports under this head in the newspapers: and as for England, almost the last significant scrap of news from that quarter was the revolution of 1649;* and if you have learned the history of her crops for an average year, you never need attend to that thing again, unless your speculations are of a merely pecuniary character. If one may judge who rarely looks into the newspapers, nothing new does ever happen in foreign parts, a French revolution not excepted.

What news! how much more important to know what that is which was never old! "Kieou-he-yu* (great dignitary of the state of Wei) sent a man to Khoung-tseu to know his news. Khoung-tseu caused the messenger to be seated near him, and questioned him in these terms: What is your master doing? The messenger answered with respect: My master desires to diminish the number of his faults, but he cannot come to the end of them.* The messenger being gone, the philosopher remarked: What a worthy messenger! What a worthy messenger!" The preacher, instead of vexing the ears of drowsy farmers on their day of rest at the end of the week,—for Sunday is the fit conclusion of an ill-spent week, and not the fresh and brave beginning of a new one,—with this one other draggletail of a sermon, should shout with thundering voice,—"Pause! Avast! Why so seeming fast, but deadly slow?"

Shams and delusions are esteemed for soundest truths, while reality is fabulous. If men would steadily observe realities only, and not allow themselves to be deluded, life, to compare it with such things as we know, would be like a fairy tale and the Arabian Nights' Entertainments. If we respected only what is inevitable and has a right to be, music and poetry would resound along the streets. When we are unhurried and wise, we perceive that only great and worthy things have any permanent and absolute existence,—that petty fears and petty pleasures are but the shadow of the reality. This is always exhilarating and sublime. By closing the eyes and slumbering, and consenting to be deceived by shows, men establish and confirm their daily life of routine and habit every where, which still is built on purely illusory foundations. Children, who play life, discern its true law and relations more clearly than men, who fail to live it worthily, but who think that they are wiser by experience, that is, by failure. I have read in a Hindoo book,* that "there was a king's son, who, being expelled in infancy from his native city, was brought up by a forester, and, growing up to maturity in that state, imagined himself to belong to the barbarous race with which he lived. One of his father's ministers having discovered him, revealed to him what he was, and the misconception of his character was removed, and he knew himself to be a prince. So soul," continues the Hindoo philosopher, "from the circumstances in which it is placed, mistakes its own character, until the truth is revealed to it by some holy teacher, and then it knows itself to be *Brahme*." I perceive that we inhabitants of New England live this mean life that we do because our vision does not penetrate the surface of things. We think that that *is* which *appears* to be. If a man should walk through this town and see only the reality, where, think you, would the "Mill-dam"* go to? If he should give us an account of the realities he beheld there, we should not recognize the place in his description. Look at a meeting-house, or a court-house, or a jail, or a shop, or a dwelling-house, and say what that thing really is before a true gaze, and they would all go to pieces in your account of them. Men esteem truth remote, in the outskirts of the system, behind the farthest star, before Adam and after the last man. In eternity there is indeed something true and sublime. But all these times and places and occasions are now and here. God himself culminates in the present moment, and will never be more divine in the lapse of all the ages. And we are enabled to apprehend at all what is sublime and noble only by the perpetual instilling and drenching of the reality that surrounds us. The universe constantly and obediently answers to our conceptions; whether we travel fast or slow, the track is laid for us. Let us spend our lives in conceiving then. The poet or the artist never yet had so fair and noble a design but some of his posterity at least could accomplish it.

Let us spend one day as deliberately as Nature, and not be thrown off the track by every nutshell and mosquito's wing that falls on the rails. Let us rise early and fast, or break fast, gently and without perturbation; let company come and let company go, let the bells ring and the children cry,— determined to make a day of it. Why should we knock under and go with the stream? Let us not be upset and overwhelmed in that terrible rapid and whirlpool called a dinner, situated in the meridian shallows. Weather this danger and you are safe, for the rest of the way is down hill. With unrelaxed nerves,* with morning vigor, sail by it, looking another way, tied to the mast like Ulysses.* If the engine whistles, let it whistle till it is hoarse for its pains. If the bell rings,* why should we run? We will consider what kind of music they are like. Let us settle ourselves, and work and wedge our feet downward through the mud and slush of opinion, and prejudice, and tradition, and delusion, and appearance, that alluvion which covers the globe, through Paris and London, through New York and Boston and Concord, through church and state, through poetry and philosophy and religion, till we come to a hard bottom and rocks in place, which we can call *reality*, and say, This is, and no mistake; and then begin, having a *point d'appui*,* below freshet and frost and fire, a place where you might found a wall or a state, or set a lamp-post safely, or perhaps a gauge, not a Nilometer,* but a Realometer, that future ages might know how deep a freshet of shams and appearances had gathered from time to time. If you stand right fronting and face to face to a fact, you will see the sun glimmer on both its surfaces, as if it were a cimeter,* and feel its sweet edge dividing you through the heart and marrow, and so you will happily conclude your mortal career. Be it life or death, we crave only reality. If we are really dying, let us hear the rattle in our throats and feel cold in the extremities; if we are alive, let us go about our business.

Time is but the stream I go a-fishing in. I drink at it; but while I drink I see the sandy bottom and detect how shallow it is. Its thin current slides away, but eternity remains. I would drink deeper; fish in the sky, whose bottom is pebbly with stars. I cannot count one. I know not the first letter of the alphabet. I have always been regretting that I was not as wise as the day I was born. The intellect is a cleaver; it discerns and rifts its way into the secret of things. I do not wish to be any more busy with my hands than is necessary. My head is hands and feet. I feel all my best faculties concentrated in it. My instinct tells me that my head is an organ for burrowing, as some creatures use their snout and fore-paws, and with it I would mine and burrow my way through these hills. I think that the richest vein is somewhere hereabouts; so by the divining rod and thin rising vapors I judge; and here I will begin to mine.

books of travel, in the intervals of my work, till that employment made me ashamed of myself, and I asked where it was
then that I lived.

READING

WITH a little more deliberation in the choice of their
pursuits, all men would perhaps become essentially students
and observers, for certainly their nature and destiny are interesting to all alike. In accumulating property for ourselves or
our posterity, in founding a family or a state, or acquiring
fame even, we are mortal; but in dealing with truth we are
immortal, and need fear no change nor accident. The oldest
Egyptian or Hindoo philosopher raised a corner of the veil
from the statue of the divinity; and still the trembling robe
remains raised, and I gaze upon as fresh a glory as he did,
since it was I in him that was then so bold, and it is he in me
that now reviews the vision. No dust has settled on that robe;
no time has elapsed since that divinity was revealed. That time
which we really improve, or which is improvable, is neither
past, present, nor future.

My residence was more favorable, not only to thought, but
to serious reading, than a university; and though I was beyond
the range of the ordinary circulating library, I had more than
ever come within the influence of those books which circulate
round the world, whose sentences were first written on bark,
and are now merely copied from time to time on to linen
paper. Says the poet Mîr Camar Uddîn Mast,* "Being seated
to run through the region of the spiritual world; I have had
this advantage in books. To be intoxicated by a single glass of
wine; I have experienced this pleasure when I have drunk the
liquor of the esoteric doctrines." I kept Homer's Iliad on my
table through the summer, though I looked at his page only
now and then. Incessant labor with my hands, at first, for I had
my house to finish and my beans to hoe at the same time,
made more study impossible. Yet I sustained myself by the
prospect of such reading in future. I read one or two shallow

books of travel in the intervals of my work, till that employment made me ashamed of myself, and I asked where it was
then that *I* lived.

The student may read Homer or Æschylus in the Greek
without danger of dissipation or luxuriousness, for it implies
that he in some measure emulate their heroes, and consecrate morning hours to their pages. The heroic books, even if
printed in the character of our mother tongue, will always be
in a language dead to degenerate times; and we must labori-
ously seek the meaning of each word and line, conjecturing a
larger sense than common use permits out of what wisdom
and valor and generosity we have. The modern cheap and
fertile press, with all its translations, has done little to bring us
nearer to the heroic writers of antiquity. They seem as solitary, and the letter in which they are printed as rare and
curious, as ever. It is worth the expense of youthful days and
costly hours, if you learn only some words of an ancient
language, which are raised out of the trivialness of the street,
to be perpetual suggestions and provocations. It is not in vain
that the farmer remembers and repeats the few Latin words
which he has heard. Men sometimes speak as if the study of
the classics would at length make way for more modern and
practical studies; but the adventurous student will always
study classics, in whatever language they may be written and
however ancient they may be. For what are the classics but
the noblest recorded thoughts of man? They are the only
oracles which are not decayed, and there are such answers to
the most modern inquiry in them as Delphi and Dodona*
never gave. We might as well omit to study Nature because
she is old. To read well, that is, to read true books in a true
spirit, is a noble exercise, and one that will task the reader
more than any exercise which the customs of the day esteem.
It requires a training such as the athletes underwent, the
steady intention almost of the whole life to this object. Books
must be read as deliberately and reservedly as they were
written. It is not enough even to be able to speak the language

of that nation by which they are written, for there is a memorable interval between the spoken and the written language, the language heard and the language read. The one is commonly transitory, a sound, a tongue, a dialect merely, almost brutish, and we learn it unconsciously, like the brutes, of our mothers. The other is the maturity and experience of that; if that is our mother tongue, this is our father tongue, a reserved and select expression, too significant to be heard by the ear, which we must be born again in order to speak. The crowds of men who merely *spoke* the Greek and Latin tongues in the middle ages were not entitled by the accident of birth to *read* the works of genius written in those languages; for these were not written in that Greek or Latin which they knew, but in the select language of literature. They had not learned the nobler dialects of Greece and Rome, but the very materials on which they were written were waste paper* to them, and they prized instead a cheap contemporary literature. But when the several nations of Europe had acquired distinct though rude written languages of their own, sufficient for the purposes of their rising literatures, then first learning revived, and scholars were enabled to discern from that remoteness the treasures of antiquity. What the Roman and Grecian multitude could not *hear*, after the lapse of ages a few scholars *read*, and a few scholars only are still reading it.

However much we may admire the orator's occasional bursts of eloquence, the noblest written words are commonly as far behind or above the fleeting spoken language as the firmament with its stars is behind the clouds. *There* are the stars, and they who can may read them. The astronomers forever comment on and observe them. They are not exhalations like our daily colloquies and vaporous breath. What is called eloquence in the forum is commonly found to be rhetoric in the study. The orator yields to the inspiration of a transient occasion, and speaks to the mob before him, to those who can *hear* him; but the writer, whose more equable life is his occasion, and who would be distracted by the event and the crowd which inspire the orator, speaks to the intellect and heart of mankind, to all in any age who can *understand* him.

No wonder that Alexander carried the Iliad with him on his expeditions in a precious casket.* A written word is the choicest of relics. It is something at once more intimate with us and more universal than any other work of art. It is the work of art nearest to life itself. It may be translated into every language, and not only be read but actually breathed from all human lips;—not be represented on canvas or in marble only, but be carved out of the breath of life itself. The symbol of an ancient man's thought becomes a modern man's speech. Two thousand summers have imparted to the monuments of Grecian literature, as to her marbles, only a maturer golden and autumnal tint, for they have carried their own serene and celestial atmosphere into all lands to protect them against the corrosion of time. Books are the treasured wealth of the world and the fit inheritance of generations and nations. Books, the oldest and best, stand naturally and rightfully on the shelves of every cottage. They have no cause of their own to plead, but while they enlighten and sustain the reader his common sense will not refuse them. Their authors are a natural and irresistible aristocracy in every society, and, more than kings or emperors, exert an influence on mankind. When the illiterate and perhaps scornful trader has earned by enterprise and industry his coveted leisure and independence, and is admitted to the circles of wealth and fashion, he turns inevitably at last to those still higher but yet inaccessible circles of intellect and genius, and is sensible only of the imperfection of his culture and the vanity and insufficiency of all his riches, and further proves his good sense by the pains which he takes to secure for his children that intellectual culture whose want he so keenly feels; and thus it is that he becomes the founder of a family.

Those who have not learned to read the ancient classics in the language in which they were written must have a very

imperfect knowledge of the history of the human race; for it is
remarkable that no transcript of them has ever been made
into any modern tongue, unless our civilization itself may
be regarded as such a transcript. Homer has never yet been
printed in English, nor Æschylus, nor Virgil even,*—works as
refined, as solidly done, and as beautiful almost as the morning itself; for later writers, say what we will of their genius,
have rarely, if ever, equalled the elaborate beauty and finish
and the lifelong and heroic literary labors of the ancients.
They only talk of forgetting them who never knew them.
It will be soon enough to forget them when we have the
learning and the genius which will enable us to attend to and
appreciate them. That age will be rich indeed when those
relics which we call Classics, and the still older and more
than classic but even less known Scriptures of the nations,
shall have still further accumulated, when the Vaticans* shall
be filled with Vedas and Zendavestas* and Bibles, with Homers and Dantes and Shakspeares, and all the centuries to come
shall have successively deposited their trophies in the forum
of the world. By such a pile we may hope to scale heaven*
at last.

The works of the great poets have never yet been read by
mankind, for only great poets can read them. They have only
been read as the multitude read the stars, at most astro-
logically, not astronomically. Most men have learned to serve
a paltry convenience, as they have learned to cipher in order
to keep accounts and not be cheated in trade; but of reading
as a noble intellectual exercise they know little or nothing; yet
this only is reading, in a high sense, not that which lulls us as
a luxury and suffers the nobler faculties to sleep the while, but
what we have to stand on tip-toe to read and devote our most
alert and wakeful hours to.

I think that having learned our letters we should read the
best that is in literature, and not be forever repeating our a b
abs, and words of one syllable, in the fourth or fifth classes,
sitting on the lowest and foremost form* all our lives. Most

men are satisfied if they read or hear read, and perchance
have been convicted by the wisdom of one good book, the
Bible, and for the rest of their lives vegetate and dissipate
their faculties in what is called easy reading. There is a work
in several volumes in our Circulating Library entitled Little
Reading,* which I thought referred to a town of that name
which I had not been to. There are those who, like cormorants
and ostriches, can digest all sorts of this, even after the fullest
dinner of meats and vegetables, for they suffer nothing to
be wasted. If others are the machines to provide this provender, they are the machines to read it. They read the nine
thousandth tale about Zebulon and Sephronia,* and how
they loved as none had ever loved before, and neither did
the course of their true love run smooth,*—at any rate, how it
did run and stumble, and get up again and go on! how some
poor unfortunate got up on to a steeple, who had better
never have gone up as far as the belfry; and then, having
needlessly got him up there, the happy novelist rings the bell
for all the world to come together and hear, O dear! how
he did get down again! For my part, I think that they had
better metamorphose all such aspiring heroes of universal
noveldom into man weathercocks, as they used to put
heroes among the constellations, and let them swing round
there till they are rusty, and not come down at all to bother
honest men with their pranks. The next time the novelist
rings the bell I will not stir* though the meeting-house*
burn down. "The Skip of the Tip-Toe-Hop, a Romance of
the Middle Ages, by the celebrated author of 'Tittle-Tol-
Tan,'* to appear in monthly parts; a great rush; don't all come
together." All this they read with saucer eyes, and erect and
primitive curiosity, and with unwearied gizzard, whose corrugations even yet need no sharpening,* just as some little
four-year-old bencher* his two-cent gilt-covered edition of
Cinderella,—without any improvement, that I can see, in
the pronunciation, or accent, or emphasis, or any more skill
in extracting or inserting the moral. The result is dulness of

sight, a stagnation of the vital circulations, and a general deliquium* and sloughing off of all the intellectual faculties. This sort of gingerbread is baked daily and more sedulously than pure wheat or rye-and-Indian in almost every oven, and finds a surer market.

The best books are not read even by those who are called good readers. What does our Concord culture amount to? There is in this town, with a very few exceptions, no taste for the best or for very good books even in English literature, whose words all can read and spell. Even the college-bred and so called liberally educated men here and elsewhere have really little or no acquaintance with the English classics;* and as for the recorded wisdom of mankind, the ancient classics and Bibles, which are accessible to all who will know of them, there are the feeblest efforts any where made to become acquainted with them. I know a woodchopper, of middle age, who takes a French paper, not for news as he says, for he is above that, but to "keep himself in practice," he being a Canadian by birth; and when I ask him what he considers the best thing he can do in this world, he says, beside this, to keep up and add to his English. This is about as much as the college bred generally do or aspire to do, and they take an English paper for the purpose. One who has just come from reading perhaps one of the best English books will find how many with whom he can converse about it? Or suppose he comes from reading a Greek or Latin classic in the original, whose praises are familiar even to the so called illiterate; he will find nobody at all to speak to, but must keep silence about it. Indeed, there is hardly the professor in our colleges, who, if he has mastered the difficulties of the language, has proportionally mastered the difficulties of the wit and poetry of a Greek poet, and has any sympathy to impart to the alert and heroic reader; and as for the sacred Scriptures, or Bibles of mankind, who in this town can tell me even their titles? Most men do not know that any nation but the Hebrews have had a scripture. A man, any man, will go considerably out of his way to pick up a silver dollar; but here are golden words, which the wisest men of antiquity have uttered, and whose worth the wise of every succeeding age have assured us of;—and yet we learn to read only as far as Easy Reading,* the primers and class-books, and when we leave school, the "Little Reading,"* and story books, which are for boys and beginners; and our reading, our conversation and thinking, are all on a very low level, worthy only of pygmies and manikins.

I aspire to be acquainted with wiser men than this our Concord soil has produced, whose names are hardly known here. Or shall I hear the name of Plato and never read his book? As if Plato were my townsman and I never saw him,—my next neighbor and I never heard him speak or attended to the wisdom of his words. But how actually is it? His Dialogues, which contain what was immortal in him, lie on the next shelf, and yet I never read them. We are under-bred and low-lived and illiterate; and in this respect I confess I do not make any very broad distinction between the illiterateness of my townsman who cannot read at all, and the illiterateness of him who has learned to read only what is for children and feeble intellects. We should be as good as the worthies of antiquity, but partly by first knowing how good they were. We are a race of tit-men,* and soar but little higher in our intellectual flights than the columns of the daily paper.

It is not all books that are as dull as their readers. There are probably words addressed to our condition exactly, which, if we could really hear and understand, would be more salutary than the morning or the spring to our lives, and possibly put a new aspect on the face of things for us. How many a man has dated a new era in his life from the reading of a book. The book exists for us perchance which will explain our miracles and reveal new ones. The at present unutterable things we may find somewhere uttered. These same questions that disturb and puzzle and confound us have in their turn occurred to all the wise men; not one has been omitted; and each has answered them, according to his ability, by his words and his life. Moreover, with wisdom we shall learn liberality. The solitary hired man on a farm in the outskirts of Concord, who has had his second birth and peculiar religious experience, and is driven as he believes into silent gravity and exclusiveness by his faith, may think it is not true; but Zoroaster, thousands of years ago, travelled the same road and had the same experience; but he, being wise, knew it to be universal, and treated his neighbors accordingly, and is even said to have invented and established worship among men.* Let him humbly commune with Zoroaster then, and, through the liberalizing influence of all the worthies, with Jesus Christ himself, and let "our church" go by the board.

We boast that we belong to the nineteenth century and are making the most rapid strides of any nation. But consider how little this village does for its own culture. I do not wish to flatter my townsmen, nor to be flattered by them, for that will not advance either of us. We need to be provoked,—goaded like oxen, as we are, into a trot. We have a comparatively decent system of common schools,* schools for infants only; but excepting the half-starved Lyceum* in the winter, and latterly the puny beginning of a library suggested by the state,* no school for ourselves. We spend more on almost any article of bodily aliment or ailment than on our mental aliment. It is time that we had uncommon schools, that we did not leave off our education when we begin to be men and women. It is time that villages were universities, and their elder inhabitants the fellows of universities, with leisure—if they are indeed so well off—to pursue liberal studies the rest of their lives. Shall the world be confined to one Paris or one Oxford forever? Cannot students be boarded here and get a liberal education under the skies of Concord? Can we not hire some Abelard* to lecture to us? Alas! what with foddering the cattle and tending the store, we are kept from school too long, and our education is sadly neglected. In this country, the village should in some respects take the place of the nobleman of Europe. It should be the patron of the fine arts. It is rich enough. It wants only the magnanimity and refinement. It can spend money enough on such things as farmers and traders value, but it is thought Utopian to propose spending money for things which more intelligent men know to be of far more worth. This town has spent seventeen thousand dollars on a town-house,* thank fortune or politics, but probably it will not spend so much on living wit, the true meat to put into that shell, in a hundred years. The one hundred and twenty-five dollars annually subscribed* for a Lyceum in the winter is better spent than any other equal sum raised in the town. If we live in the nineteenth century, why should we not enjoy the advantages which the nineteenth century offers? Why should our life be in any respect provincial? If we will read newspapers, why not skip the gossip of Boston and take the best newspaper in the world at once?—not be sucking the pap of "neutral family" papers,* or browsing "Olive-Branches"* here in New England. Let the reports of all the learned societies come to us, and we will see if they know any thing. Why should we leave it to Harper & Brothers and Redding & Co.* to select our reading? As the nobleman of cultivated taste surrounds himself with whatever conduces to his culture,—genius—learning—wit—books—paintings—statuary—music—philosophical instruments,* and the like; so let the village do,—not stop short at a pedagogue, a parson, a sexton, a parish library, and three selectmen, because our pilgrim forefathers got through a cold winter once on a bleak rock with these. To act collectively is according to the spirit of our institutions; and I am confident that, as our circumstances are more flourishing, our means are greater than the nobleman's. New England can hire all the wise men in the world to come and teach her, and board them round the while, and not be provincial at all. That is the *uncommon* school we

want. Instead of noblemen, let us have noble villages of men.
If it is necessary, omit one bridge over the river, go round a
little there, and throw one arch at least over the darker gulf of
ignorance which surrounds us.

SOUNDS

BUT while we are confined to books, though the most select and classic, and read only particular written languages, which are themselves but dialects and provincial, we are in danger of forgetting the language which all things and events speak without metaphor, which alone is copious and standard. Much is published, but little printed. The rays which stream through the shutter will be no longer remembered when the shutter is wholly removed. No method nor discipline can supersede the necessity of being forever on the alert. What is a course of history, or philosophy, or poetry, no matter how well selected, or the best society, or the most admirable routine of life, compared with the discipline of looking always at what is to be seen? Will you be a reader, a student merely, or a seer? Read your fate, see what is before you, and walk on into futurity.

I did not read books the first summer; I hoed beans. Nay, I often did better than this. There were times when I could not afford to sacrifice the bloom of the present moment to any work, whether of the head or hands. I love a broad margin to my life. Sometimes, in a summer morning, having taken my accustomed bath, I sat in my sunny doorway from sunrise till noon, rapt in a revery, amidst the pines and hickories and sumachs, in undisturbed solitude and stillness, while the birds sang around or flitted noiseless through the house, until by the sun falling in at my west window, or the noise of some traveller's wagon on the distant highway, I was reminded of the lapse of time. I grew in those seasons like corn in the night,* and they were far better than any work of the hands would have been. They were not time subtracted from my life, but so much over and above my usual allowance. I realized what the Orientals mean by contemplation and the forsaking of works.

For the most part, I minded not how the hours went. The day advanced as if to light some work of mine; it was morning, and lo, now it is evening, and nothing memorable is accomplished. Instead of singing like the birds, I silently smiled at my incessant good fortune. As the sparrow had its trill, sitting on the hickory before my door, so had I my chuckle or suppressed warble which he might hear out of my nest. My days were not days of the week, bearing the stamp of any heathen deity,* nor were they minced into hours and fretted by the ticking of a clock; for I lived like the Puri Indians, of whom it is said that "for yesterday, to-day, and to-morrow they have only one word, and they express the variety of meaning by pointing backward for yesterday, forward for to-morrow, and overhead for the passing day."* This was sheer idleness to my fellow-townsmen, no doubt; but if the birds and flowers had tried me by their standard, I should not have been found wanting. A man must find his occasions in himself, it is true. The natural day is very calm, and will hardly reprove his indolence.

I had this advantage, at least, in my mode of life, over those who were obliged to look abroad for amusement, to society and the theatre, that my life itself was become my amusement and never ceased to be novel. It was a drama of many scenes and without an end. If we were always indeed getting our living, and regulating our lives according to the last and best mode we had learned, we should never be troubled with ennui. Follow your genius closely enough, and it will not fail to show you a fresh prospect every hour. Housework was a pleasant pastime. When my floor was dirty, I rose early, and, setting all my furniture out of doors on the grass, bed and bedstead making but one budget, dashed water on the floor, and sprinkled white sand from the pond on it, and then with a broom scrubbed it clean and white; and by the time the villagers had broken their fast the morning sun had dried my house sufficiently to allow me to move in again, and my meditations were almost uninterrupted. It was pleasant to see my whole household effects out on the grass, making a little pile like a gypsy's pack, and my three-legged table, from which I did not remove the books and pen and ink, standing amid the pines and hickories. They seemed glad to get out themselves, and as if unwilling to be brought in. I was sometimes tempted to stretch an awning over them and take my seat there. It was worth the while to see the sun shine on these things, and hear the free wind blow on them; so much more interesting most familiar objects look out of doors than in the house. A bird sits on the next bough, life-everlasting grows under the table, and blackberry vines run round its legs; pine cones, chestnut burs, and strawberry leaves are strewn about. It looked as if this was the way these forms came to be transferred to our furniture, to tables, chairs, and bedsteads,—because they once stood in their midst.

My house was on the side of a hill, immediately on the edge of the larger wood, in the midst of a young forest of pitch pines and hickories, and half a dozen rods* from the pond, to which a narrow footpath led down the hill. In my front yard grew the strawberry, blackberry, and life-everlasting, johnswort and goldenrod, shrub-oaks and sand-cherry, blueberry and ground-nut. Near the end of May, the sand-cherry, (*cerasus pumila*,) adorned the sides of the path with its delicate flowers arranged in umbels cylindrically about its short stems, which last, in the fall, weighed down with good sized and handsome cherries, fell over in wreaths like rays on every side. I tasted them out of compliment to Nature, though they were scarcely palatable. The sumach, (*rhus glabra*,) grew luxuriantly about the house, pushing up through the embankment which I had made, and growing five or six feet the first season. Its broad pinnate tropical leaf was pleasant though strange to look on. The large buds, suddenly pushing out late in the spring from dry sticks which had seemed to be dead, developed themselves as by magic into graceful green and tender boughs, an inch in diameter; and sometimes, as I sat at my window, so heedlessly did they grow and tax their weak joints, I heard a fresh and tender bough suddenly fall like a fan to the ground, when there was not a breath of air stirring, broken off by its own weight. In August, the large masses of berries, which, when in flower, had attracted many wild bees, gradually assumed their bright velvety crimson hue, and by their weight again bent down and broke the tender limbs.

As I sit at my window this summer afternoon, hawks are circling about my clearing; the tantivy* of wild pigeons,* flying by twos and threes athwart my view, or perching restless on the white-pine boughs behind my house, gives a voice to the air; a fishhawk dimples the glassy surface of the pond and brings up a fish; a mink steals out of the marsh before my door and seizes a frog by the shore; the sedge is bending under the weight of the reed-birds flitting hither and thither; and for the last half hour I have heard the rattle of railroad cars, now dying away and then reviving like the beat of a partridge,* conveying travellers from Boston to the country. For I did not live so out of the world as that boy, who, as I hear, was put out to a farmer in the east part of the town, but ere long ran away and came home again, quite down at the heel and homesick. He had never seen such a dull and out-of-the-way place; the folks were all gone off; why, you couldn't even hear the whistle! I doubt if there is such a place in Massachusetts now:—

> "In truth, our village has become a butt
> For one of those fleet railroad shafts, and o'er
> Our peaceful plain its soothing sound is—Concord."*

The Fitchburg Railroad touches the pond about a hundred rods south of where I dwell. I usually go to the village along its causeway, and am, as it were, related to society by this link. The men on the freight trains, who go over the whole length of the road, bow to me as to an old acquaintance, they pass me so often, and apparently they take me for an employee; and so

I am. I too would fain be a track-repairer somewhere in the orbit of the earth.

The whistle of the locomotive penetrates my woods summer and winter, sounding like the scream of a hawk sailing over some farmer's yard, informing me that many restless city merchants are arriving within the circle of the town, or adventurous country traders from the other side. As they come under one horizon, they shout their warning to get off the track to the other, heard sometimes through the circles of two towns. Here come your groceries, country; your rations, countrymen! Nor is there any man so independent on his farm that he can say them nay. And here's your pay for them! screams the countryman's whistle; timber like long battering rams going twenty miles an hour against the city's walls, and chairs enough to seat all the weary and heavy laden* that dwell within them. With such huge and lumbering civility the country hands a chair to the city. All the Indian huckleberry hills are stripped, all the cranberry meadows are raked* into the city. Up comes the cotton, down goes the woven cloth; up comes the silk, down goes the woollen; up come the books, but down goes the wit that writes them.

When I meet the engine with its train of cars moving off with planetary motion,—or, rather, like a comet, for the beholder knows not if with that velocity and with that direction it will ever revisit this system, since its orbit does not look like a returning curve,—with its steam cloud like a banner streaming behind in golden and silver wreaths, like many a downy cloud which I have seen, high in the heavens, unfolding its masses to the light,—as if this travelling demigod, this cloud-compeller,* would ere long take the sunset sky for the livery of his train; when I hear the iron horse make the hills echo with his snort like thunder, shaking the earth with his feet, and breathing fire and smoke from his nostrils, (what kind of winged horse or fiery dragon they will put into the new Mythology I don't know,) it seems as if the earth had got a race now worthy to inhabit it. If all were as it seems, and men made the elements their servants for noble ends! If the cloud that hangs over the engine were the perspiration of heroic deeds, or as beneficent as that which floats over the farmer's fields, then the elements and Nature herself would cheerfully accompany men on their errands and be their escort.

I watch the passage of the morning cars with the same feeling that I do the rising of the sun, which is hardly more regular. Their train of clouds stretching far behind and rising higher and higher, going to heaven while the cars are going to Boston, conceals the sun for a minute and casts my distant field into the shade, a celestial train* beside which the petty train of cars which hugs the earth is but the barb of the spear. The stabler of the iron horse was up early this winter morning by the light of the stars amid the mountains, to fodder and harness his steed. Fire, too, was awakened thus early to put the vital heat in him and get him off. If the enterprise were as innocent as it is early! If the snow lies deep, they strap on his snow-shoes, and with the giant plough plough a furrow from the mountains to the seaboard, in which the cars, like a following drill-barrow,* sprinkle all the restless men and floating merchandise in the country for seed. All day the fire-steed flies over the country, stopping only that his master may rest, and I am awakened by his tramp and defiant snort at midnight, when in some remote glen in the woods he fronts the elements incased in ice and snow; and he will reach his stall only with the morning star, to start once more on his travels without rest or slumber. Or perchance, at evening, I hear him in his stable blowing off the superfluous energy of the day, that he may calm his nerves and cool his liver and brain for a few hours of iron slumber. If the enterprise were as heroic and commanding as it is protracted and unwearied!

Far through unfrequented woods on the confines of towns, where once only the hunter penetrated by day, in the darkest night dart these bright saloons without the knowledge of their inhabitants; this moment stopping at some brilliant station-house in town or city, where a social crowd is gathered, the next in the Dismal Swamp,* scaring the owl and fox. The startings and arrivals of the cars are now the epochs in the village day. They go and come with such regularity and precision, and their whistle can be heard so far, that the farmers set their clocks* by them, and thus one well conducted institution regulates a whole country. Have not men improved somewhat in punctuality since the railroad was invented? Do they not talk and think faster in the depot* than they did in the stage-office? There is something electrifying in the atmosphere of the former place. I have been astonished at the miracles it has wrought; that some of my neighbors, who, I should have prophesied, once for all, would never get to Boston by so prompt a conveyance, are on hand when the bell rings.* To do things "railroad fashion" is now the by-word; and it is worth the while to be warned so often and so sincerely by any power to get off its track. There is no stopping to read the riot act,* no firing over the heads of the mob, in this case. We have constructed a fate, an *Atropos,** that never turns aside. (Let that be the name of your engine.) Men are advertised that at a certain hour and minute these bolts* will be shot toward particular points of the compass; yet it interferes with no man's business, and the children go to school on the other track. We live the steadier for it. We are all educated thus to be sons of Tell.* The air is full of invisible bolts. Every path but your own is the path of fate. Keep on your own track, then.

What recommends commerce to me is its enterprise and bravery. It does not clasp its hands and pray to Jupiter. I see these men every day go about their business with more or less courage and content, doing more even than they suspect, and perchance better employed than they could have consciously devised. I am less affected by their heroism who stood up for half an hour in the front line at Buena Vista,* than by the steady and cheerful valor of the men who inhabit the snow-plough for their winter quarters; who have not merely the three-o'-clock in the morning courage, which Bonaparte thought was the rarest,* but whose courage does not go to rest so early, who go to sleep only when the storm sleeps or the sinews of their iron steed are frozen. On this morning of the Great Snow,* perchance, which is still raging and chilling men's blood, I hear the muffled tone of their engine bell from out the fog bank of their chilled breath, which announces that the cars *are coming*, without long delay, notwithstanding the veto* of a New England north-east snow storm, and I behold the ploughmen covered with snow and rime, their heads peering above the mould-board* which is turning down other than daisies and the nests of field-mice, like bowlders of the Sierra Nevada, that occupy an outside place in the universe.

Commerce is unexpectedly confident and serene, alert, adventurous, and unwearied. It is very natural in its methods withal, far more so than many fantastic enterprises and senti-mental experiments, and hence its singular success. I am refreshed and expanded when the freight train rattles past me, and I smell the stores which go dispensing their odors all the way from Long Wharf to Lake Champlain,* reminding me of foreign parts, of coral reefs, and Indian oceans, and tropical climes, and the extent of the globe. I feel more like a citizen of the world at the sight of the palm-leaf* which will cover so many flaxen New England heads the next summer, the Manilla hemp and cocoa-nut husks,* the old junk, gunny* bags, scrap iron, and rusty nails. This car-load of torn sails* is more legible and interesting now than if they should be wrought into paper and printed books. Who can write so graphically the history of the storms they have weathered as these rents have done? They are proof-sheets which need no correction. Here goes lumber from the Maine woods, which did not go out to sea in the last freshet,* risen four dollars on the thousand because of what did go out or was split up; pine, spruce, cedar,—first, second, third and fourth qualities, so lately all of one quality, to wave over the bear, and moose, and caribou. Next rolls Thomaston* lime, a prime lot, which will

get far among the hills before it gets slacked.* These rags in
bales, of all hues and qualities, the lowest condition to which
cotton and linen descend, the final result of dress,—of patterns which are now no longer cried up, unless it be in
Milwaukie,* as those splendid articles, English, French, or
American prints, ginghams, muslins, &c., gathered from all
quarters both of fashion and poverty, going to become paper
of one color or a few shades only, on which forsooth will be
written tales of real life, high and low, and founded on fact!
This closed car smells of salt fish, the strong New England and
commercial scent, reminding me of the Grand Banks* and
the fisheries. Who has not seen a salt fish, thoroughly cured
for this world, so that nothing can spoil it, and putting the
perseverance of the saints to the blush? with which you may
sweep or pave the streets, and split your kindlings, and the
teamster shelter himself and his lading against sun, wind and
rain behind it,—and the trader, as a Concord trader once did,
hang it up by his door for a sign when he commences business, until at last his oldest customer cannot tell surely
whether it be animal, vegetable, or mineral, and yet it shall be
as pure as a snowflake, and if it be put into a pot and boiled,
will come out an excellent dun* fish for a Saturday's dinner.
Next Spanish hides, with the tails still preserving their twist
and the angle of elevation they had when the oxen that wore
them were careering over the pampas of the Spanish main,—
a type of all obstinacy, and evincing how almost hopeless and
incurable are all constitutional vices. I confess, that practically
speaking, when I have learned a man's real disposition, I have
no hopes of changing it for the better or worse in this state of
existence. As the Orientals say, "A cur's tail may be warmed,
and pressed, and bound round with ligatures, and after a
twelve years' labor bestowed upon it, still it will retain its
natural form."* The only effectual cure for such inveteracies
as these tails exhibit is to make glue of them, which I believe
is what is usually done with them, and then they will stay
put and stick. Here is a hogshead of molasses or of brandy

directed to John Smith, Cuttingsville, Vermont, some trader
among the Green Mountains,* who imports for the farmers
near his clearing, and now perchance stands over his bulk-
head and thinks of the last arrivals on the coast, how they may
affect the price for him, telling his customers this moment, as
he has told them twenty times before this morning, that he
expects some by the next train of prime quality. It is advertised in the Cuttingsville Times.

While these things go up other things come down. Warned
by the whizzing sound, I look up from my book and see some
tall pine, hewn on far northern hills, which has winged its way
over the Green Mountains and the Connecticut,* shot like an
arrow through the township within ten minutes, and scarce
another eye beholds it; going

> "to be the mast
> Of some great ammiral."*

And hark! here comes the cattle-train bearing the cattle of a
thousand hills,* sheepcots, stables, and cow-yards in the air,
drovers with their sticks, and shepherd boys in the midst of
their flocks, all but the mountain pastures, whirled along like
leaves blown from the mountains by the September gales.
The air is filled with the bleating of calves and sheep, and the
hustling of oxen, as if a pastoral valley were going by. When
the old bell-weather at the head rattles his bell, the mountains
do indeed skip like rams and the little hills like lambs.* A car-
load of drovers, too, in the midst, on a level with their droves
now, their vocation gone, but still clinging to their useless
sticks as their badge of office. But their dogs, where are they?
It is a stampede to them; they are quite thrown out; they have
lost the scent. Methinks I hear them barking behind the
Peterboro' Hills,* or panting up the western slope of the
Green Mountains. They will not be in at the death. Their
vocation, too, is gone. Their fidelity and sagacity are below par
now. They will slink back to their kennels in disgrace, or
perchance run wild and strike a league with the wolf and the

fox. So is your pastoral life whirled past and away. But the bell rings, and I must get off the track and let the cars go by;—

> What's the railroad to me?
> I never go to see
> Where it ends.
> It fills a few hollows,
> And makes banks for the swallows,
> It sets the sand a-blowing,
> And the blackberries a-growing,*

but I cross it like a cart-path in the woods. I will not have my eyes put out and my ears spoiled by its smoke and steam and hissing.

Now that the cars are gone by and all the restless world with them, and the fishes in the pond no longer feel their rumbling, I am more alone than ever. For the rest of the long afternoon, perhaps, my meditations are interrupted only by the faint rattle of a carriage or team along the distant highway.

Sometimes, on Sundays, I heard the bells, the Lincoln, Acton, Bedford, or Concord* bell, when the wind was favorable, a faint, sweet, and, as it were, natural melody, worth importing into the wilderness. At a sufficient distance over the woods this sound acquires a certain vibratory hum, as if the pine needles in the horizon were the strings of a harp which it swept. All sound heard at the greatest possible distance produces one and the same effect, a vibration of the universal lyre, just as the intervening atmosphere makes a distant ridge of earth interesting to our eyes by the azure tint it imparts to it. There came to me in this case a melody which the air had strained, and which had conversed with every leaf and needle of the wood, that portion of the sound which the elements had taken up and modulated and echoed from vale to vale. The echo is, to some extent, an original sound, and therein is the magic and charm of it. It is not merely a repetition of what was worth repeating in the bell, but partly the voice of the wood; the same trivial words and notes sung by a wood-nymph.

At evening, the distant lowing of some cow in the horizon beyond the woods sounded sweet and melodious, and at first I would mistake it for the voices of certain minstrels by whom I was sometimes serenaded, who might be straying over hill and dale; but soon I was not unpleasantly disappointed when it was prolonged into the cheap and natural music of the cow. I do not mean to be satirical, but to express my appreciation of those youths' singing, when I state that I perceived clearly that it was akin to the music of the cow, and they were at length one articulation of Nature.

Regularly at half past seven, in one part of the summer, after the evening train had gone by, the whippoorwills chanted their vespers for half an hour, sitting on a stump by my door, or upon the ridge pole of the house. They would begin to sing almost with as much precision as a clock, within five minutes of a particular time, referred to the setting of the sun, every evening. I had a rare opportunity to become acquainted with their habits. Sometimes I heard four or five at once in different parts of the wood, by accident one a bar behind another, and so near me that I distinguished not only the cluck after each note, but often that singular buzzing sound like a fly in a spider's web, only proportionally louder. Sometimes one would circle round and round me in the woods a few feet distant as if tethered by a string, when probably I was near its eggs. They sang at intervals throughout the night, and were again as musical as ever just before and about dawn.

When other birds are still the screech owls take up the strain, like mourning women their ancient u-lu-lu.* Their dismal scream is truly Ben Jonsonian.* Wise midnight hags! It is no honest and blunt tu-whit tu-who* of the poets, but, without jesting, a most solemn graveyard ditty, the mutual consolations of suicide lovers remembering the pangs and the delights of supernal love in the infernal groves. Yet I love to hear their wailing, their doleful responses, trilled along the wood-side; reminding me sometimes of music and singing birds; as if it were the dark and tearful side of music, the regrets and sighs that would fain be sung. They are the spirits, the low spirits and melancholy forebodings, of fallen souls that once in human shape nightwalked the earth and did the deeds of darkness, now expiating their sins with their wailing hymns or threnodies in the scenery of their transgressions. They give me a new sense of the variety and capacity of that nature which is our common dwelling. *Oh-o-o-o-o that I never had been bor-r-r-r-n!* sighs one on this side of the pond, and circles with the restlessness of despair to some new perch on the gray oaks. Then—*that I never had been bor-r-r-r-n!* echoes another on the farther side with tremulous sincerity, and—*bor-r-r-r-n!* comes faintly from far in the Lincoln woods.

I was also serenaded by a hooting owl. Near at hand you could fancy it the most melancholy sound in Nature, as if she meant by this to stereotype and make permanent in her choir the dying moans of a human being,—some poor weak relic of mortality who has left hope behind, and howls like an animal, yet with human sobs, on entering the dark valley, made more awful by a certain gurgling melodiousness,—I find myself beginning with the letters gl when I try to imitate it,—expressive of a mind which has reached the gelatinous mildewy stage in the mortification of all healthy and coura-geous thought. It reminded me of ghouls and idiots and in-sane howlings. But now one answers from far woods in a strain made really melodious by distance,—*Hoo hoo hoo, hoorer hoo*; and indeed for the most part it suggested only pleasing associations, whether heard by day or night, summer or winter.

I rejoice that there are owls. Let them do the idiotic and maniacal hooting for men. It is a sound admirably suited to swamps and twilight woods which no day illustrates, suggesting a vast and undeveloped nature which men have not recognized. They represent the stark twilight and unsatisfied thoughts which all have. All day the sun has shone on the surface of some savage swamp, where the single spruce* stands hung with usnea lichens, and small hawks circulate above, and the chicadee lisps amid the evergreens, and the partridge and rabbit skulk beneath; but now a more dismal and fitting day dawns, and a different race of creatures awakes to express the meaning of Nature there.

Late in the evening I heard the distant rumbling of wagons over bridges,—a sound heard farther than almost any other at night,—the baying of dogs, and sometimes again the lowing of some disconsolate cow in a distant barn-yard. In the mean while all the shore rang with the trump of bullfrogs, the sturdy spirits of ancient winebibbers and wassailers,* still unrepentant, trying to sing a catch* in their Stygian* lake,—if the Walden nymphs will pardon the comparison, for though there are almost no weeds, there are frogs there,—who would fain keep up the hilarious rules of their old festal tables, though their voices have waxed hoarse and solemnly grave, mocking at mirth, and the wine has lost its flavor, and become only liquor to distend their paunches, and sweet intoxication never comes to drown the memory of the past, but mere saturation and waterloggedness and distention. The most aldermanic, with his chin upon a heart-leaf, which serves for a napkin to his drooling chaps, under this northern shore quaffs a deep draught of the once scorned water, and passes round the cup with the ejaculation *tr-r-r-oonk*, *tr-r-r-oonk*, *tr-r-r-oonk!* and straightway comes over the water from some distant cove the same password repeated, where the next in seniority and girth has gulped down to his mark; and when this observance has made the circuit of the shores, then ejaculates the master of ceremonies, with satisfaction, *tr-r-r-oonk!* and each in his turn repeats the same down to the least distended, leakiest, and flabbiest paunched, that there be no mistake; and then the bowl goes round again and again, until the sun disperses the morning mist, and only the patriarch is not under the pond, but vainly bellowing *troonk* from time to time, and pausing for a reply.

I am not sure that I ever heard the sound of cock-crowing from my clearing, and I thought that it might be worth the while to keep a cockerel for his music merely, as a singing bird. The note of this once wild Indian pheasant* is certainly the most remarkable of any bird's, and if they could be naturalized without being domesticated, it would soon become the most famous sound in our woods, surpassing the clangor of the goose and the hooting of the owl; and then imagine the cackling of the hens to fill the pauses when their lords' clarions rested! No wonder that man added this bird to his tame stock,—to say nothing of the eggs and drumsticks. To walk in a winter morning in a wood where these birds abounded, their native woods, and hear the wild cockerels crow on the trees, clear and shrill for miles over the resounding earth, drowning the feebler notes of other birds,—think of it! It would put nations on the alert. Who would not be early to rise,* and rise earlier and earlier every successive day of his life, till he became unspeakably healthy, wealthy, and wise? This foreign bird's note is celebrated by the poets of all countries along with the notes of their native songsters. All climates agree with brave Chanticleer. He is more indigenous even than the natives. His health is ever good, his lungs are sound, his spirits never flag. Even the sailor* on the Atlantic and Pacific is awakened by his voice; but its shrill sound never roused me from my slumbers. I kept neither dog, cat, cow, pig, nor hens, so that you would have said there was a deficiency of domestic sounds; neither the churn, nor the spinning wheel, nor even the singing of the kettle, nor the hissing of the urn, nor children crying, to comfort one. An old-fashioned man would have lost his senses or died of ennui before this. Not even rats in the wall, for they were starved out, or rather were never baited in,—only squirrels on the roof and under the floor, a whipporwill on the ridge pole, a blue-jay screaming beneath the window, a hare or woodchuck under the house, a screech-owl or a cat-owl behind it, a flock of wild geese or a laughing loon on the pond, and a fox to bark in the night. Not even a lark or an oriole, those mild plantation birds, ever visited my clearing. No cockerels to crow nor hens to cackle in the yard. No yard! but unfenced Nature reaching up to your very sills. A young forest growing up under your windows, and wild sumachs and blackberry vines breaking through into your cellar; sturdy pitch-pines rubbing and creaking against the shingles for want of room, their roots reaching quite under the house. Instead of a scuttle or a blind blown off in the gale,— a pine tree snapped off or torn up by the roots behind your house for fuel. Instead of no path to the front-yard gate in the Great Snow,*—no gate—no front-yard,—and no path to the civilized world!

SOLITUDE*

THIS is a delicious evening, when the whole body is one sense, and imbibes delight through every pore. I go and come with a strange liberty in Nature, a part of herself. As I walk along the stony shore of the pond in my shirt sleeves, though it is cool as well as cloudy and windy, and I see nothing special to attract me, all the elements are unusually congenial to me. The bullfrogs trump to usher in the night, and the note of the whippoorwill is borne on the rippling wind from over the water. Sympathy with the fluttering alder and poplar leaves almost takes away my breath; yet, like the lake, my serenity is rippled but not ruffled. These small waves raised by the evening wind are as remote from storm as the smooth reflecting surface. Though it is now dark, the wind still blows and roars in the wood, the waves still dash, and some creatures lull the rest with their notes. The repose is never complete. The wildest animals do not repose, but seek their prey now; the fox, and skunk, and rabbit, now roam the fields and woods without fear. They are Nature's watchmen—links which connect the days of animated life.

When I return to my house I find that visitors have been there and left their cards, either a bunch of flowers, or a wreath of evergreen, or a name in pencil on a yellow walnut leaf or a chip. They who come rarely to the woods take some little piece of the forest into their hands to play with by the way, which they leave, either intentionally or accidentally. One has peeled a willow wand, woven it into a ring, and dropped it on my table. I could always tell if visitors had called in my absence, either by the bended twigs or grass, or the print of their shoes, and generally of what sex or age or quality they were by some slight trace left, as a flower dropped, or a bunch of grass plucked and thrown away, even as far off as the railroad, half a mile distant, or by the lingering odor of a cigar or pipe. Nay, I was frequently notified of the passage of a traveller along the highway sixty rods off by the scent of his pipe.

There is commonly sufficient space about us. Our horizon is never quite at our elbows. The thick wood is not just at our door, nor the pond, but somewhat is always clearing, familiar and worn by us, appropriated and fenced in some way, and reclaimed from Nature. For what reason have I this vast range and circuit, some square miles of unfrequented forest, for my privacy, abandoned to me by men?* My nearest neighbor is a mile distant, and no house is visible from any place but the hill-tops within half a mile of my own. I have my horizon bounded by woods all to myself; a distant view of the railroad where it touches the pond on the one hand, and of the fence which skirts the woodland road on the other. But for the most part it is as solitary where I live as on the prairies. It is as much Asia or Africa as New England. I have, as it were, my own sun and moon and stars, and a little world all to myself. At night there was never a traveller passed my house, or knocked at my door, more than if I were the first or last man; unless it were in the spring, when at long intervals some came from the village to fish for pouts,—they plainly fished much more in the Walden Pond of their own natures, and baited their hooks with darkness,—but they soon retreated, usually with light baskets, and left "the world to darkness and to me,"* and the black kernel of the night was never profaned by any human neighborhood. I believe that men are generally still a little afraid of the dark, though the witches are all hung, and Christianity and candles have been introduced.

Yet I experienced sometimes that the most sweet and tender, the most innocent and encouraging society may be found in any natural object, even for the poor misanthrope and most melancholy man. There can be no very black melancholy* to him who lives in the midst of Nature and has his senses still. There was never yet such a storm but it was Æolian music* to a healthy and innocent ear. Nothing can rightly compel a simple and brave man to a vulgar sadness. While I enjoy the friendship of the seasons I trust that nothing can make life a burden to me. The gentle rain which waters my beans and keeps me in the house to-day is not drear and melancholy, but good for me too. Though it prevents my hoeing them, it is of far more worth than my hoeing. If it should continue so long as to cause the seeds to rot in the ground and destroy the potatoes in the low lands, it would still be good for the grass on the uplands, and, being good for the grass, it would be good for me. Sometimes, when I compare myself with other men, it seems as if I were more favored by the gods than they, beyond any deserts that I am conscious of; as if I had a warrant and surety at their hands which my fellows have not, and were especially guided and guarded. I do not flatter myself, but if it be possible they flatter me. I have never felt lonesome, or in the least oppressed by a sense of solitude, but once, and that was a few weeks after I came to the woods, when, for an hour, I doubted if the near neighborhood of man was not essential to a serene and healthy life. To be alone was something unpleasant. But I was at the same time conscious of a slight insanity in my mood, and seemed to foresee my recovery. In the midst of a gentle rain while these thoughts prevailed, I was suddenly sensible of such sweet and beneficent society in Nature, in the very pattering of the drops, and in every sound and sight around my house, an infinite and unaccountable friendliness all at once like an atmosphere sustaining me, as made the fancied advantages of human neighborhood insignificant, and I have never thought of them since. Every little pine needle expanded and swelled with sympathy and befriended me. I was so distinctly made aware of the presence of something kindred to me, even in scenes which we are accustomed to call wild and dreary, and also that the nearest of blood to me and humanest was not a person nor a villager, that I thought no place could ever be strange to me again.—

"Mourning untimely consumes the sad;
Few are their days in the land of the living,
Beautiful daughter of Toscar."*

Some of my pleasantest hours were during the long rain
storms in the spring or fall, which confined me to the house
for the afternoon as well as the forenoon, soothed by their
ceaseless roar and pelting; when an early twilight ushered in a
long evening in which many thoughts had time to take root
and unfold themselves. In those driving north-east rains*
which tried the village houses so, when the maids stood ready
with mop and pail in front entries to keep the deluge out, I sat
behind my door in my little house, which was all entry, and
thoroughly enjoyed its protection. In one heavy thunder
shower the lightning struck a large pitch-pine across the pond,
making a very conspicuous and perfectly regular spiral groove
from top to bottom, an inch or more deep, and four or five
inches wide, as you would groove a walking-stick. I passed it
again the other day, and was struck with awe on looking up
and beholding that mark, now more distinct than ever, where
a terrific and resistless bolt came down out of the harmless sky
eight years ago. Men frequently say to me, "I should think you
would feel lonesome down there, and want to be nearer to
folks, rainy and snowy days and nights especially." I am
tempted to reply to such,—This whole earth which we inhabit
is but a point in space. How far apart, think you, dwell the two
most distant inhabitants of yonder star, the breadth of whose
disk cannot be appreciated by our instruments? Why should I
feel lonely? is not our planet in the Milky Way? This which
you put seems to me not to be the most important question.
What sort of space is that which separates a man from his
fellows and makes him solitary? I have found that no exertion
of the legs can bring two minds much nearer to one another.
What do we want most to dwell near to? Not to many men
surely, the depot, the post-office, the bar-room, the meeting-
house, the school-house, the grocery, Beacon Hill,* or the

Five Points,* where men most congregate, but to the perennial source of our life, whence in all our experience we have
found that to issue, as the willow stands near the water and
sends out its roots in that direction. This will vary with different natures, but this is the place where a wise man will dig his
cellar. . . . I one evening overtook one of my townsmen, who
has accumulated what is called "a handsome property,"—
though I never got a *fair* view of it,—on the Walden road,*
driving a pair of cattle to market, who inquired of me how I
could bring my mind to give up so many of the comforts of
life. I answered that I was very sure I liked it passably well; I
was not joking. And so I went home to my bed, and left him
to pick his way through the darkness and the mud to
Brighton,*—or Bright-town,*—which place he would reach
some time in the morning.

Any prospect of awakening or coming to life to a dead man
makes indifferent all times and places. The place where that
may occur is always the same, and indescribably pleasant to all
our senses. For the most part we allow only outlying and
transient circumstances to make our occasions. They are, in
fact, the cause of our distraction. Nearest to all things is that
power which fashions their being. *Next* to us the grandest laws
are continually being executed. *Next* to us is not the workman
whom we have hired, with whom we love so well to talk, but
the workman whose work we are.

"How vast and profound is the influence of the subtile
powers of Heaven and of Earth!"

"We seek to perceive them, and we do not see them; we
seek to hear them, and we do not hear them; identified
with the substance of things, they cannot be separated from
them."

"They cause that in all the universe men purify and sanctify
their hearts, and clothe themselves in their holiday garments
to offer sacrifices and oblations to their ancestors. It is an
ocean of subtile intelligences. They are every where, above us,
on our left, on our right; they environ us on all sides."*

We are the subjects of an experiment which is not a little interesting to me. Can we not do without the society of our gossips a little while under these circumstances,—have our own thoughts to cheer us? Confucius says truly, "Virtue does not remain as an abandoned orphan; it must of necessity have neighbors."*

With thinking we may be beside ourselves in a sane sense. By a conscious effort of the mind we can stand aloof from actions and their consequences; and all things, good and bad, go by us like a torrent. We are not wholly involved in Nature. I may be either the driftwood in the stream, or Indra* in the sky looking down on it. I *may* be affected by a theatrical exhibition; on the other hand, I *may not* be affected by an actual event which appears to concern me much more. I only know myself as a human entity; the scene, so to speak, of thoughts and affections; and am sensible of a certain doubleness by which I can stand as remote from myself as from another. However intense my experience, I am conscious of the presence and criticism of a part of me, which, as it were, is not a part of me, but spectator, sharing no experience, but taking note of it; and that is no more I than it is you. When the play, it may be the tragedy, of life is over, the spectator goes his way. It was a kind of fiction, a work of the imagination only, so far as he was concerned. This doubleness may easily make us poor neighbors and friends sometimes.

I find it wholesome to be alone the greater part of the time. To be in company, even with the best, is soon wearisome and dissipating. I love to be alone. I never found the companion that was so companionable as solitude. We are for the most part more lonely when we go abroad among men than when we stay in our chambers. A man thinking or working is always alone, let him be where he will. Solitude is not measured by the miles of space that intervene between a man and his fellows. The really diligent student in one of the crowded hives of Cambridge College is as solitary as a dervis in the desert. The farmer can work alone in the field or the woods all day, hoeing or chopping, and not feel lonesome, because he is employed; but when he comes home at night he cannot sit down in a room alone, at the mercy of his thoughts, but must be where he can "see the folks," and recreate, and as he thinks remunerate, himself for his day's solitude; and hence he wonders how the student can sit alone in the house all night and most of the day without ennui and "the blues;" but he does not realize that the student, though in the house, is still at work in *his* field, and chopping in *his* woods, as the farmer in his, and in turn seeks the same recreation and society that the latter does, though it may be a more condensed form of it.

Society* is commonly too cheap. We meet at very short intervals, not having had time to acquire any new value for each other. We meet at meals three times a day, and give each other a new taste of that old musty cheese that we are. We have had to agree on a certain set of rules, called etiquette and politeness, to make this frequent meeting tolerable and that we need not come to open war. We meet at the post-office, and at the sociable, and about the fireside every night; we live thick and are in each other's way, and stumble over one another, and I think that we thus lose some respect for one another. Certainly less frequency would suffice for all important and hearty communications. Consider the girls in a factory,*—never alone, hardly in their dreams. It would be better if there were but one inhabitant to a square mile, as where I live. The value of a man is not in his skin, that we should touch him.

I have heard of a man lost in the woods and dying of famine and exhaustion at the foot of a tree, whose loneliness was relieved by the grotesque visions with which, owing to bodily weakness, his diseased imagination surrounded him, and which he believed to be real. So also, owing to bodily and mental health and strength, we may be continually cheered by a like but more normal and natural society, and come to know that we are never alone.

I have a great deal of company in my house; especially in the morning, when nobody calls. Let me suggest a few comparisons, that some one may convey an idea of my situation. I am no more lonely than the loon in the pond that laughs so loud, or than Walden Pond itself. What company has that lonely lake, I pray? And yet it has not the blue devils,* but the blue angels in it, in the azure tint of its waters. The sun is alone, except in thick weather, when there sometimes appear to be two, but one is a mock sun.* God is alone,—but the devil, he is far from being alone; he sees a great deal of company; he is legion.* I am no more lonely than a single mullein or dandelion in a pasture, or a bean leaf, or sorrel, or a horse-fly, or a humble-bee. I am no more lonely than the Mill Brook,* or a weathercock, or the north star, or the south wind, or an April shower, or a January thaw, or the first spider in a new house.

I have occasional visits in the long winter evenings, when the snow falls fast and the wind howls in the wood, from an old settler* and original proprietor, who is reported to have dug Walden Pond, and stoned it, and fringed it with pine woods; who tells me stories of old time and of new eternity; and between us we manage to pass a cheerful evening with social mirth and pleasant views of things, even without apples or cider,—a most wise and humorous friend, whom I love much, who keeps himself more secret than ever did Goffe or Whalley;* and though he is thought to be dead, none can show where he is buried. An elderly dame,* too, dwells in my neighborhood, invisible to most persons, in whose odorous herb garden I love to stroll sometimes, gathering simples and listening to her fables; for she has a genius of unequalled fertility, and her memory runs back farther than mythology, and she can tell me the original of every fable, and on what fact every one is founded, for the incidents occurred when she was young. A ruddy and lusty old dame, who delights in all weathers and seasons, and is likely to outlive all her children yet.

The indescribable innocence and beneficence of Nature,—of sun and wind and rain, of summer and winter,—such health, such cheer, they afford forever! and such sympathy have they ever with our race, that all Nature would be affected, and the sun's brightness fade, and the winds would sigh humanely, and the clouds rain tears, and the woods shed their leaves and put on mourning in midsummer, if any man should ever for a just cause grieve. Shall I not have intelligence with the earth? Am I not partly leaves and vegetable mould myself?

What is the pill which will keep us well, serene, contented? Not my or thy great-grandfather's, but our great-grandmother Nature's universal, vegetable, botanic medicines, by which she has kept herself young always, outlived so many old Parrs* in her day, and fed her health with their decaying fatness. For my panacea, instead of one of those quack vials of a mixture dipped from Acheron* and the Dead Sea, which come out of those long shallow black-schooner looking wagons which we sometimes see made to carry bottles, let me have a draught of undiluted morning air. Morning air! If men will not drink of this at the fountain-head of the day, why, then, we must even bottle up some and sell it in the shops, for the benefit of those who have lost their subscription ticket to morning time in this world. But remember, it will not keep quite till noon-day even in the coolest cellar, but drive out the stopples* long ere that and follow westward the steps of Aurora.* I am no worshipper of Hygeia,* who was the daughter of that old herb-doctor Æsculapius,* and who is represented on monuments holding a serpent in one hand, and in the other a cup out of which the serpent sometimes drinks; but rather of Hebe,* cupbearer to Jupiter, who was the daughter of Juno and wild lettuce, and who had the power of restoring gods and men to the vigor of youth. She was probably the only thoroughly sound-conditioned, healthy, and robust young lady that ever walked the globe, and wherever she came it was spring.

VISITORS

I think that I love society as much as most, and am ready enough to fasten myself like a bloodsucker for the time to any full-blooded man that comes in my way. I am naturally no hermit, but might possibly sit out the sturdiest frequenter of the bar-room, if my business called me thither.

I had three chairs in my house; one for solitude, two for friendship, three for society. When visitors came in larger and unexpected numbers there was but the third chair for them all, but they generally economized the room by standing up. It is surprising how many great men and women a small house will contain. I have had twenty-five or thirty souls, with their bodies, at once under my roof, and yet we often parted without being aware that we had come very near to one another. Many of our houses, both public and private, with their almost innumerable apartments, their huge halls and their cellars for the storage of wines and other munitions of peace, appear to me extravagantly large for their inhabitants. They are so vast and magnificent that the latter seem to be only vermin which infest them. I am surprised when the herald blows his summons before some Tremont or Astor or Middlesex House,* to see come creeping out over the piazza for all inhabitants a ridiculous mouse, which soon again slinks into some hole in the pavement.

One inconvenience I sometimes experienced in so small a house, the difficulty of getting to a sufficient distance from my guest when we began to utter the big thoughts in big words. You want room for your thoughts to get into sailing trim and run a course or two before they make their port. The bullet of your thought must have overcome its lateral and ricochet motion and fallen into its last and steady course before it reaches the ear of the hearer, else it may plough out again through the side of his head. Also, our sentences wanted room to unfold and form their columns in the interval. Individuals, like nations, must have suitable broad and natural boundaries, even a considerable neutral ground, between them. I have found it a singular luxury to talk across the pond to a companion on the opposite side. In my house we were so near that we could not begin to hear,—we could not speak low enough to be heard; as when you throw two stones into a calm water so near that they break each other's undulations. If we are merely loquacious and loud talkers, then we can afford to stand very near together, cheek by jowl, and feel each other's breath; but if we speak reservedly and thoughtfully, we want to be farther apart, that all animal heat and moisture may have a chance to evaporate. If we would enjoy the most intimate society with that in each of us which is without, or above, being spoken to, we must not only be silent, but commonly so far apart bodily that we cannot possibly hear each other's voice in any case. Referred to this standard, speech is for the convenience of those who are hard of hearing; but there are many fine things which we cannot say if we have to shout. As the conversation began to assume a loftier and grander tone, we gradually shoved our chairs farther apart till they touched the wall in opposite corners, and then commonly there was not room enough.*

My "best" room, however, my withdrawing room,* always ready for company, on whose carpet the sun rarely fell,* was the pine wood behind my house. Thither in summer days, when distinguished guests came, I took them, and a priceless domestic swept the floor and dusted the furniture and kept the things in order.

If one guest came he sometimes partook of my frugal meal, and it was no interruption to conversation to be stirring a hasty-pudding,* or watching the rising and maturing of a loaf of bread in the ashes, in the mean while. But if twenty came and sat in my house there was nothing said about dinner, though there might be bread enough for two, more than if eating were a forsaken habit; but we naturally practised abstinence; and this was never felt to be an offense against hospitality, but the most proper and considerate course. The waste and decay of physical life, which so often needs repair, seemed miraculously retarded in such a case, and the vital vigor stood its ground. I could entertain thus a thousand as well as twenty; and if any ever went away disappointed or hungry from my house when they found me at home, they may depend upon it that I sympathized with them at least. So easy is it, though many housekeepers doubt it, to establish new and better customs in the place of old. You need not rest your reputation on the dinners you give. For my own part, I was never so effectually deterred from frequenting a man's house, by any kind of Cerberus* whatever, as by the parade one made about dining me, which I took to be a very polite and roundabout hint never to trouble him so again. I think I shall never revisit those scenes. I should be proud to have for the motto of my cabin those lines of Spenser which one of my visitors inscribed on a yellow walnut leaf for a card:—

> "Arrivéd there, the little house they fill,
> Ne looke for entertainment where none was;
> Rest is their feast, and all things at their will:
> The noblest mind the best contentment has."*

When Winslow,* afterward governor of the Plymouth Colony, went with a companion on a visit of ceremony to Massassoit* on foot through the woods, and arrived tired and hungry at his lodge, they were well received by the king, but nothing was said about eating that day. When the night arrived, to quote their own words,—"He laid us on a bed with himself and his wife, they at the one end and we at the other, it being only plank, laid a foot from the ground, and a thin mat upon them. Two more of his chief men, for want of room, pressed by and upon us; so that we were worse weary of our lodging than of our journey." At one o'clock the next day

Massassoit "brought two fishes that he had shot," about thrice as big as bream; "these being boiled, there were at least forty looked for a share in them. The most ate of them. This meal only we had in two nights and a day; and had not one of us bought a partridge, we had taken our journey fasting." Fearing that they would be light-headed for want of food and also sleep, owing to "the savages' barbarous singing, (for they used to sing themselves asleep,)"* and that they might get home while they had strength to travel, they departed. As for lodging, it is true they were but poorly entertained, though what they found an inconvenience was no doubt intended for an honor; but as far as eating was concerned, I do not see how the Indians could have done better. They had nothing to eat themselves, and they were wiser than to think that apologies could supply the place of food to their guests; so they drew their belts tighter and said nothing about it. Another time when Winslow visited them, it being a season of plenty with them, there was no deficiency in this respect.

As for men, they will hardly fail one any where. I had more visitors while I lived in the woods than at any other period of my life; I mean that I had some. I met several there under more favorable circumstances than I could any where else. But fewer came to see me upon trivial business. In this respect, my company was winnowed by my mere distance from town. I had withdrawn so far within the great ocean of solitude, into which the rivers of society empty, that for the most part, so far as my needs were concerned, only the finest sediment was deposited around me. Beside, there were wafted to me evidences of unexplored and uncultivated continents on the other side.

Who should come to my lodge this morning but a true Homeric or Paphlagonian man,*—he had so suitable and poetic a name that I am sorry I cannot print it here,—a Canadian, a wood-chopper and post-maker, who can hole fifty posts in a day, who made his last supper on a woodchuck which his dog caught. He, too, has heard of Homer, and, "if it

were not for books," would "not know what to do rainy days,"
though perhaps he has not read one wholly through for many
rainy seasons. Some priest who could pronounce the Greek
itself taught him to read his verse in the testament in his
native parish far away; and now I must translate to him, while
he holds the book, Achilles' reproof to Patroclus for his sad
countenance.—"Why are you in tears, Patroclus, like a young
girl?"—

    "Or have you alone heard some news from Phthia?
    They say that Menœtius lives yet, son of Actor,
    And Peleus lives, son of Æacus, among the Myrmidons,
    Either of whom having died, we should greatly grieve."*

He says, "That's good." He has a great bundle of white-oak
bark* under his arm for a sick man, gathered this Sunday
morning. "I suppose there's no harm in going after such a
thing to-day," says he. To him Homer was a great writer,
though what his writing was about he did not know. A more
simple and natural man it would be hard to find. Vice and
disease, which cast such a sombre moral hue over the world,
seemed to have hardly any existence for him. He was about
twenty-eight years old, and had left Canada and his father's
house a dozen years before to work in the States, and earn
money to buy a farm with at last, perhaps in his native country.
He was cast in the coarsest mould; a stout but sluggish body,
yet gracefully carried, with a thick sunburnt neck, dark bushy
hair, and dull sleepy blue eyes, which were occasionally lit up
with expression. He wore a flat gray cloth cap, a dingy wool-
colored greatcoat, and cowhide boots. He was a great con-
sumer of meat, usually carrying his dinner to his work a couple
of miles past my house,—for he chopped all summer,—in a
tin pail; cold meats, often cold woodchucks, and coffee in a
stone bottle which dangled by a string from his belt; and
sometimes he offered me a drink. He came along early, crossing my bean-field, though without anxiety or haste to get to his
work, such as Yankees exhibit. He wasn't a-going to hurt

himself. He didn't care if he only earned his board. Frequently he would leave his dinner in the bushes, when his dog had caught a woodchuck by the way, and go back a mile and a half to dress it and leave it in the cellar of the house where he boarded, after deliberating first for half an hour whether he could not sink it in the pond safely till nightfall,—loving to dwell long upon these themes. He would say, as he went by in the morning, "How thick the pigeons are! If working every day were not my trade, I could get all the meat I should want by hunting,—pigeons, woodchucks, rabbits, partridges,—by gosh! I could get all I should want for a week in one day."

He was a skilful chopper, and indulged in some flourishes and ornaments in his art. He cut his trees level and close to the ground, that the sprouts which came up afterward might be more vigorous and a sled might slide over the stumps; and instead of leaving a whole tree to support his corded wood, he would pare it away to a slender stake or splinter which you could break off with your hand at last.

He interested me because he was so quiet and solitary and so happy withal; a well of good humor and contentment which overflowed at his eyes. His mirth was without alloy. Sometimes I saw him at his work in the woods, felling trees, and he would greet me with a laugh of inexpressible satisfaction, and a salutation in Canadian French, though he spoke English as well. When I approached him he would suspend his work, and with half-suppressed mirth lie along the trunk of a pine which he had felled, and, peeling off the inner bark, roll it up into a ball and chew it while he laughed and talked. Such an exuberance of animal spirits had he that he sometimes tumbled down and rolled on the ground with laughter at any thing which made him think and tickled him. Looking round upon the trees he would exclaim,—"By George! I can enjoy myself well enough here chopping; I want no better sport." Sometimes, when at leisure, he amused himself all day in the woods with a pocket pistol, firing salutes to himself at regular intervals as he walked. In the winter he had a fire by which at noon he warmed his coffee in a kettle; and as he sat on a log to eat his dinner the chicadees would sometimes come round and alight on his arm and peck at the potato in his fingers; and he said that he "liked to have the little *fellers* about him."

In him the animal man chiefly was developed. In physical endurance and contentment he was cousin to the pine and rock. I asked him once if he was not sometimes tired at night, after working all day; and he answered, with a sincere and serious look, "Gorrappit, I never was tired in my life." But the intellectual and what is called spiritual man in him were slumbering as in an infant. He had been instructed only in that innocent and ineffectual way in which the Catholic priests teach the aborigines, by which the pupil is never educated to the degree of consciousness, but only to the degree of trust and reverence, and a child is not made a man, but kept a child. When Nature made him, she gave him a strong body and contentment for his portion, and propped him on every side with reverence and reliance, that he might live out his threescore* years and ten a child. He was so genuine and unsophisticated that no introduction would serve to introduce him, more than if you introduced a woodchuck to your neighbor. He had got to find him out as you did. He would not play any part. Men paid him wages for work, and so helped to feed and clothe him; but he never exchanged opinions with them. He was so simply and naturally humble—if he can be called humble who never aspires—that humility was no distinct quality in him, nor could he conceive of it. Wiser men were demigods to him. If you told him that such a one was coming, he did as if he thought that any thing so grand would expect nothing of himself, but take all the responsibility on itself, and let him be forgotten still. He never heard the sound of praise. He particularly reverenced the writer and the preacher. Their performances were miracles. When I told him that I wrote considerably, he thought for a long time that it was merely the handwriting which I meant, for he could write a remarkably good hand himself. I sometimes found the name of his native parish handsomely written in the snow by the highway, with the proper French accent, and knew that he had passed. I asked him if he ever wished to write his thoughts. He said that he had read and written letters for those who could not, but he never tried to write thoughts,— no, he could not, he could not tell what to put first, it would kill him, and then there was spelling to be attended to at the same time!

I heard that a distinguished wise man and reformer asked him if he did not want the world to be changed; but he answered with a chuckle of surprise in his Canadian accent, not knowing that the question had ever been entertained before, "No, I like it well enough." It would have suggested many things to a philosopher to have dealings with him. To a stranger he appeared to know nothing of things in general; yet I sometimes saw in him a man whom I had not seen before, and I did not know whether he was as wise as Shakspeare or as simply ignorant as a child, whether to suspect him of a fine poetic consciousness or of stupidity. A townsman told me that when he met him sauntering through the village in his small close-fitting cap, and whistling to himself, he reminded him of a prince in disguise.

His only books were an almanac and an arithmetic, in which last he was considerably expert. The former was a sort of cyclopædia to him, which he supposed to contain an abstract of human knowledge, as indeed it does to a considerable extent. I loved to sound him on the various reforms of the day, and he never failed to look at them in the most simple and practical light. He had never heard of such things before. Could he do without factories? I asked. He had worn the home-made Vermont gray,* he said, and that was good. Could he dispense with tea and coffee? Did this country afford any beverage beside water? He had soaked hemlock leaves in water and drank it, and thought that was better than water in warm weather. When I asked him if he could do without money, he showed the convenience of money in such a way as to suggest and coincide with the most philosophical accounts of the origin of this institution, and the very derivation of the word *pecunia*.* If an ox were his property, and he wished to get needles and thread at the store, he thought it would be inconvenient and impossible soon to go on mortgaging some portion of the creature each time to that amount. He could defend many institutions better than any philosopher, because, in describing them as they concerned him, he gave the true reason for their prevalence, and speculation had not suggested to him any other. At another time, hearing Plato's definition of a man,*—a biped without feathers,—and that one exhibited a cock plucked and called it Plato's man, he thought it an important difference that the *knees* bent the wrong way. He would sometimes exclaim, "How I love to talk! By George, I could talk all day!" I asked him once, when I had not seen him for many months, if he had got a new idea this summer. "Good Lord," said he, "a man that has to work as I do, if he does not forget the ideas he has had, he will do well. May be the man you hoe with is inclined to race; then, by gorry, your mind must be there; you think of weeds." He would sometimes ask me first on such occasions, if I had made any improvement. One winter day I asked him if he was always satisfied with himself, wishing to suggest a substitute within him for the priest without, and some higher motive for living. "Satisfied!" said he; "some men are satisfied with one thing, and some with another. One man, perhaps, if he has got enough, will be satisfied to sit all day with his back to the fire and his belly to the table, by George!" Yet I never, by any manœuvring, could get him to take the spiritual view of things; the highest that he appeared to conceive of was a simple expediency, such as you might expect an animal to appreciate; and this, practically, is true of most men. If I suggested any improvement in his mode of life, he merely answered, without expressing any regret, that it was too late. Yet he thoroughly believed in honesty and the like virtues.

There was a certain positive originality, however slight, to be detected in him, and I occasionally observed that he was thinking for himself and expressing his own opinion, a phenomenon so rare that I would any day walk ten miles to observe it, and it amounted to the re-origination of many of the institutions of society. Though he hesitated, and perhaps failed to express himself distinctly, he always had a presentable thought behind. Yet his thinking was so primitive and immersed in his animal life, that, though more promising than a merely learned man's, it rarely ripened to any thing which can be reported. He* suggested that there might be men of genius in the lowest grades of life, however permanently humble and illiterate, who take their own view always, or do not pretend to see at all; who are as bottomless even as Walden Pond was thought to be, though they may be dark and muddy.

Many a traveller came out of his way to see me and the inside of my house, and, as an excuse for calling, asked for a glass of water. I told them that I drank at the pond, and pointed thither, offering to lend them a dipper.* Far off as I lived, I was not exempted from that annual visitation which occurs, methinks, about the first of April, when every body is on the move; and I had my share of good luck, though there were some curious specimens among my visitors. Half-witted men from the almshouse* and elsewhere came to see me; but I endeavored to make them exercise all the wit they had, and make their confessions to me; in such cases making wit the theme of our conversation; and so was compensated. Indeed, I found some of them to be wiser than the so called *overseers** of the poor and selectmen* of the town, and thought it was time that the tables were turned. With respect to wit, I learned that there was not much difference between the half and the whole. One day, in particular, an inoffensive, simple-minded pauper, whom with others I had often seen used as fencing stuff, standing or sitting on a bushel in the fields to keep cattle and himself from straying, visited me, and expressed a wish to live as I did. He told me, with the utmost simplicity and truth, quite superior, or rather *inferior*, to any thing that is called humility, that he was "deficient in intellect." These were his words. The Lord had made him so, yet he supposed the Lord cared as much for him as for another. "I have always been so," said he, "from my childhood; I never had much mind; I was not like other children; I am weak in the head. It was the Lord's will, I suppose." And there he was to prove the truth of his words. He was a metaphysical puzzle to me. I have rarely met a fellow-man on such promising ground,—it was so simple and sincere and so true all that he said. And, true enough, in proportion as he appeared to humble himself was he exalted.* I did not know at first but it was the result of a wise policy. It seemed that from such a basis of truth and frankness as the poor weak-headed pauper had laid, our intercourse might go forward to something better than the intercourse of sages.

I had some guests from those not reckoned commonly among the town's poor, but who should be; who are among the world's poor, at any rate; guests who appeal, not to your hospitality, but to your *hospitalality*;* who earnestly wish to be helped, and preface their appeal with the information that they are resolved, for one thing, never to help themselves. I require of a visitor that he be not actually starving, though he may have the very best appetite in the world, however he got it. Objects of charity are not guests. Men who did not know when their visit had terminated, though I went about my business again, answering from greater and greater remoteness. Men of almost every degree of wit called on me in the migrating season. Some who had more wits than they knew what to do with; runaway slaves with plantation manners, who listened from time to time, like the fox in the fable,* as if they heard the hounds a-baying on their track, and looked at me beseechingly, as much as to say,—

"O Christian, will you send me back?"*

One real runaway slave,* among the rest, whom I helped to forward toward the northstar. Men of one idea, like a hen with one chicken, and that a duckling; men of a thousand ideas, and unkempt heads, like those hens which are made to take charge of a hundred chickens, all in pursuit of one bug, a score of them lost in every morning's dew,—and become frizzled and mangy in consequence; men of ideas instead of legs, a sort of intellectual centipede that made you crawl all over. One man proposed a book in which visitors should write their names, as at the White Mountains;* but, alas! I have too good a memory to make that necessary.

I could not but notice some of the peculiarities of my visitors. Girls and boys and young women generally seemed glad to be in the woods. They looked in the pond and at the flowers, and improved their time. Men of business, even farmers, thought only of solitude and employment, and of the great distance at which I dwelt from something or other; and though they said that they loved a ramble in the woods occasionally, it was obvious that they did not. Restless committed men, whose time was all taken up in getting a living or keeping it; ministers who spoke of God as if they enjoyed a monopoly of the subject, who could not bear all kinds of opinions; doctors, lawyers, uneasy housekeepers who pried into my cupboard and bed when I was out—how came Mrs.—to know that my sheets were not as clean as hers?—young men who had ceased to be young, and had concluded that it was safest to follow the beaten track of the professions,—all these generally said that it was not possible to do so much good in my position. Ay! there was the rub.* The old and infirm and the timid, of whatever age or sex, thought most of sickness, and sudden accident and death; to them life seemed full of danger,—what danger is there if you don't think of any?—and they thought that a prudent man would carefully select the safest position, where Dr. B.* might be on hand at a moment's warning. To them the village was literally a *com-munity*,* a league for mutual defence, and you would suppose that they would not go a-huckleberrying without a medicine chest. The amount of it is, if a man is alive, there is always *danger* that he may die, though the danger must be allowed to be less in proportion as he is dead-and-alive to begin with. A man sits as many risks as he runs. Finally, there were the self-styled reformers, the greatest bores of all, who thought that I was forever singing,—

> This is the house I built;
> This is the man that lives in the house I built;*

but they did not know that the third line was,—

> These are the folks that worry the man
> That lives in the house that I built.

I did not fear the hen-harriers, for I kept no chickens; but I feared the men-harriers rather.

I had more cheering visitors than the last. Children come a-berrying, railroad men taking a Sunday morning walk in clean shirts, fishermen and hunters, poets and philosophers, in short, all honest pilgrims, who came out to the woods for freedom's sake, and really left the village behind, I was ready to greet with,—"Welcome, Englishmen! welcome, Englishmen!"* for I had had communication with that race.

THE BEAN-FIELD

MEANWHILE my beans, the length of whose rows, added together, was seven miles already planted, were impatient to be hoed, for the earliest had grown considerably before the latest were in the ground; indeed they were not easily to be put off. What was the meaning of this so steady and self-respecting, this small Herculean labor,* I knew not. I came to love my rows, my beans, though so many more than I wanted. They attached me to the earth, and so I got strength like Antæus.* But why should I raise them?* Only Heaven knows. This was my curious labor all summer,—to make this portion of the earth's surface, which had yielded only cinquefoil, blackberries, johnswort, and the like, before, sweet wild fruits and pleasant flowers, produce instead this pulse. What shall I learn of beans or beans of me? I cherish them, I hoe them, early and late I have an eye to them; and this is my day's work. It is a fine broad leaf to look on. My auxiliaries are the dews and rains which water this dry soil, and what fertility is in the soil itself, which for the most part is lean and effete. My enemies are worms, cool days, and most of all woodchucks. The last have nibbled for me a quarter of an acre clean. But what right had I to oust johnswort and the rest, and break up their ancient herb garden? Soon, however, the remaining beans will be too tough for them, and go forward to meet new foes.

When I was four years old, as I well remember, I was brought from Boston to this my native town, through these very woods and this field, to the pond. It is one of the oldest scenes stamped on my memory. And now to-night my flute has waked the echoes over that very water. The pines still stand here older than I; or, if some have fallen, I have cooked my supper with their stumps, and a new growth is rising all around, preparing another aspect for new infant eyes. Almost the same johnswort springs from the same perennial root in this pasture, and even I have at length helped to clothe that fabulous landscape of my infant dreams, and one of the results of my presence and influence is seen in these bean leaves, corn blades, and potato vines.

I planted about two acres and a half of upland; and as it was only about fifteen years since the land was cleared, and I myself had got out two or three cords* of stumps, I did not give it any manure; but in the course of the summer it appeared by the arrow-heads which I turned up in hoeing, that an extinct nation had anciently dwelt here and planted corn and beans ere white men came to clear the land, and so, to some extent, had exhausted the soil for this very crop.

Before yet any woodchuck or squirrel had run across the road, or the sun had got above the shrub-oaks, while all the dew was on, though the farmers warned me against it,—I would advise you to do all your work if possible while the dew is on,—I began to level the ranks of haughty weeds in my bean-field and throw dust upon their heads.* Early in the morning I worked barefooted, dabbling like a plastic artist* in the dewy and crumbling sand, but later in the day the sun blistered my feet. There the sun lighted me to hoe beans, pacing slowly backward and forward over that yellow gravelly upland, between the long green rows, fifteen rods, the one end terminating in a shrub oak copse where I could rest in the shade, the other in a blackberry field where the green berries deepened their tints by the time I had made another bout. Removing the weeds, putting fresh soil about the bean stems, and encouraging this weed which I had sown, making the yellow soil express its summer thought in bean leaves and blossoms rather than in wormwood and piper and millet grass, making the earth say beans instead of grass,—this was my daily work. As I had little aid from horses or cattle, or hired men or boys, or improved implements of husbandry, I was much slower, and became much more intimate with my beans than usual. But labor of the hands, even when pursued to the verge of drudgery, is perhaps never the worst form of idleness. It has a constant and imperishable moral, and to the scholar it yields a classic result. A very *agricola laboriosus** was I to travellers bound westward through Lincoln and Wayland* to nobody knows where; they sitting at their ease in gigs,* with elbows on knees, and reins loosely hanging in festoons; I the home-staying, laborious native of the soil. But soon my homestead was out of their sight and thought. It was the only open and cultivated field for a great distance on either side of the road; so they made the most of it; and sometimes the man in the field heard more of travellers' gossip and comment than was meant for his ear: "Beans so late! peas so late!"—for I continued to plant when others had began to hoe,*—the ministerial husbandman* had not suspected it. "Corn, my boy, for fodder; corn for fodder." "Does he *live* there?" asks the black bonnet of the gray coat; and the hard-featured farmer reins up his grateful dobbin to inquire what you are doing where he sees no manure in the furrow, and recommends a little chip dirt, or any little waste stuff, or it may be ashes or plaster. But here were two acres and a half of furrows, and only a hoe for cart and two hands to draw it,— there being an aversion to other carts and horses,—and chip dirt far away. Fellow-travellers as they rattled by compared it aloud with the fields which they had passed, so that I came to know how I stood in the agricultural world. This was one field not in Mr. Coleman's report.* And, by the way, who estimates the value of the crop which Nature yields in the still wilder fields unimproved by man? The crop of *English* hay* is carefully weighed, the moisture calculated, the silicates and potash; but in all dells and pond holes in the woods and pastures and swamps grows a rich and various crop only unreaped by man. Mine was, as it were, the connecting link between wild and cultivated fields; as some states are civilized, and others half-civilized, and others savage or barbarous, so my field was, though not in a bad sense, a half-cultivated field. They were

beans cheerfully returning to their wild and primitive state
that I cultivated, and my hoe played the *Rans des Vaches** for
them.

Near at hand, upon the topmost spray of a birch, sings the
brown-thrasher—or red mavis, as some love to call him—all
the morning, glad of your society, that would find out another
farmer's field if yours were not here. While you are planting
the seed, he cries—"Drop it, drop it,—cover it up,—pull it
up, pull it up, pull it up." But this was not corn, and so it was
safe from such enemies as he. You may wonder what his
rigmarole, his amateur Paganini* performances on one string
or on twenty, have to do with your planting, and yet prefer it
to leached ashes or plaster. It was a cheap sort of top dressing
in which I had entire faith.

As I drew a still fresher soil about the rows with my hoe, I
disturbed the ashes of unchronicled nations who in primeval
years lived under these heavens, and their small implements
of war and hunting were brought to the light of this modern
day. They lay mingled with other natural stones, some of
which bore the marks of having been burned by Indian fires,
and some by the sun, and also bits of pottery and glass brought
hither by the recent cultivators of the soil. When my hoe
tinkled against the stones, that music echoed to the woods and
the sky, and was an accompaniment to my labor which yielded
an instant and immeasurable crop. It was no longer beans that
I hoed, nor I that hoed beans; and I remembered with as
much pity as pride, if I remembered at all, my acquaintances
who had gone to the city to attend the oratorios. The night-
hawk circled overhead in the sunny afternoons—for I sometimes made a day of it—like a mote in the eye, or in heaven's
eye, falling from time to time with a swoop and a sound as if
the heavens were rent, torn at last to very rags and tatters, and
yet a seamless cope remained; small imps that fill the air and
lay their eggs on the ground on bare sand or rocks on the tops
of hills, where few have found them; graceful and slender like
ripples caught up from the pond, as leaves are raised by the

wind to float in the heavens; such kindredship is in Nature. The hawk is aerial brother of the wave which he sails over and surveys, those his perfect air-inflated wings answering to the elemental unfledged pinions of the sea. Or sometimes I watched a pair of hen-hawks circling high in the sky, alternately soaring and descending, approaching and leaving one another, as if they were the imbodiment of my own thoughts. Or I was attracted by the passage of wild pigeons from this wood to that, with a slight quivering winnowing sound and carrier haste; or from under a rotten stump my hoe turned up a sluggish portentous and outlandish spotted salamander, a trace of Egypt and the Nile, yet our contemporary. When I paused to lean on my hoe, these sounds and sights I heard and saw any where in the row, a part of the inexhaustible entertainment which the country offers.

On gala days the town fires its great guns,* which echo like popguns to these woods, and some waifs of martial music occasionally penetrate thus far. To me, away there in my bean-field at the other end of the town, the big guns sounded as if a puff ball had burst; and when there was a military turnout of which I was ignorant, I have sometimes had a vague sense all the day of some sort of itching and disease in the horizon, as if some eruption would break out there soon, either scarlatina or canker-rash, until at length some more favorable puff of wind, making haste over the fields and up the Wayland road, brought me information of the "trainers."* It seemed by the distant hum as if somebody's bees had swarmed, and that the neighbors, according to Virgil's advice, by a faint *tintinnabulum** upon the most sonorous of their domestic utensils, were endeavoring to call them down into the hive again. And when the sound died quite away, and the hum had ceased, and the most favorable breezes told no tale, I knew that they had got the last drone of them all safely into the Middlesex* hive, and that now their minds were bent on the honey with which it was smeared.

I felt proud to know that the liberties of Massachusetts and of our fatherland were in such safe keeping; and as I turned to my hoeing again I was filled with an inexpressible confidence, and pursued my labor cheerfully with a calm trust in the future.

When there were several bands of musicians, it sounded as if all the village was a vast bellows, and all the buildings expanded and collapsed alternately with a din. But sometimes it was a really noble and inspiring strain that reached these woods, and the trumpet that sings of fame,* and I felt as if I could spit a Mexican* with a good relish,—for why should we always stand for trifles?—and looked round for a woodchuck or a skunk to exercise my chivalry upon. These martial strains seemed as far away as Palestine, and reminded me of a march of crusaders in the horizon, with a slight tantivy and tremulous motion of the elm-tree tops which overhang the village. This was one of the *great* days; though the sky had from my clearing only the same everlastingly great look that it wears daily, and I saw no difference in it.

It was a singular experience that long acquaintance which I cultivated with beans, what with planting, and hoeing, and harvesting, and threshing, and picking over, and selling them,—the last was the hardest of all,—I might add eating, for I did taste. I was determined to know beans.* When they were growing, I used to hoe from five o'clock in the morning till noon, and commonly spent the rest of the day about other affairs. Consider the intimate and curious acquaintance one makes with various kinds of weeds,—it will bear some iteration in the account, for there was no little iteration in the labor,—disturbing their delicate organization so ruthlessly, and making such invidious distinctions with his hoe, levelling whole ranks of one species, and sedulously cultivating another. That's Roman wormwood,—that's pigweed,—that's sorrel,—that's piper-grass,—have at him, chop him up, turn his roots upward to the sun, don't let him have a fibre in the shade, if you do he'll turn himself t'other side up and be as green as a leek in two days. A long war, not with cranes,* but with weeds, those Trojans who had sun and rain and dews on their side. Daily the beans saw me come to their rescue armed with a hoe, and thin the ranks of their enemies, filling up the trenches with weedy dead. Many a lusty crest-waving Hector,* that towered a whole foot above his crowding comrades, fell before my weapon and rolled in the dust.

Those summer days which some of my contemporaries devoted to the fine arts in Boston or Rome, and others to contemplation in India, and others to trade in London or New York, I thus, with the other farmers of New England, devoted to husbandry. Not that I wanted beans to eat, for I am by nature a Pythagorean,* so far as beans are concerned, whether they mean porridge or voting,* and exchanged them for rice; but, perchance, as some must work in fields if only for the sake of tropes and expression, to serve a parable-maker one day. It was on the whole a rare amusement, which, continued too long, might have become a dissipation. Though I gave them no manure, and did not hoe them all once, I hoed them unusually well as far as I went, and was paid for it in the end, "there being in truth," as Evelyn* says, "no compost or lætation whatsoever comparable to this continual motion, repastination, and turning of the mould with the spade." "The earth," he adds elsewhere, "especially if fresh, has a certain magnetism in it, by which it attracts the salt, power, or virtue (call it either) which gives it life, and is the logic of all the labor and stir we keep about it, to sustain us; all dungings and other sordid temperings being but the vicars succedaneous to this improvement." Morever, this being one of those "worn-out and exhausted lay fields which enjoy their sabbath," had perchance, as Sir Kenelm Digby* thinks likely, attracted "vital spirits" from the air. I harvested twelve bushels of beans.

But to be more particular, for it is complained that Mr. Coleman* has reported chiefly the expensive experiments of gentlemen farmers, my outgoes were,—

For a hoe,	$ 0 54	
Ploughing, harrowing, and furrowing,	7 50,	Too much.
Beans for seed,	3 12½	
Potatoes "	1 33	
Peas "	0 40	
Turnip seed,	0 06	
White line for crow fence,	0 02	
Horse cultivator and boy three hours,	1 00	
Horse and cart to get crop,	0 75	
In all,	$14 72½	

My income* was, (patrem familias vendacem, non emacem esse oportet,)* from

Nine bushels and twelve quarts of beans sold,	$16 94
Five " large potatoes,	2 50
Nine " small,	2 25
Grass,	1 00
Stalks,	0 75
In all,	$23 44
Leaving a pecuniary profit, as I have elsewhere said, of	$8 71½.

This is the result of my experience in raising beans. Plant the common small white bush bean about the first of June, in rows three feet by eighteen inches apart, being careful to select fresh round and unmixed seed. First look out for worms, and supply vacancies by planting anew. Then look out for woodchucks, if it is an exposed place, for they will nibble off the earliest tender leaves almost clean as they go; and again, when the young tendrils make their appearance, they have notice of it, and will shear them off with both buds and young pods, sitting erect like a squirrel. But above all harvest as early as possible, if you would escape frosts and have a fair and salable crop; you may save much loss by this means.

This further experience also I gained. I said to myself, I will not plant beans and corn with so much industry another summer, but such seeds, if the seed is not lost, as sincerity,

truth, simplicity, faith, innocence, and the like, and see if they will not grow in this soil, even with less toil and manurance, and sustain me, for surely it has not been exhausted for these crops. Alas! I said this to myself; but now another summer is gone, and another, and another, and I am obliged to say to you, Reader, that the seeds which I planted, if indeed they *were* the seeds of virtues, were wormeaten or had lost their vitality, and so did not come up. Commonly men will only be brave as their fathers were brave, or timid. This generation is very sure to plant corn and beans each new year precisely as the Indians did centuries ago and taught the first settlers to do, as if there were a fate in it. I saw an old man the other day, to my astonishment, making the holes with a hoe for the seventieth time at least, and not for himself to lie down in! But why should not the New Englander try new adventures, and not lay so much stress on his grain, his potato and grass crop, and his orchards,—raise other crops than these? Why concern ourselves so much about our beans for seed, and not be concerned at all about a new generation of men? We should really be fed and cheered if when we met a man we were sure to see that some of the qualities which I have named, which we all prize more than those other productions, but which are for the most part broadcast and floating in the air, had taken root and grown in him. Here comes such a subtile and ineffable quality, for instance, as truth or justice, though the slightest amount or new variety of it, along the road. Our ambassadors should be instructed to send home such seeds as these, and Congress help to distribute them over all the land.*
We should never stand upon ceremony with sincerity. We should never cheat and insult and banish one another by our meanness, if there were present the kernel of worth and friendliness. We should not meet thus in haste. Most men I do not meet at all, for they seem not to have time; they are busy about their beans. We would not deal with a man thus plodding ever, leaning on a hoe or a spade as a staff between his work, not as a mushroom, but partially risen out of the earth,

something more than erect, like swallows alighted and walking on the ground:—

> "And as he spake, his wings would now and then
> Spread, as he meant to fly, then close again,"*

so that we should suspect that we might be conversing with an angel. Bread may not always nourish us; but it always does us good, it even takes stiffness out of our joints, and makes us supple and buoyant, when we knew not what ailed us, to recognize any generosity in man or Nature, to share any unmixed and heroic joy.

Ancient poetry and mythology suggest, at least, that husbandry was once a sacred art; but it is pursued with irreverent haste and heedlessness by us, our object being to have large farms and large crops merely. We have no festival, nor procession, nor ceremony, not excepting our Cattle-shows* and so called Thanksgivings, by which the farmer expresses a sense of the sacredness of his calling, or is reminded of its sacred origin. It is the premium and the feast which tempt him. He sacrifices not to Ceres* and the Terrestrial Jove,* but to the infernal Plutus* rather. By avarice and selfishness, and a grovelling habit, from which none of us is free, of regarding the soil as property, or the means of acquiring property chiefly, the landscape is deformed, husbandry is degraded with us, and the farmer leads the meanest of lives. He knows Nature but as a robber. Cato says that the profits of agriculture are particularly pious or just, (*maximeque pius quaestus,*) and according to Varro* the old Romans "called the same earth Mother and Ceres, and thought that they who cultivated it led a pious and useful life, and that they alone were left of the race of King Saturn."*

We are wont to forget that the sun looks on our cultivated fields and on the prairies and forests without distinction. They all reflect and absorb his rays alike, and the former make but a small part of the glorious picture which he beholds in his daily course. In his view the earth is all equally cultivated like a garden. Therefore we should receive the benefit of his light and heat with a corresponding trust and magnanimity. What though I value the seed of these beans, and harvest that in the fall of the year? This broad field which I have looked at so long looks not to me as the principal cultivator, but away from me to influences more genial to it, which water and make it green. These beans have results which are not harvested by me. Do they not grow for woodchucks partly? The ear of wheat, (in Latin *spica*, obsoletely *speca*, from *spe*, hope,) should not be the only hope of the husbandman; its kernel or grain (*granum*, from *gerendo*, bearing,)* is not all that it bears. How, then, can our harvest fail? Shall I not rejoice also at the abundance of the weeds whose seeds are the granary of the birds? It matters little comparatively whether the fields fill the farmer's barns. The true husbandman will cease from anxiety, as the squirrels manifest no concern whether the woods will bear chestnuts this year or not, and finish his labor with every day, relinquishing all claim to the produce of his fields, and sacrificing in his mind not only his first but his last fruits also.

THE VILLAGE

AFTER hoeing, or perhaps reading and writing, in the forenoon, I usually bathed again in the pond, swimming across
one of its coves for a stint,* and washed the dust of labor from
my person, or smoothed out the last wrinkle which study had
made, and for the afternoon was absolutely free. Every day or
two I strolled to the village to hear some of the gossip which
is incessantly going on there, circulating either from mouth to
mouth, or from newspaper to newspaper, and which, taken in
homœopathic doses,* was really as refreshing in its way as the
rustle of leaves and the peeping of frogs. As I walked in the
woods to see the birds and squirrels, so I walked in the village
to see the men and boys; instead of the wind among the pines
I heard the carts rattle. In one direction from my house there
was a colony of muskrats in the river meadows; under the
grove of elms and buttonwoods in the other horizon was a
village of busy men, as curious to me as if they had been
prairie dogs, each sitting at the mouth of its burrow, or running over to a neighbor's to gossip. I went there frequently to
observe their habits. The village appeared to me a great news
room; and on one side, to support it, as once at Redding &
Company's* on State Street, they kept nuts and raisins, or salt
and meal and other groceries. Some have such a vast appetite
for the former commodity, that is, the news, and such sound
digestive organs, that they can sit forever in public avenues
without stirring, and let it simmer and whisper through them
like the Etesian winds,* or as if inhaling ether, it only producing numbness and insensibility to pain,—otherwise it would
often be painful to hear,—without affecting the consciousness. I hardly ever failed, when I rambled through the village,
to see a row of such worthies, either sitting on a ladder
sunning themselves, with their bodies inclined forward and

their eyes glancing along the line this way and that, from time to time, with a voluptuous expression, or else leaning against a barn with their hands in their pockets, like caryatides,* as if to prop it up. They, being commonly out of doors, heard whatever was in the wind. These are the coarsest mills, in which all gossip is first rudely digested or cracked up before it is emptied into finer and more delicate hoppers within doors. I observed that the vitals of the village were the grocery, the bar-room, the post-office, and the bank; and, as a necessary part of the machinery, they kept a bell, a big gun, and a fire-engine, at convenient places; and the houses were so arranged as to make the most of mankind, in lanes and fronting one another, so that every traveller had to run the gantlet, and every man, woman, and child might get a lick at him. Of course, those who were stationed nearest to the head of the line, where they could most see and be seen, and have the first blow at him, paid the highest prices for their places; and the few straggling inhabitants in the outskirts, where long gaps in the line began to occur, and the traveller could get over walls or turn aside into cow paths, and so escape, paid a very slight ground or window tax. Signs were hung out on all sides to allure him; some to catch him by the appetite, as the tavern and victualling cellar; some by the fancy, as the dry goods store and the jeweller's; and others by the hair or the feet or the skirts, as the barber, the shoemaker, or the tailor. Besides, there was a still more terrible standing invitation to call at every one of these houses, and company expected about these times. For the most part I escaped wonderfully from these dangers, either by proceeding at once boldly and without deliberation to the goal, as is recommended to those who run the gantlet, or by keeping my thoughts on high things, like Orpheus, who, "loudly singing the praises of the gods to his lyre, drowned the voices of the Sirens,* and kept out of danger." Sometimes I bolted suddenly, and nobody could tell my whereabouts, for I did not stand much about gracefulness, and never hesitated at a gap in a fence. I was even accustomed to make an irruption into some houses, where I was well entertained, and after learning the kernels and very last sieve-ful of news, what had subsided, the prospects of war and peace, and whether the world was likely to hold together much longer, I was let out through the rear avenues,* and so escaped to the woods again.

It was very pleasant, when I staid late in town, to launch myself into the night, especially if it was dark and tempestu-ous, and set sail from some bright village parlor or lecture room, with a bag of rye or Indian meal upon my shoulder, for my snug harbor in the woods, having made all tight without and withdrawn under hatches with a merry crew of thoughts, leaving only my outer man at the helm, or even tying up the helm when it was plain sailing. I had many a genial thought by the cabin fire "as I sailed."* I was never cast away nor dis-tressed in any weather, though I encountered some severe storms. It is darker in the woods, even in common nights, than most suppose. I frequently had to look up at the opening between the trees above the path in order to learn my route, and, where there was no cart-path, to feel with my feet the faint track which I had worn, or steer by the known relation of particular trees which I felt with my hands, passing between two pines for instance, not more than eighteen inches apart, in the midst of the woods, invariably in the darkest night. Sometimes, after coming home thus late in a dark and muggy night, when my feet felt the path which my eyes could not see, dreaming and absent-minded all the way, until I was aroused by having to raise my hand to lift the latch, I have not been able to recall a single step of my walk, and I have thought that perhaps my body would find its way home if its master should forsake it, as the hand finds its way to the mouth without assistance. Several times, when a visitor chanced to stay into evening, and it proved a dark night, I was obliged to conduct him to the cart-path in the rear of the house, and then point out to him the direction he was to pursue, and in keeping which he was to be guided rather by his feet than his eyes.

One very dark night I directed thus on their way two young men who had been fishing in the pond. They lived about a mile off through the woods, and were quite used to the route. A day or two after one of them told me that they wandered about the greater part of the night, close by their own premises, and did not get home till toward morning, by which time, as there had been several heavy showers in the mean while, and the leaves were very wet, they were drenched to their skins. I have heard of many going astray even in the village streets, when the darkness was so thick that you could cut it with a knife, as the saying is. Some who live in the outskirts, having come to town a-shopping in their wagons, have been obliged to put up for the night; and gentlemen and ladies making a call have gone half a mile out of their way, feeling the sidewalk only with their feet, and not knowing when they turned. It is a surprising and memorable, as well as valuable experience, to be lost in the woods any time. Often in a snow storm, even by day, one will come out upon a well-known road and yet find it impossible to tell which way leads to the village. Though he knows that he has travelled it a thousand times, he cannot recognize a feature in it, but it is as strange to him as if it were a road in Siberia. By night, of course, the perplexity is infinitely greater. In our most trivial walks, we are constantly, though unconsciously, steering like pilots by certain well-known beacons and headlands, and if we go beyond our usual course we still carry in our minds the bearing of some neighboring cape; and not till we are completely lost, or turned round,—for a man needs only to be turned round once with his eyes shut in this world to be lost,—do we appreciate the vastness and strangeness of Nature. Every man has to learn the points of compass again as often as he awakes, whether from sleep or any abstraction. Not till we are lost, in other words, not till we have lost the world, do we begin to find ourselves, and realize where we are and the infinite extent of our relations.

One afternoon, near the end of the first summer, when I went to the village to get a shoe from the cobbler's, I was seized and put into jail, because, as I have elsewhere related,* I did not pay a tax to, or recognize the authority of, the state which buys and sells men, women, and children, like cattle at the door of its senate-house. I had gone down to the woods for other purposes. But, wherever a man goes, men will pursue and paw him with their dirty institutions, and, if they can, constrain him to belong to their desperate odd-fellow society.* It is true, I might have resisted forcibly with more or less effect, might have run "amok" against society; but I preferred that society should run "amok" against me, it being the desperate party. However, I was released the next day, obtained my mended shoe, and returned to the woods in season to get my dinner of huckleberries on Fair-Haven Hill.* I was never molested by any person but those who represented the state. I had no lock nor bolt but for the desk which held my papers, not even a nail to put over my latch or windows. I never fastened my door night or day, though I was to be absent several days; not even when the next fall I spent a fortnight in the woods of Maine. And yet my house was more respected than if it had been surrounded by a file of soldiers. The tired rambler could rest and warm himself by my fire, the literary amuse himself with the few books on my table, or the curious, by opening my closet door, see what was left of my dinner, and what prospect I had of a supper. Yet, though many people of every class came this way to the pond, I suffered no serious inconvenience from these sources, and I never missed any thing but one small book, a volume of Homer, which perhaps was improperly gilded,* and this I trust a soldier of our camp* has found by this time. I am convinced, that if all men were to live as simply as I then did, thieving and robbery would be unknown. These take place only in communities where some have got more than is sufficient while others have not enough. The Pope's Homers would soon get properly distributed.—

"Nec bella fuerunt,
Faginus astabat dum scyphus ante dapes."
"Nor wars did men molest,
When only beechen bowls were in request."*

"You who govern public affairs, what need have you to employ punishments? Love virtue, and the people will be virtuous. The virtues of a superior man are like the wind; the virtues of a common man are like the grass; the grass, when the wind passes over it, bends."*

THE PONDS

SOMETIMES, having had a surfeit of human society and gossip, and worn out all my village friends, I rambled still farther westward than I habitually dwell, into yet more unfrequented parts of the town, "to fresh woods and pastures new,"* or, while the sun was setting, made my supper of huckleberries and blueberries on Fair Haven Hill, and laid up a store for several days. The fruits do not yield their true flavor to the purchaser of them, nor to him who raises them for the market. There is but one way to obtain it, yet few take that way. If you would know the flavor of huckleberries, ask the cow-boy* or the partridge. It is a vulgar error to suppose that you have tasted huckleberries who never plucked them. A huckleberry never reaches Boston; they have not been known there since they grew on her three hills.* The ambrosial and essential part of the fruit is lost with the bloom* which is rubbed off in the market cart, and they become mere provender. As long as Eternal Justice reigns, not one innocent huckleberry can be transported thither from the country's hills.

Occasionally, after my hoeing was done for the day, I joined some impatient companion who had been fishing on the pond since morning, as silent and motionless as a duck or a floating leaf, and, after practising various kinds of philosophy, had concluded commonly, by the time I arrived, that he belonged to the ancient sect of Cœnobites.* There was one older man, an excellent fisher and skilled in all kinds of woodcraft, who was pleased to look upon my house as a building erected for the convenience of fishermen; and I was equally pleased when he sat in my doorway to arrange his lines. Once in a while we sat together on the pond, he at one end of the boat, and I at the other; but not many words passed between us, for he had grown deaf in his later years, but he occasionally hummed a psalm, which harmonized well enough with my philosophy. Our intercourse was thus altogether one of unbroken harmony, far more pleasing to remember than if it had been carried on by speech. When, as was commonly the case, I had none to commune with, I used to raise the echoes by striking with a paddle on the side of my boat, filling the surrounding woods with circling and dilating sound, stirring them up as the keeper of a menagerie his wild beasts, until I elicited a growl from every wooded vale and hill-side.

In warm evenings I frequently sat in the boat playing the flute, and saw the perch, which I seemed to have charmed, hovering around me, and the moon travelling over the ribbed bottom, which was strewed with the wrecks of the forest. Formerly I had come to this pond adventurously, from time to time, in dark summer nights, with a companion, and making a fire close to the water's edge, which we thought attracted the fishes, we caught pouts with a bunch of worms strung on a thread; and when we had done, far in the night, threw the burning brands high into the air like skyrockets, which, coming down into the pond, were quenched with a loud hissing, and we were suddenly groping in total darkness. Through this, whistling a tune, we took our way to the haunts of men again. But now I had made my home by the shore.

Sometimes, after staying in a village parlor till the family had all retired, I have returned to the woods, and, partly with a view to the next day's dinner, spent the hours of midnight fishing from a boat by moonlight, serenaded by owls and foxes, and hearing, from time to time, the creaking note of some unknown bird close at hand. These experiences were very memorable and valuable to me,—anchored in forty feet of water, and twenty or thirty rods* from the shore, surrounded sometimes by thousands of small perch and shiners, dimpling the surface with their tails in the moonlight, and communicating by a long flaxen line with mysterious nocturnal fishes which had their dwelling forty feet below, or sometimes dragging sixty feet of line about the pond as I drifted in the gentle night breeze, now and then feeling a slight vibration along it, indicative of some life prowling about its extremity, of dull uncertain blundering purpose there, and slow to make up its mind. At length you slowly raise, pulling hand over hand, some horned pout squeaking and squirming to the upper air. It was very queer, especially in dark nights, when your thoughts had wandered to vast and cosmogonal themes in other spheres, to feel this faint jerk, which came to interrupt your dreams and link you to Nature again. It seemed as if I might next cast my line upward into the air, as well as downward into this element which was scarcely more dense. Thus I caught two fishes as it were with one hook.

The scenery of Walden is on a humble scale, and, though very beautiful, does not approach to grandeur, nor can it much concern one who has not long frequented it or lived by its shore; yet this pond is so remarkable for its depth and purity as to merit a particular description. It is a clear and deep green well, half a mile long and a mile and three quarters in circumference, and contains about sixty-one and a half acres; a perennial spring* in the midst of pine and oak woods, without any visible inlet or outlet except by the clouds and evaporation. The surrounding hills rise abruptly from the water to the height of forty to eighty feet, though on the south-east and east they attain to about one hundred and one hundred and fifty feet respectively, within a quarter and a third of a mile. They are exclusively woodland. All our Concord waters have two colors at least, one when viewed at a distance, and another, more proper, close at hand. The first depends more on the light, and follows the sky. In clear weather, in summer, they appear blue at a little distance, especially if agitated, and at a great distance all appear alike. In stormy weather they are sometimes of a dark slate color. The sea, however, is said to be blue one day and green another without any perceptible change in the atmosphere. I have seen our river, when, the landscape being covered with snow, both water and ice were almost as green as grass. Some consider blue "to be the color of pure water, whether liquid or solid."* But, looking directly down into our waters from a boat, they are seen to be of very different colors. Walden is blue at one time and green at another, even from the same point of view. Lying between the earth and the heavens, it partakes of the color of both. Viewed from a hill-top it reflects the color of the sky, but near at hand it is of a yellowish tint next the shore where you can see the sand, then a light green, which gradually deepens to a uniform dark green in the body of the pond. In some lights, viewed even from a hill-top, it is of a vivid green next the shore. Some have referred this to the reflection of the verdure; but it is equally green there against the railroad sand-bank, and in the spring, before the leaves are expanded, and it may be simply the result of the prevailing blue mixed with the yellow of the sand. Such is the color of its iris. This is that portion, also, where in the spring, the ice being warmed by the heat of the sun reflected from the bottom, and also transmitted through the earth, melts first and forms a narrow canal about the still frozen middle. Like the rest of our waters, when much agitated, in clear weather, so that the surface of the waves may reflect the sky at the right angle, or because there is more light mixed with it, it appears at a little distance of a darker blue than the sky itself; and at such a time, being on its surface, and looking with divided vision, so as to see the reflection, I have discerned a matchless and indescribable light blue, such as watered or changeable silks* and sword blades suggest, more cerulean than the sky itself, alternating with the original dark green on the opposite sides of the waves, which last appeared but muddy in comparison. It is a vitreous greenish blue, as I remember it, like those patches of the winter sky seen through cloud vistas in the west before sundown. Yet a single glass of its water held up to the light is as colorless as an equal quantity of air. It is well known that a large plate of glass will have a green tint, owing, as the makers say, to its "body," but a small piece of the same will be colorless. How large a body of Walden water would be required to reflect a green tint I have never proved. The water of our river is black or a very dark brown to one looking directly down on it, and, like that of most ponds, imparts to the body of one bathing in it a yellowish tinge; but this water is of such crystalline purity that the body of the bather appears of an alabaster whiteness, still more unnatural, which, as the limbs are magnified and distorted withal, produces a monstrous effect, making fit studies for a Michael Angelo.*

The water is so transparent that the bottom can easily be discerned at the depth of twenty-five or thirty feet. Paddling over it, you may see many feet beneath the surface the schools of perch and shiners, perhaps only an inch long, yet the former easily distinguished by their transverse bars, and you think that they must be ascetic fish that find a subsistence there. Once, in the winter, many years ago, when I had been cutting holes through the ice in order to catch pickerel, as I stepped ashore I tossed my axe back on to the ice, but, as if some evil genius had directed it, it slid four or five rods directly into one of the holes, where the water was twenty-five feet deep. Out of curiosity, I lay down on the ice and looked through the hole, until I saw the axe a little on one side, standing on its head, with its helve erect and gently swaying to and fro with the pulse of the pond; and there it might have stood erect and swaying till in the course of time the handle rotted off, if I had not disturbed it. Making another hole directly over it with an ice chisel which I had, and cutting down the longest birch which I could find in the neighorhood* with my knife, I made a slip-noose, which I attached to its end, and, letting it down carefully, passed it over the knob of the handle, and drew it by a line along the birch, and so pulled the axe out again.

The shore is composed of a belt of smooth rounded white stones like paving stones, excepting one or two short sand beaches, and is so steep that in many places a single leap will carry you into water over your head; and were it not for its remarkable transparency, that would be the last to be seen of its bottom till it rose on the opposite side. Some think it is bottomless.* It is nowhere muddy, and a casual observer would say that there were no weeds at all in it; and of notice-able plants, except in the little meadows recently over-flowed, which do not properly belong to it, a closer scrutiny does not detect a flag nor a bulrush, nor even a lily, yellow or white, but only a few small heart-leaves and potamogetons, and perhaps a water-target or two; all which however a bather might not perceive; and these plants are clean and bright like the ele-ment they grow in. The stones extend a rod or two into the water, and then the bottom is pure sand, except in the deepest parts, where there is usually a little sediment, probably from the decay of the leaves which have been wafted on to it so many successive falls, and a bright green weed is brought up on anchors even in midwinter.

We have one other pond just like this, White Pond in Nine Acre Corner, about two and a half miles westerly; but, though I am acquainted with most of the ponds within a dozen miles of this centre, I do not know a third of this pure and well-like character. Successive nations perchance have drank at, ad-mired, and fathomed it, and passed away, and still its water is green and pellucid as ever. Not an intermitting spring! Per-haps on that spring morning when Adam and Eve were driven out of Eden Walden Pond was already in existence, and even then breaking up in a gentle spring rain accompanied with mist and a southerly wind, and covered with myriads of ducks and geese, which had not heard of the fall, when still such pure lakes sufficed them. Even then it had commenced to rise and fall, and had clarified its waters and colored them of the hue they now wear, and obtained a patent of heaven to be the only Walden Pond in the world and distiller of celestial dews. Who knows in how many unremembered nations' literatures this has been the Castalian Fountain?* or what nymphs presided over it in the Golden Age?* It is a gem of the first water which Concord wears in her coronet.

Yet perchance the first who came to this well have left some trace of their footsteps. I have been surprised to detect encircling the pond, even where a thick wood has just been cut down on the shore, a narrow shelf-like path* in the steep hillside, alternately rising and falling, approaching and receding from the water's edge, as old probably as the race of man here, worn by the feet of aboriginal hunters, and still from time to time unwittingly trodden by the present occupants of the land. This is particularly distinct to one standing on the middle of the pond in winter, just after a light snow has fallen, appearing as a clear undulating white line, unobscured by weeds and twigs, and very obvious a quarter of a mile off in many places where in summer it is hardly distinguishable close at hand. The snow reprints it, as it were, in clear white type alto-relievo.* The ornamented grounds of villas which will one day be built here may still preserve some trace of this.

The pond rises and falls,* but whether regularly or not, and within what period, nobody knows, though, as usual, many pretend to know. It is commonly higher in the winter and lower in the summer, though not corresponding to the general wet and dryness. I can remember when it was a foot or two lower, and also when it was at least five feet higher, than when I lived by it. There is a narrow sand-bar running into it, with very deep water on one side, on which I helped boil a kettle of chowder,* some six rods from the main shore, about the year 1824, which it has not been possible to do for twenty-five years; and on the other hand, my friends used to listen with incredulity when I told them, that a few years later I was accustomed to fish from a boat in a secluded cove in the woods, fifteen rods from the only shore they knew, which place was long since converted into a meadow. But the pond has risen steadily for two years, and now, in the summer of '52, is just five feet higher than when I lived there, or as high as it was thirty years ago, and fishing goes on again in the meadow. This makes a difference of level, at the outside, of six or seven feet; and yet the water shed by the surrounding hills is insignificant in amount, and this overflow must be referred to causes which affect the deep springs. This same summer the pond has begun to fall again. It is remarkable that this fluctuation, whether periodical or not, appears thus to require many years for its accomplishment. I have observed one rise and a part of two falls, and I expect that a dozen or fifteen years hence the water will again be as low as I have ever known it. Flints' Pond, a mile eastward, allowing for the disturbance occasioned by its inlets and outlets, and the smaller intermediate ponds also, sympathize with Walden,* and recently attained their greatest height at the same time with the latter. The same is true, as far as my observation goes, of White Pond.*

This rise and fall of Walden at long intervals serves this use at least; the water standing at this great height for a year or more, though it makes it difficult to walk round it, kills the shrubs and trees which have sprung up about its edge since the last rise, pitch-pines, birches, alders, aspens, and others, and, falling again, leaves an unobstructed shore; for, unlike many ponds and all waters which are subject to a daily tide, its shore is cleanest when the water is lowest. On the side of the pond next to my house, a row of pitch pines fifteen feet high has been killed and tipped over as if by a lever, and thus a stop put to their encroachments; and their size indicates how many years have elapsed since the last rise to this height. By this fluctuation the pond asserts its title to a shore, and thus the *shore* is *shorn*, and the trees cannot hold it by right of possession. These are the lips of the lake on which no beard grows. It licks its chaps from time to time. When the water is at its height, the alders, willows, and maples send forth a mass of fibrous red roots several feet long from all sides of their stems in the water, and to the height of three or four feet from the ground, in the effort to maintain themselves; and I have known the high-blueberry bushes about the shore, which commonly produce no fruit, bear an abundant crop under these circumstances.

Some have been puzzled to tell how the shore became so regularly paved. My townsmen have all heard the tradition,* the oldest people tell me that they heard it in their youth, that anciently the Indians were holding a pow-wow upon a hill here, which rose as high into the heavens as the pond now sinks deep into the earth, and they used much profanity, as the story goes, though this vice is one of which the Indians were never guilty, and while they were thus engaged the hill shook and suddenly sank, and only one old squaw, named Walden, escaped, and from her the pond was named. It has been conjectured that when the hill shook these stones rolled down its side and became the present shore. It is very certain, at any rate, that once there was no pond here, and now there is one; and this Indian fable does not in any respect conflict with the account of that ancient settler* whom I have mentioned, who remembers so well when he first came here with his divining rod, saw a thin vapor rising from the sward, and the hazel pointed steadily downward, and he concluded to dig a well here. As for the stones, many still think that they are hardly to be accounted for by the action of the waves on these hills; but I observe that the surrounding hills are remarkably full of the same kind of stones, so that they have been obliged to pile them up in walls on both sides of the railroad cut nearest the pond; and, moreover, there are most stones where the shore is most abrupt; so that, unfortunately, it is no longer a mystery to me. I detect the paver.* If the name was not derived from that of some English locality,—Saffron Walden, for instance,—one might suppose that it was called, originally, *Walled-in* Pond.

The pond was my well ready dug. For four months in the year its water is as cold as it is pure at all times; and I think that it is then as good as any, if not the best, in the town. In the winter, all water which is exposed to the air is colder than springs and wells which are protected from it. The temperature of the pond water which had stood in the room where I sat from five o'clock in the afternoon till noon the next day, the sixth of March, 1846, the thermometer having been up to 65° or 70° some of the time, owing partly to the sun on the roof, was 42°, or one degree colder than the water of one of the coldest wells in the village just drawn. The temperature of the Boiling Spring* the same day was 45°, or the warmest of any water tried, though it is the coldest that I know of in summer, when, beside, shallow and stagnant surface water is not mingled with it. Moreover, in summer, Walden never becomes so warm as most water which is exposed to the sun, on account of its depth. In the warmest weather I usually placed a pailful in my cellar, where it became cool in the night, and remained so during the day; though I also resorted to a spring in the neighborhood. It was as good when a week old as the day it was dipped, and had no taste of the pump. Whoever camps for a week in summer by the shore of a pond, needs only bury a pail of water a few feet deep in the shade of his camp to be independent on the luxury of ice.

There have been caught in Walden, pickerel, one weighing seven pounds, to say nothing of another which carried off a reel with great velocity, which the fisherman safely set down at eight pounds because he did not see him, perch and pouts, some of each weighing over two pounds, shiners, chivins or roach, (*Leuciscus pulchellus*,) a very few breams, and* a couple of eels, one weighing four pounds,—I am thus particular because the weight of a fish is commonly its only title to fame, and these are the only eels I have heard of here; also, I have a faint recollection of a little fish some five inches long, with silvery sides and a greenish back, somewhat dace-like in its character, which I mention here chiefly to link my facts to fable. Nevertheless, this pond is not very fertile in fish. Its pickerel, though not abundant, are its chief boast. I have seen at one time lying on the ice pickerel of at least three different kinds; a long and shallow one, steel-colored, most like those caught in the river; a bright golden kind, with greenish reflections and remarkably deep, which is the most common here; and another, golden-colored, and shaped like the last, but peppered on the sides with small dark brown or black spots, intermixed with a few faint blood-red ones, very much like a trout. The specific name *reticulatus** would not apply to this; it should be *guttatus** rather. These are all very firm fish, and weigh more than their size promises. The shiners, pouts, and perch also, and indeed all the fishes which inhabit this pond, are much cleaner, handsomer, and firmer fleshed than those in the river and most other ponds, as the water is purer, and they can easily be distinguished from them. Probably many ichthyologists would make new varieties of some of them. There are also a clean race of frogs and tortoises, and a few muscles in it; muskrats and minks leave their traces about it, and occasionally a travelling mud-turtle visits it. Sometimes, when I pushed off my boat in the morning, I disturbed a great mud-turtle which had secreted himself under the boat in the night. Ducks and geese frequent it in the spring and fall, the white-bellied swallows (*Hirundo bicolor*) skim over it, and the peetweets* (*Totanus macularius*) "teter" along its stony shores all summer. I have sometimes disturbed a fishhawk sitting on a white-pine over the water; but I doubt if it is ever profaned by the wing of a gull, like Fair Haven.* At most, it tolerates one annual loon. These are all the animals of consequence which frequent it now.

You may see from a boat, in calm weather, near the sandy eastern shore, where the water is eight or ten feet deep, and also in some other parts of the pond, some circular heaps half a dozen feet in diameter by a foot in height, consisting of small stones less than a hen's egg in size, where all around is bare sand. At first you wonder if the Indians could have formed them on the ice for any purpose, and so, when the ice melted, they sank to the bottom; but they are too regular and some of them plainly too fresh for that. They are similar to those found in rivers; but as there are no suckers nor lampreys here, I know not by what fish they could be made. Perhaps they are the nests of the chivin.* These lend a pleasing mystery to the bottom.

The shore is irregular enough not to be monotonous. I have in my mind's eye the western indented with deep bays, the bolder northern, and the beautifully scolloped southern shore, where successive capes overlap each other and suggest unexplored coves between. The forest has never so good a setting, nor is so distinctly beautiful, as when seen from the middle of a small lake amid hills which rise from the water's edge; for the water in which it is reflected not only makes the best foreground in such a case, but, with its winding shore, the most natural and agreeable boundary to it. There is no rawness nor imperfection in its edge there, as where the axe has cleared a part, or a cultivated field abuts on it. The tress have ample room to expand on the water side, and each sends forth its most vigorous branch in that direction. There Nature has woven a natural selvage, and the eye rises by just gradations from the low shrubs of the shore to the highest tress. There are few traces of man's hand to be seen. The water laves the shore as it did a thousand years ago.

A lake is the landscape's most beautiful and expressive feature. It is earth's eye; looking into which the beholder measures the depth of his own nature. The fluviatile* trees next the shore are the slender eyelashes which fringe it, and the wooded hills and cliffs around are its overhanging brows.

Standing on the smooth sandy beach at the east end of the pond, in a calm September afternoon, when a slight haze makes the opposite shore line indistinct, I have seen whence came the expression, "the glassy surface of a lake." When you invert your head, it looks like a thread of finest gossamer stretched across the valley, and gleaming against the distant pine woods, separating one stratum of the atmosphere from another. You would think that you could walk dry under it to the opposite hills, and that the swallows which skim over might perch on it. Indeed, they sometimes dive below the line, as it were by mistake, and are undeceived. As you look over the pond westward you are obliged to employ both your hands to defend your eyes against the reflected as well as true sun, for they are equally bright; and if, between the two, you survey its surface critically, it is literally as smooth as glass, except where the skater insects, at equal intervals* scattered over its whole extent, by their motions in the sun produce the finest imaginable sparkle on it, or, perchance, a duck plumes itself, or, as I have said, a swallow skims so low as to touch it. It may be that in the distance a fish describes an arc of three or four feet in the air, and there is one bright flash where it emerges, and another where it strikes the water; sometimes the whole silvery arc is revealed; or here and there, perhaps, is a thistle-down floating on its surface, which the fishes dart at and so dimple it again. It is like molten glass cooled but not congealed, and the few motes in it are pure and beautiful like the imperfections in glass. You may often detect a yet smoother and darker water, separated from the rest as if by an invisible cobweb, boom of the water nymphs, resting on it. From a hill-top you can see a fish leap in almost any part; for not a pickerel or shiner picks an insect from this smooth surface but it manifestly disturbs the equilibrium of the whole lake. It is wonderful with what elaborateness this simple fact is advertised,—this piscine* murder will out,*—and from my distant perch I distinguish the circling undulations when they are half a dozen rods in diameter. You can even detect a water-bug, (*Gyrinus*) ceaselessly progressing over the smooth surface a quarter of a mile off; for they furrow the water slightly, making a conspicuous ripple bounded by two diverging lines, but the skaters glide over it without rippling it perceptibly. When the surface is considerably agitated there are no skaters nor water-bugs on it, but apparently, in calm days, they leave their havens and adventurously glide forth from the shore by short impulses till they completely cover it. It is a soothing employment, on one of those fine days in the fall when all the warmth of the sun is fully appreciated, to sit on a stump on such a height as this, overlooking the pond, and study the dimpling circles which are incessantly inscribed on its otherwise invisible surface amid the reflected skies and trees. Over this great expanse there is no disturbance but it is thus at once gently smoothed away and assuaged, as, when a vase of water is jarred, the trembling circles seek the shore and all is smooth again. Not a fish can leap or an insect fall on the pond but it is thus reported in circling dimples, in lines of beauty, as it were the constant welling up of its fountain, the gentle pulsing of its life, the heaving of its breast. The thrills of joy and thrills of pain are undistinguishable. How peaceful the phenomena of the lake! Again the works of man shine as in the spring. Ay, every leaf and twig and stone and cobweb sparkles now at mid-afternoon as when covered with dew in a spring morning. Every motion of an oar or an insect produces a flash of light; and if an oar falls, how sweet the echo!

In such a day, in September or October, Walden is a perfect forest mirror, set round with stones as precious to my eye as if fewer or rarer. Nothing so fair, so pure, and at the same time so large, as a lake, perchance, lies on the surface of the earth. Sky water. It needs no fence. Nations come and go without defiling it. It is a mirror which no stone can crack, whose quicksilver* will never wear off, whose gilding Nature continually repairs; no storms, no dust, can dim its surface ever fresh;—a mirror in which all impurity presented to it sinks, swept and dusted by the sun's hazy brush,—this the light dust-cloth,—which retains no breath that is breathed on it, but sends its own to float as clouds high above its surface, and be reflected in its bosom still.

A field of water betrays the spirit that is in the air. It is continually receiving new life and motion from above. It is intermediate in its nature between land and sky. On land only the grass and trees wave, but the water itself is rippled by the wind. I see where the breeze dashes across it by the streaks or flakes of light. It is remarkable that we can look down on its surface. We shall, perhaps, look down thus on the surface of air at length, and mark where a still subtler spirit sweeps over it.

The skaters and water-bugs finally disappear in the latter part of October, when the severe frosts have come; and then and in November, usually, in a calm day, there is absolutely nothing to ripple the surface. One November afternoon, in the calm at the end of a rain storm of several days' duration, when the sky was still completely overcast and the air was full of mist, I observed that the pond was remarkably smooth, so that it was difficult to distinguish its surface; though it no longer reflected the bright tints of October, but the sombre November colors of the surrounding hills. Though I passed over it as gently as possible, the slight undulations produced by my boat extended almost as far as I could see, and gave a ribbed appearance to the reflections. But, as I was looking over the surface, I saw here and there at a distance a faint glimmer, as if some skater insects which had escaped the frosts might be collected there, or, perchance, the surface, being so smooth, betrayed where a spring welled up from the bottom. Paddling gently to one of these places, I was surprised to find myself surrounded by myriads of small perch, about five inches long, of a rich bronze color in the green water, sporting there and constantly rising to the surface and dimpling it, sometimes leaving bubbles on it. In such transparent and seemingly bottomless water, reflecting the clouds, I seemed to be floating through the air as in a balloon, and their swimming impressed me as a kind of flight or hovering, as if they were a compact flock of birds passing just beneath my level on the right or left, their fins, like sails, set all around them. There were many such schools in the pond, apparently improving the short season before winter would draw an icy shutter over their broad skylight, sometimes giving to the surface an appearance as if a slight breeze struck it, or a few rain-drops fell there. When I approached carelessly and alarmed them, they made a sudden plash and rippling with their tails, as if one had struck the water with a brushy bough, and instantly took refuge in the depths. At length the wind rose, the mist increased, and the waves began to run, and the perch leaped much higher than before, half out of water, a hundred black points, three inches long, at once above the surface. Even as late as the fifth of December, one year, I saw some dimples on the surface, and thinking it was going to rain hard immediately, the air being full of mist, I made haste to take my place at the oars and row homeward; already the rain seemed rapidly increasing, though I felt none on my cheek, and I anticipated a thorough soaking. But suddenly the dimples ceased, for they were produced by the perch, which the noise of my oars had scared into the depths, and I saw their schools dimly disappearing; so I spent a dry afternoon after all.

An old man who used to frequent this pond nearly sixty years ago, when it was dark with surrounding forests, tells me in those days he sometimes saw it all alive with ducks and other water fowl, and that there were many eagles about it. He came here a-fishing, and used an old log canoe which he found on the shore. It was made of two white-pine logs dug out and pinned together, and was cut off square at the ends. It was very clumsy, but lasted a great many years before it became water-logged and perhaps sank to the bottom. He did not know whose it was; it belonged to the pond. He used to make a cable for his anchor of strips of hickory bark tied together. An old man, a potter, who lived by the pond before the Revolution, told him once that there was an iron chest at the bottom, and that he had seen it. Sometimes it would come floating up to the shore; but when you went toward it, it would go back into deep water and disappear. I was pleased to hear of the old log canoe, which took the place of an Indian one of the same material but more graceful construction, which perchance had first been a tree on the bank, and then, as it were, fell into the water, to float there for a generation, the most proper vessel for the lake. I remember that when I first looked into these depths there were many large trunks to be seen indistinctly lying on the bottom, which had either been blown over formerly, or left on the ice at the last cutting, when wood was cheaper; but now they have mostly disappeared.

When I first paddled a boat on Walden, it was completely surrounded by thick and lofty pine and oak woods, and in some of its coves grape vines had run over the trees next the water and formed bowers under which a boat could pass. The hills which form its shores are so steep, and the woods on them were then so high, that, as you looked down from the west end, it had the appearance of an amphitheatre for some kind of sylvan spectacle. I have spent many an hour, when I was younger, floating over its surface as the zephyr willed, having paddled my boat to the middle, and lying on my back across the seats, in a summer forenoon, dreaming awake, until I was aroused by the boat touching the sand, and I arose to see what shore my fates had impelled me to; days when idleness was the most attractive and productive industry. Many a forenoon have I stolen away, preferring to spend thus the most valued part of the day; for I was rich, if not in money, in sunny hours and summer days, and spent them lavishly; nor do I regret that I did not waste more of them in the workshop or the teacher's desk.* But since I left those shores the woodchoppers have still further laid them waste, and now for many a year there will be no more rambling through the aisles of the wood, with occasional vistas through which you see the water. My Muse may be excused if she is silent henceforth. How can you expect the birds to sing when their groves are cut down?

Now the trunks of trees on the bottom, and the old log canoe, and the dark surrounding woods, are gone, and the villagers, who scarcely know where it lies, instead of going to the pond to bathe or drink,* are thinking to bring its water, which should be as sacred as the Ganges* at least, to the village in a pipe,* to wash their dishes with!—to earn their

Walden by the turning of a cock or drawing of a plug! That devilish Iron Horse, whose ear-rending neigh is heard throughout the town, has muddied the Boiling Spring with his foot, and he it is that has browsed off all the woods on Walden shore;* that Trojan horse,* with a thousand men in his belly, introduced by mercenary Greeks! Where is the country's champion, the Moore of Moore Hall,* to meet him at the Deep Cut* and thrust an avenging lance* between the ribs of the bloated pest?

Nevertheless, of all the characters I have known, perhaps Walden wears best, and best preserves its purity. Many men have been likened to it, but few deserve that honor. Though the woodchoppers have laid bare first this shore and then that, and the Irish have built their sties by it, and the railroad has infringed on its border, and the ice-men have skimmed it once, it is itself unchanged, the same water which my youthful eyes fell on; all the change is in me. It has not acquired one permanent wrinkle after all its ripples. It is perennially young, and I may stand and see a swallow dip apparently to pick an insect from its surface as of yore. It struck me again to-night, as if I had not seen it almost daily for more than twenty years,—Why, here is Walden, the same woodland lake that I discovered so many years ago; where a forest was cut down last winter another is springing up by its shore as lustily as ever; the same thought is welling up to its surface that was then; it is the same liquid joy and happiness to itself and its Maker, ay, and it *may* be to me. It is the work of a brave man surely, in whom there was no guile! He rounded this water with his hand, deepened and clarified it in his thought, and in his will bequeathed it to Concord. I see by its face that it is visited by the same reflection; and I can almost say, Walden, is it you?

> It is no dream of mine,
> To ornament a line;
> I cannot come nearer to God and Heaven
> Than I live to Walden even.

> I am its stony shore,
> And the breeze that passes o'er;
> In the hollow of my hand
> Are its water and its sand,
> And its deepest resort
> Lies high in my thought.*

The cars never pause to look at it; yet I fancy that the engineers and firemen and brakemen,* and those passengers who have a season ticket and see it often, are better men for the sight. The engineer does not forget at night, or his nature does not, that he has beheld this vision of serenity and purity once at least during the day. Though seen but once, it helps to wash out State-street* and the engine's soot. One proposes that it be called "God's Drop."*

I have said that Walden has no visible inlet nor outlet, but it is on the one hand distantly and indirectly related to Flints' Pond, which is more elevated, by a chain of small ponds coming from that quarter, and on the other directly and manifestly to Concord River, which is lower, by a similar chain of ponds through which in some other geological period it may have flowed, and by a little digging, which God forbid, it can be made to flow thither again. If by living thus reserved and austere, like a hermit in the woods, so long, it has acquired such wonderful purity, who would not regret that the comparatively impure water of Flints' Pond should be mingled with it, or itself should ever go to waste its sweetness* in the ocean wave?

Flints', or Sandy Pond, in Lincoln, our greatest lake and inland sea, lies about a mile east of Walden. It is much larger, being said to contain one hundred and ninety-seven acres, and is more fertile in fish; but it is comparatively shallow, and not remarkably pure. A walk through the woods thither was often my recreation. It was worth the while, if only to feel the wind blow on your cheek freely, and see the waves run, and remember the life of mariners. I went a-chestnutting there in the fall, on windy days, when the nuts were dropping into the water and were washed to my feet; and one day, as I crept along its sedgy shore, the fresh spray blowing in my face, I came upon the mouldering wreck of a boat, the sides gone, and hardly more than the impression of its flat bottom left amid the rushes; yet its model was sharply defined, as if it were a large decayed pad,* with its veins. It was as impressive a wreck as one could imagine on the sea-shore, and had as good a moral. It is by this time mere vegetable mould and undistinguishable pond shore, through which rushes and flags have pushed up. I used to admire the ripple marks on the sandy bottom, at the north end of this pond, made firm and hard to the feet of the wader by the pressure of the water, and the rushes which grew in Indian file,* in waving lines, corresponding to these marks, rank behind rank, as if the waves had planted them. There also I have found, in considerable quantities, curious balls, composed apparently of fine grass or roots, of pipewort perhaps, from half an inch to four inches in diameter, and perfectly spherical. These wash back and forth in shallow water on a sandy bottom, and are sometimes cast on the shore. They are either solid grass, or have a little sand in the middle. At first you would say that they were formed by the action of the waves, like a pebble; yet the smallest are made of equally coarse materials, half an inch long, and they are produced only at one season of the year. Moreover, the waves, I suspect, do not so much construct as wear down a material which has already acquired consistency. They preserve their form when dry for an indefinite period.

Flints' Pond! Such is the poverty of our nomenclature. What right had the unclean and stupid farmer, whose farm abutted on this sky water, whose shores he has ruthlessly laid bare, to give his name to it? Some skin-flint,* who loved better the reflecting surface of a dollar, or a bright cent, in which he could see his own brazen face; who regarded even the wild ducks which settled in it as trespassers; his fingers grown into crooked and horny talons from the long habit of grasping harpy-like;—so it is not named for me. I go not there to see him nor to hear of him; who never *saw* it, who never bathed in it, who never loved it, who never protected it, who never spoke a good word for it, nor thanked God that he had made it. Rather let it be named from the fishes that swim in it, the wild fowl or quadrupeds which frequent it, the wild flowers which grow by its shores, or some wild man or child the thread of whose history is interwoven with its own; not from him who could show no title to it but the deed which a like-minded neighbor or legislature gave him,—him who thought only of its money value; whose presence perchance cursed all the shore; who exhausted the land around it, and would fain have exhausted the waters within it; who regretted only that it was not English hay* or cranberry meadow,—there was nothing to redeem it, forsooth, in his eyes,—and would have drained and sold it for the mud at its bottom. It did not turn his mill, and it was no *privilege** to him to behold it. I respect not his labors, his farm where every thing has its price; who would carry the landscape,* who would carry his God, to market, if he could get any thing for him; who goes to market *for* his god as it is; on whose farm nothing grows free, whose fields bear no crops, whose meadows no flowers, whose trees no fruits, but dollars; who loves not the beauty of his fruits, whose fruits are not ripe for him till they are turned to dollars. Give me the poverty that enjoys true wealth. Farmers are respectable and interesting to me in proportion as they are poor,—poor farmers. A model farm!* where the house stands like a fungus in a muck-heap, chambers for men, horses, oxen, and swine, cleansed and uncleansed, all contiguous to one another! Stocked with men! A great grease-spot, redolent of manures and buttermilk! Under a high state of cultivation, being manured with the hearts and brains of men! As if you were to raise your potatoes in the church-yard! Such is a model farm.

No, no; if the fairest features of the landscape are to be named after men, let them be the noblest and worthiest men alone. Let our lakes receive as true names at least as the Icarian Sea,* where "still the shore" a "brave attempt resounds."*

Goose Pond, of small extent, is on my way* to Flints'; Fair-Haven, an expansion of Concord River,* said to contain some seventy acres, is a mile south-west; and White Pond, of about forty acres, is a mile and a half beyond Fair-Haven. This is my lake country.* These, with Concord River, are my water privileges; and night and day, year in year out, they grind such grist as I carry to them.

Since the woodcutters, and the railroad, and I myself have profaned Walden,* perhaps the most attractive, if not the most beautiful, of all our lakes, the gem of the woods, is White Pond;—a poor name from its commonness, whether derived from the remarkable purity of its waters or the color of its sands. In these as in other respects, however, it is a lesser twin of Walden. They are so much alike that you would say they must be connected under ground. It has the same stony shore, and its waters are of the same hue. As at Walden, in sultry dog-day* weather, looking down through the woods on some of its bays which are not so deep but that the reflection from the bottom tinges them, its waters are of a misty bluish-green or glaucous color. Many years since I used to go there to collect the sand by cart-loads, to make sand-paper* with, and I have continued to visit it ever since. One who frequents it proposes to call it Virid* Lake. Perhaps it might be called Yellow-Pine Lake, from the following circumstance. About fifteen years ago you could see the top of a pitch-pine, of the kind called yellow-pine hereabouts, though it is not a distinct species, projecting above the surface in deep water, many rods from the shore. It was even supposed by some that the pond had sunk, and this was one of the primitive forest that formerly stood there. I find that even so long ago as 1792, in a "Topographical Description of the Town of Concord,"* by one of its citizens, in the Collections of the Massachusetts Historical Society, the author, after speaking of Walden and

White Ponds, adds: "In the middle of the latter may be seen, when the water is very low, a tree which appears as if it grew in the place where it now stands, although the roots are fifty feet below the surface of the water; the top of this tree is broken off, and at that place measures fourteen inches in diameter." In the spring of '49 I talked with the man who lives nearest the pond in Sudbury, who told me that it was he who got out this tree ten or fifteen years before. As near as he could remember, it stood twelve or fifteen rods from the shore, where the water was thirty or forty feet deep. It was in the winter, and he had been getting out ice in the forenoon, and had resolved that in the afternoon, with the aid of his neighbors, he would take out the old yellow-pine. He sawed a channel in the ice toward the shore, and hauled it over and along and out on to the ice with oxen; but, before he had gone far in his work, he was surprised to find that it was wrong end upward, with the stumps of the branches pointing down, and the small end firmly fastened in the sandy bottom. It was about a foot in diameter at the big end, and he had expected to get a good saw-log, but it was so rotten as to be fit only for fuel, if for that. He had some of it in his shed then. There were marks of an axe and of woodpeckers on the but. He thought that it might have been a dead tree on the shore, but was finally blown over into the pond, and after the top had become water-logged, while the but-end was still dry and light, had drifted out and sunk wrong end up. His father, eighty years old, could not remember when it was not there. Several pretty large logs may still be seen lying on the bottom, where, owing to the undulation of the surface, they look like huge snakes in motion.

This pond has rarely been profaned by a boat, for there is little in it to tempt a fisherman. Instead of the white lily, which requires mud, or the common sweet flag, the blue flag (*Iris versicolor*) grows thinly in the pure water, rising from the stony bottom all around the shore, where it is visited by humming birds in June, and the color both of its bluish blades and its flowers, and especially their reflections, are in singular harmony with the glaucous water.

White Pond and Walden are great crystals on the surface of the earth, Lakes of Light. If they were permanently congealed, and small enough to be clutched, they would, perchance, be carried off by slaves, like precious stones, to adorn the heads of emperors; but being liquid, and ample, and secured to us and our successors forever, we disregard them, and run after the diamond of Kohinoor.* They are too pure to have a market value; they contain no muck. How much more beautiful than our lives, how much more transparent than our characters, are they! We never learned meanness of them. How much fairer than the pool before the farmer's door, in which his ducks swim! Hither the clean wild ducks come. Nature has no human inhabitant who appreciates her. The birds with their plumage and their notes are in harmony with the flowers, but what youth or maiden conspires with the wild luxuriant beauty of Nature? She flourishes most alone, far from the towns where they reside. Talk of heaven! ye disgrace earth.

BAKER FARM*

SOMETIMES I rambled to pine groves, standing like temples, or like fleets at sea, full-rigged, with wavy boughs, and rippling with light, so soft and green and shady that the Druids* would have forsaken their oaks to worship in them; or to the cedar wood beyond Flints' Pond, where the trees, covered with hoary blue berries, spiring higher and higher, are fit to stand before Valhalla,* and the creeping juniper covers the ground with wreaths full of fruit; or to swamps where the usnea lichen hangs in festoons from the white-spruce trees,* and toad-stools, round tables of the swamp gods, cover the ground, and more beautiful fungi adorn the stumps, like butterflies or shells, vegetable winkles; where the swamp-pink and dogwood grow, the red alder-berry glows like eyes of imps, the waxwork grooves and crushes the hardest woods in its folds, and the wild-holly berries make the beholder forget his home with their beauty, and he is dazzled and tempted by nameless other wild forbidden fruits, too fair for mortal taste. Instead of calling on some scholar, I paid many a visit to particular trees, of kinds which are rare in this neighborhood, standing far away in the middle of some pasture, or in the depths of a wood or swamp, or on a hill-top; such as the black-birch, of which we have some handsome specimens two feet in diameter; its cousin the yellow-birch, with its loose golden vest, per-fumed like the first; the beech, which has so neat a bole and beautifully lichen-painted, perfect in all its details, of which, excepting scattered specimens, I know but one small grove of sizable trees left in the township, supposed by some to have been planted by the pigeons that were once baited with beech nuts near by; it is worth the while to see the silver grain sparkle when you split this wood; the bass; the hornbeam; the *celtis occidentalis*, or false elm, of which we have but one well-grown; some taller mast of a pine, a shingle tree, or a more perfect hemlock than usual, standing like a pagoda in the midst of the woods; and many others I could mention. These were the shrines I visited both summer and winter.

Once it chanced that I stood in the very abutment of a rainbow's arch, which filled the lower stratum of the atmosphere, tinging the grass and leaves around, and dazzling me as if I looked through colored crystal. It was a lake of rainbow light, in which, for a short while, I lived like a dolphin. If it had lasted longer it might have tinged my employments and life. As I walked on the railroad causeway, I used to wonder at the halo of light around my shadow,* and would fain fancy myself one of the elect. One who visited me declared that the shadows of some Irishmen before him had no halo about them, that it was only natives* that were so distinguished. Benvenuto Cellini* tells us in his memoirs, that, after a certain terrible dream or vision which he had during his confinement in the castle of St. Angelo,* a resplendent light appeared over the shadow of his head at morning and evening, whether he was in Italy or France, and it was particularly conspicuous when the grass was moist with dew. This was probably the same phenomenon to which I have referred, which is especially observed in the morning, but also at other times, and even by moonlight. Though a constant one, it is not commonly noticed, and, in the case of an excitable imagination like Cellini's, it would be basis enough for superstition. Beside, he tells us that he showed it to very few. But are they not indeed distinguished who are conscious that they are regarded at all?

I set out one afternoon to go a-fishing to Fair-Haven, through the woods, to eke out my scanty fare of vegetables. My way led through Pleasant Meadow, an adjunct of the Baker Farm,* that retreat of which a poet has since sung, beginning,—

> "Thy entry is a pleasant field,
> Which some mossy fruit trees yield
> Partly to a ruddy brook,
> By gliding musquash undertook,
> And mercurial trout,
> Darting about."*

I thought of living there before I went to Walden. I "hooked"
the apples, leaped the brook, and scared the musquash and
the trout. It was one of those afternoons which seem indefinitely long before one, in which many events may happen, a
large portion of our natural life, though it was already half
spent when I started. By the way there came up a shower,
which compelled me to stand half an hour under a pine, piling
boughs over my head, and wearing my handkerchief for a
shed; and when at length I had made one cast over the
pickerel-weed, standing up to my middle in water, I found
myself suddenly in the shadow of a cloud, and the thunder
began to rumble with such emphasis that I could do no more
than listen to it. The gods must be proud, thought I, with such
forked flashes to rout a poor unarmed fisherman. So I made
haste for shelter to the nearest hut, which stood half a mile
from any road, but so much the nearer to the pond, and had
long been uninhabited:—

> "And here a poet builded,
> In the completed years,
> For behold a trivial cabin
> That to destruction steers."*

So the Muse fables. But therein, as I found, dwelt now John
Field, an Irishman, and his wife, and several children, from
the broad-faced boy who assisted his father at his work, and
now came running by his side from the bog to escape the rain,
to the wrinkled, sibyl-like,* cone-headed infant that sat upon
its father's knee as in the palaces of nobles, and looked out
from its home in the midst of wet and hunger inquisitively
upon the stranger, with the privilege of infancy, not knowing

but it was the last of a noble line, and the hope and cynosure of the world, instead of John Field's poor starveling brat. There we sat together under that part of the roof which leaked the least, while it showered and thundered without. I had sat there many times of old before the ship was built that floated this family to America. An honest, hard-working, but shiftless man plainly was John Field; and his wife, she too was brave to cook so many successive dinners in the recesses of that lofty stove; with round greasy face and bare breast, still thinking to improve her condition one day; with the never absent mop in one hand, and yet no effects of it visible any where. The chickens, which had also taken shelter here from the rain, stalked about the room like members of the family, too humanized methought to roast well. They stood and looked in my eye or pecked at my shoe significantly. Meanwhile my host told me his story, how hard he worked "bogging"* for a neighboring farmer, turning up a meadow with a spade or bog hoe at the rate of ten dollars an acre* and the use of the land with manure for one year, and his little broad-faced son worked cheerfully at his father's side the while, not knowing how poor a bargain the latter had made. I tried to help him with my experience, telling him that he was one of my nearest neighbors, and that I too, who came a-fishing here, and looked like a loafer, was getting my living like himself; that I lived in a tight, light, clean house, which hardly cost more than the annual rent of such a ruin as his commonly amounts to; and how, if he chose, he might in a month or two build himself a palace of his own; that I did not use tea, nor coffee, nor butter, nor milk, nor fresh meat, and so did not have to work to get them; again, as I did not work hard, I did not have to eat hard, and it cost me but a trifle for my food; but as he began with tea, and coffee, and butter, and milk, and beef, he had to work hard to pay for them, and when he had worked hard he had to eat hard again to repair the waste of his system,—and so it was as broad as it was long, indeed it was broader than it was long, for he was discontented and wasted his life into the bargain; and yet he had rated it as a gain in coming to America, that here you could get tea, and coffee, and meat every day. But the only true America is that country where you are at liberty to pursue such a mode of life as may enable you to do without these, and where the state does not endeavor to compel you to sustain the slavery and war and other superfluous expenses which directly or indirectly result from the use of such things. For I purposely talked to him as if he were a philosopher, or desired to be one. I should be glad if all the meadows on the earth were left in a wild state, if that were the consequence of men's beginning to redeem themselves. A man will not need to study history to find out what is best for his own culture. But alas! the culture of an Irishman is an enterprise to be undertaken with a sort of moral bog hoe. I told him, that as he worked so hard at bogging, he required thick boots and stout clothing, which yet were soon soiled and worn out, but I wore light shoes and thin clothing, which cost not half so much, though he might think that I was dressed like a gentleman, (which, however, was not the case,) and in an hour or two, without labor, but as a recreation, I could, if I wished, catch as many fish as I should want for two days, or earn enough money to support me a week. If he and his family would live simply, they might all go a-huckleberrying in the summer for their amusement. John heaved a sigh at this, and his wife stared with arms a-kimbo, and both appeared to be wondering if they had capital enough to begin such a course with, or arithmetic enough to carry it through. It was sailing by dead reckoning* to them, and they saw not clearly how to make their port so; therefore I suppose they still take life bravely, after their fashion, face to face, giving it tooth and nail, not having skill to split its massive columns with any fine entering wedge, and rout* it in detail;—thinking to deal with it roughly, as one should handle a thistle. But they fight at an overwhelming disadvantage,—living, John Field, alas! without arithmetic, and failing so.

"Do you ever fish?" I asked "O yes, I catch a mess now and then when I am lying by; good perch I catch." "What's your bait?" "I catch shiners with fish-worms, and bait the perch with them." "You'd better go now, John," said his wife with glistening and hopeful face; but John demurred.

The shower was now over, and a rainbow above the eastern woods promised a fair evening; so I took my departure. When I had got without I asked for a dish, hoping to get a sight of the well bottom, to complete my survey of the premises; but there, alas! are shallows and quicksands, and rope broken withal, and bucket irrecoverable. Meanwhile the right culinary vessel was selected, water was seemingly distilled, and after consultation and long delay passed out to the thirsty one,—not yet suffered to cool, not yet to settle. Such gruel sustains life here, I thought; so, shutting my eyes, and excluding the motes by a skilfully directed under-current, I drank to genuine hospitality the heartiest draught I could. I am not squeamish in such cases when manners are concerned.

As I was leaving the Irishman's roof after the rain, bending my steps again to the pond, my haste to catch pickerel, wading in retired meadows, in sloughs and bog-holes, in forlorn and savage places, appeared for an instant trivial to me who had been sent to school and college; but as I ran down the hill toward the reddening west, with the rainbow over my shoulder, and some faint tinkling sounds borne to my ear through the cleansed air, from I know not what quarter, my Good Genius seemed to say,—Go fish and hunt far and wide day by day,—farther and wider,—and rest thee by many brooks and hearth-sides without misgiving. Remember thy Creator in the days of thy youth.* Rise free from care before the dawn, and seek adventures. Let the noon find thee by other lakes, and the night overtake thee every where at home. There are no larger fields than these, no worthier games than may here be played. Grow wild according to thy nature, like these sedges and brakes, which will never become English hay.* Let the thunder rumble; what if it threaten ruin to farmers crops? that is not its errand to thee. Take shelter under the cloud, while they flee to carts and sheds. Let not to get a living be thy trade, but thy sport. Enjoy the land, but own it not. Through want of enterprise and faith men are where they are, buying and selling, and spending their lives like serfs.

O Baker Farm!

> "Landscape where the richest element
> Is a little sunshine innocent." ° °

> "No one runs to revel
> On thy rail-fenced lea." ° °

> "Debate with no man hast thou,
>   With questions art never perplexed,
> As tame at the first sight as now,
>   In thy plain russet gabardine dressed." ° °

> "Come ye who love,
>   And ye who hate,
> Children of the Holy Dove,
>   And Guy Faux of the state,
> And hang conspiracies
> From the tough rafters of the trees!"*

Men come tamely home at night only from the next field or street, where their household echoes haunt, and their life pines because it breathes its own breath over again; their shadows morning and evening reach farther than their daily steps. We should come home from far, from adventures, and perils, and discoveries every day, with new experience and character.

Before I had reached the pond some fresh impulse had brought out John Field, with altered mind, letting go "bogging" ere this sunset. But he, poor man, disturbed only a couple of fins while I was catching a fair string, and he said it was his luck; but when we changed seats in the boat luck changed seats too. Poor John Field!—I trust he does not read this, unless he will improve by it,—thinking to live by some derivative old country mode in this primitive new country,—

to catch perch with shiners. It is good bait sometimes, I allow. With his horizon all his own, yet he a poor man, born to be poor, with his inherited Irish poverty or poor life, his Adam's grandmother and boggy ways, not to rise in this world, he nor his posterity, till their wading webbed bog-trotting feet get *talaria** to their heels.

HIGHER LAWS*

As I came home through the woods with my string of fish, trailing my pole, it being now quite dark, I caught a glimpse of a woodchuck stealing across my path, and felt a strange thrill of savage delight, and was strongly tempted to seize and devour him raw; not that I was hungry then, except for that wildness which he represented. Once or twice, however, while I lived at the pond, I found myself ranging the woods, like a half-starved hound, with a strange abandonment, seeking some kind of vension which I might devour, and no morsel could have been too savage for me. The wildest scenes had become unaccountably familiar. I found in myself, and still find, an instinct toward a higher, or, as it is named, spiritual life, as do most men, and another toward a primitive rank and savage one, and I reverence them both. I love the wild not less than the good. The wildness and adventure that are in fishing still recommended it to me. I like sometimes to take rank hold on life and spend my day more as the animals do. Perhaps I have owed to this employment and to hunting, when quite young, my closest acquaintance with Nature. They early introduce us to and detain us in scenery with which otherwise, at that age, we should have little acquaintance. Fishermen, hunters, woodchoppers, and others, spending their lives in the fields and woods, in a peculiar sense a part of Nature themselves, are often in a more favorable mood for observing her, in the intervals of their pursuits, than philosophers or poets even, who approach her with expectation. She is not afraid to exhibit herself to them. The traveller on the prairie is naturally a hunter, on the head waters of the Missouri and Columbia a trapper, and at the Falls of St. Mary* a fisherman. He who is only a traveller learns things at second-hand and by the halves, and is poor authority. We are most interested when science reports what those men already know practically or instinctively, for that alone is a true *humanity*, or account of human experience.

They mistake who assert that the Yankee has few amusements, because he has not so many public holidays,* and men and boys do not play so many games as they do in England, for here the more primitive but solitary amusements of hunting fishing and the like have not yet given place to the former. Almost every New England boy among my contemporaries shouldered a fowling piece* between the ages of ten and fourteen; and his hunting and fishing grounds were not limited like the preserves of an English nobleman, but were more boundless even than those of a savage. No wonder, then, that he did not oftener stay to play on the common.* But already a change is taking place, owing, not to an increased humanity, but to an increased scarcity of game, for perhaps the hunter is the greatest friend of the animals hunted, not excepting the Humane Society.

Moreover, when at the pond, I wished sometimes to add fish to my fare for variety. I have actually fished from the same kind of necessity that the first fishers did. Whatever humanity I might conjure up against it was all factitious, and concerned my philosophy more than my feelings. I speak of fishing only now, for I had long felt differently about fowling, and sold my gun before I went to the woods. Not that I am less humane than others, but I did not perceive that my feelings were much affected. I did not pity the fishes nor the worms. This was habit. As for fowling, during the last years that I carried a gun my excuse was that I was studying ornithology, and sought only new or rare birds. But I confess that I am now inclined to think that there is a finer way of studying ornithology than this. It requires so much closer attention to the habits of the birds, that, if for that reason only, I have been willing to omit the gun. Yet notwithstanding the objection on the score of humanity, I am compelled to doubt if equally valuable sports are ever substituted for these; and when some

of my friends have asked me anxiously about their boys,
whether they should let them hunt, I have answered, yes,—
remembering that it was one of the best parts of my education,—*make* them hunters, though sportsmen only at first, if
possible, mighty hunters at last, so that they shall not find
game large enough for them in this or any vegetable wilderness,—hunters as well as fishers of men.* Thus far I am of the
opinion of Chaucer's nun, who

> "yave not of the text a pulled hen
> That saith that hunters ben not holy men."*

There is a period in the history of the individual, as of the race,
when the hunters are the "best men," as the Algonquins*
called them. We cannot but pity the boy who has never fired
a gun; he is no more humane, while his education has been
sadly neglected. This was my answer with respect to those
youths who were bent on this pursuit, trusting that they would
soon outgrow it. No humane being, past the thoughtless age
of boyhood, will wantonly murder any creature, which holds
its life by the same tenure that he does. The hare in its
extremity cries like a child. I warn you, mothers, that my
sympathies do not always make the usual phil*anthropic**
distinctions.

Such is oftenest the young man's introduction to the forest,
and the most original part of himself. He goes thither at first
as a hunter and fisher, until at last, if he has the seeds of a
better life in him, he distinguishes his proper objects, as a poet
or naturalist it may be, and leaves the gun and fish-pole
behind. The mass of men are still and always young in this
respect. In some countries a hunting parson is no uncommon
sight. Such a one might make a good shepherd's dog, but is far
from being the Good Shepherd.* I have been surprised to
consider that the only obvious employment, except wood-
chopping, ice-cutting, or the like business, which ever to my
knowledge detained at Walden Pond for a whole half day any
of my fellow-citizens, whether fathers or children of the town,

with just one exception, was fishing. Commonly they did not think that they were lucky, or well paid for their time, unless they got a long string of fish, though they had the opportunity of seeing the pond all the while. They might go there a thousand times before the sediment of fishing would sink to the bottom and leave their purpose pure; but no doubt such a clarifying process would be going on all the while. The governor and his council* faintly remember the pond, for they went a-fishing there when they were boys; but now they are too old and dignified to go a-fishing, and so they know it no more forever. Yet even they expect to go to heaven at last. If the legislature* regards it, it is chiefly to regulate the number of hooks to be used there; but they know nothing about the hook of hooks with which to angle for the pond itself, impaling the legislature for a bait. Thus, even in civilized communities, the embryo man passes through the hunter stage of development.

I have found repeatedly, of late years, that I cannot fish without falling a little in self-respect. I have tried it again and again. I have skill at it, and, like many of my fellows, a certain instinct for it, which revives from time to time, but always when I have done I feel that it would have been better if I had not fished. I think that I do not mistake. It is a faint intimation, yet so are the first streaks of morning. There is unquestionably this instinct in me which belongs to the lower orders of creation; yet with every year I am less a fisherman, though without more humanity or even wisdom; at present I am no fisherman at all. But I see that if I were to live in a wilderness I should again be tempted to become a fisher and hunter in earnest. Beside, there is something essentially unclean about this diet and all flesh, and I began to see where house-work commences, and whence the endeavor, which costs so much, to wear a tidy and respectable appearance each day, to keep the house sweet and free from all ill odors and sights. Having been my own butcher and scullion* and cook, as well as the gentleman for whom the dishes were served up, I can speak from an unusually complete experience. The practical objection to animal food in my case was its uncleanness; and, besides, when I had caught and cleaned and cooked and eaten my fish, they seemed not to have fed me essentially. It was insignificant and unnecessary, and cost more than it came to. A little bread or a few potatoes would have done as well, with less trouble and filth. Like many of my contemporaries,* I had rarely for many years used animal food, or tea, or coffee, &c.; not so much because of any ill effects which I had traced to them, as because they were not agreeable to my imagination. The repugnance to animal food is not the effect of experience, but is an instinct. It appeared more beautiful to live low and fare hard in many respects; and though I never did so, I went far enough to please my imagination. I believe that every man who has ever been earnest to preserve his higher or poetic faculties in the best condition has been particularly inclined to abstain from animal food, and from much food of any kind. It is a significant fact, stated by entomologists, I find it in Kirby and Spence,* that "some insects in their perfect state, though furnished with organs of feeding, make no use of them;" and they lay it down as "a general rule, that almost all insects in this state eat much less than in that of larvæ. The voracious caterpillar when transformed into a butterfly," . . . "and the gluttonous maggot when become a fly," content themselves with a drop or two of honey or some other sweet liquid. The abdomen under the wings of the butterfly still represents the larva. This is the tid-bit which tempts his insectivorous fate. The gross feeder is a man in the larva state; and there are whole nations in that condition, nations without fancy or imagination, whose vast abdomens betray them.

It is hard to provide and cook so simple and clean a diet as will not offend the imagination; but this, I think, is to be fed when we feed the body; they should both sit down at the same table. Yet perhaps this may be done. The fruits eaten temperately need not make us ashamed of our appetites, nor interrupt the worthiest pursuits. But put an extra condiment into your dish, and it will poison you. It is not worth the while to live by rich cookery. Most men would feel shame if caught preparing with their own hands precisely such a dinner, whether of animal or vegetable food, as is every day prepared for them by others. Yet till this is otherwise we are not civilized, and, if gentlemen and ladies, are not true men and women. This certainly suggests what change is to be made. It may be vain to ask why the imagination will not be reconciled to flesh and fat. I am satisfied that it is not. Is it not a reproach that man is a carniverous* animal? True, he can and does live, in a great measure, by preying on other animals; but this is a miserable way,—as any one who will go to snaring rabbits, or slaughtering lambs, may learn,—and he will be regarded as a benefactor of his race who shall teach man to confine himself to a more innocent and wholesome diet. Whatever my own practice may be, I have no doubt that it is a part of the destiny of the human race, in its gradual improvement, to leave off eating animals, as surely as the savage tribes have left off eating each other when they came in contact with the more civilized.

If one listens to the faintest but constant suggestions of his genius,* which are certainly true, he sees not to what extremes, or even insanity, it may lead him; and yet that way, as he grows more resolute and faithful, his road lies. The faintest assured objection which one healthy man feels will at length prevail over the arguments and customs of mankind. No man ever followed his genius till it misled him. Though the result were bodily weakness, yet perhaps no one can say that the consequences were to be regretted, for these were a life in conformity to higher principles. If the day and the night are such that you greet them with joy, and life emits a fragrance like flowers and sweet-scented herbs, is more elastic, more starry, more immortal,—that is your success. All nature is your congratulation, and you have cause momentarily to bless yourself. The greatest gains and values are farthest from being appreciated. We easily come to doubt if they exist. We soon forget them. They are the highest reality. Perhaps the facts most astounding and most real are never communicated by man to man. The true harvest of my daily life is somewhat as intangible and indescribable as the tints of morning or evening. It is a little star-dust caught, a segment of the rain-bow which I have clutched.

Yet, for my part, I was never unusually squeamish; I could sometimes eat a fried rat with a good relish, if it were neces-sary. I am glad to have drunk water so long, for the same reason that I prefer the natural sky to an opium-eater's heaven. I would fain keep sober always; and there are infinite degrees of drunkenness. I believe that water is the only drink for a wise man; wine is not so noble a liquor; and think of dashing the hopes of a morning with a cup of warm coffee, or of an evening with a dish of tea! Ah, how low I fall when I am tempted by them! Even music may be intoxicating. Such apparently slight causes destroyed Greece and Rome, and will destroy England and America. Of all ebriosity,* who does not prefer to be intoxicated by the air he breathes? I have found it to be the most serious objection to coarse labors long con-tinued, that they compelled me to eat and drink coarsely also. But to tell the truth, I find myself at present somewhat less particular in these respects. I carry less religion to the table, ask no blessing; not because I am wiser than I was, but, I am obliged to confess, because, however much it is to be regret-ted, with years I have grown more coarse and indifferent. Perhaps these questions are entertained only in youth, as most believe of poetry. My practice is "nowhere," my opinion is here. Nevertheless I am far from regarding myself as one of those privileged ones to whom the Ved refers when it says, that "he who has true faith in the Omnipresent Supreme Being may eat all that exists," that is, is not bound to inquire what is his food, or who prepares it; and even in their case it is to be observed, as a Hindoo commentator has remarked, that the Vedant limits this privilege to "the time of distress."*

Who has not sometimes derived an inexpressible satisfaction from his food in which appetite had no share? I have been thrilled to think that I owed a mental perception to the commonly gross sense of taste, that I have been inspired through the palate, that some berries which I had eaten on a hill-side had fed my genius. "The soul not being mistress of herself," says Thseng-tseu, "one looks, and one does not see; one listens, and one does not hear; one eats, and one does not know the savor of food."* He who distinguishes the true savor of his food can never be a glutton; he who does not cannot be otherwise. A puritan may go to his brown-bread crust with as gross an appetite as ever an alderman to his turtle. Not that food which entereth into the mouth defileth a man,* but the appetite with which it is eaten. It is neither the quality nor the quantity, but the devotion to sensual savors; when that which is eaten is not a viand to sustain our animal, or inspire our spiritual life, but food for the worms that possess us. If the hunter has a taste for mud-turtles, muskrats, and other such savage tid-bits, the fine lady indulges a taste for jelly made of a calf's foot, or for sardines from over the sea, and they are even. He goes to the mill-pond, she to her preserve-pot. The wonder is how they, how you and I, can live this slimy beastly life, eating and drinking.

Our whole life is startlingly moral. There is never an instant's truce between virtue and vice. Goodness is the only investment that never fails. In the music of the harp which trembles round the world it is the insisting on this which thrills us. The harp is the travelling patterer for the Universe's Insurance Company, recommending its laws, and our little goodness is all the assessment that we pay. Though the youth at last grows indifferent, the laws of the universe are not indifferent, but are forever on the side of the most sensitive. Listen to every zephyr for some reproof, for it is surely there, and he is unfortunate who does not hear it. We cannot touch a string or move a stop but the charming moral transfixes us.

Many an irksome noise, go a long way off, is heard as music, a proud sweet satire on the meanness of our lives.

We are conscious of an animal in us, which awakens in proportion as our higher nature slumbers. It is reptile* and sensual, and perhaps cannot be wholly expelled; like the worms which, even in life and health, occupy our bodies. Possibly we may withdraw from it, but never change its nature. I fear that it may enjoy a certain health of its own; that we may be well, yet not pure. The other day I picked up the lower jaw of a hog, with white and sound teeth and tusks, which suggested that there was an animal health and vigor distinct from the spiritual. This creature succeeded by other means than temperance and purity. "That in which men differ from brute beasts," says Mencius, "is a thing very inconsiderable; the common herd lose it very soon; superior men preserve it carefully."* Who knows what sort of life would result if we had attained to purity? If I knew so wise a man as could teach me purity I would go to seek him forthwith. "A command over our passions, and over the external senses of the body, and good acts, are declared by the Ved to be indispensable in the mind's approximation to God."* Yet the spirit can for the time pervade and control every member and function of the body, and transmute what in form is the grossest sensuality into purity and devotion. The generative energy, which, when we are loose, dissipates and makes us unclean, when we are continent invigorates and inspires us. Chastity is the flowering of man; and what are called Genius, Heroism, Holiness, and the like, are but various fruits which succeed it. Man flows at once to God when the channel of purity is open. By turns our purity inspires and our impurity casts us down. He is blessed who is assured that the animal is dying out in him day by day, and the divine being established. Perhaps there is none but has cause for shame on account of the inferior and brutish nature to which he is allied. I fear that we are such gods or demigods only as fauns and satyrs,* the divine allied to beasts, the creatures of appetite, and that, to some extent, our very life is our disgrace.—

> "How happy's he who hath due place assigned
> To his beasts and disaforested his mind!
>
> Can use his horse, goat, wolf, and ev'ry beast,
> And is not ass himself to all the rest!
> Else man not only is the herd of swine,
> But he's those devils too which did incline
> Them to a headlong rage, and made them worse."*

All sensuality is one, though it takes many forms; all purity is one. It is the same whether a man eat, or drink, or cohabit, or sleep sensually. They are but one appetite, and we only need to see a person do any one of these things to know how great a sensualist he is. The impure can neither stand nor sit with purity. When the reptile is attacked at one mouth of his burrow, he shows himself at another. If you would be chaste, you must be temperate. What is chastity? How shall a man know if he is chaste? He shall not know it. We have heard of this virtue, but we know not what it is. We speak conformably to the rumor which we have heard. From exertion come wisdom and purity; from sloth ignorance and sensuality. In the student sensuality is a sluggish habit of mind. An unclean person is universally a slothful one, one who sits by a stove, whom the sun shines on prostrate, who reposes without being fatigued. If you would avoid uncleanness, and all the sins, work earnestly, though it be at cleaning a stable. Nature is hard to be overcome, but she must be overcome. What avails it that you are Christian, if you are not purer than the heathen, if you deny yourself no more, if you are not more religious? I know of many systems of religion esteemed heathenish whose precepts fill the reader with shame, and provoke him to new endeavors, though it be to the performance of rites merely.

I hesitate to say these things, but it is not because of the

subject,—I care not how obscene my *words* are,—but because I cannot speak of them without betraying my impurity.
We discourse freely without shame of one form of sensuality,
and are silent about another. We are so degraded that we
cannot speak simply of the necessary functions of human
nature. In earlier ages, in some countries, every function was
reverently spoken of and regulated by law. Nothing was too
trivial for the Hindoo lawgiver,* however offensive it may be
to modern taste. He teaches how to eat, drink, cohabit, void
excrement and urine, and the like, elevating what is mean,
and does not falsely excuse himself by calling these things
trifles.

Every man is the builder of a temple, called his body, to the
god he worships, after a style purely his own, nor can he get
off by hammering marble instead. We are all sculptors and
painters, and our material is our own flesh and blood and
bones. Any nobleness begins at once to refine a man's features, any meanness or sensuality to imbrute them.

John Farmer sat at his door one September evening, after a
hard day's work, his mind still running on his labor more or
less. Having bathed he sat down to recreate his intellectual
man. It was a rather cool evening, and some of his neighbors
were apprehending a frost. He had not attended to the train
of his thoughts long when he heard some one playing on a
flute, and that sound harmonized with his mood. Still he
thought of his work; but the burden of his thought was, that
though this kept running in his head, and he found himself
planning and contriving it against his will, yet it concerned
him very little. It was no more than the scurf of his skin, which
was constantly shuffled off. But the notes of the flute came
home to his ears out of a different sphere from that he worked
in, and suggested work for certain faculties which slumbered
in him. They gently did away with the street, and the village,
and the state in which he lived. A voice said to him,—Why do
you stay here and live this mean moiling life, when a glorious
existence is possible for you? Those same stars twinkle over

other fields than these.—But how to come out of this condition and actually migrate thither? All that he could think of was to practise some new austerity, to let his mind descend into his body and redeem it, and treat himself with ever increasing respect.

BRUTE NEIGHBORS

SOMETIMES I had a companion* in my fishing, who came through the village to my house from the other side of the town, and the catching of the dinner was as much a social exercise as the eating of it.

Hermit. I wonder what the world is doing now. I have not heard so much as a locust over the sweet-fern these three hours. The pigeons are all asleep upon their roosts,— no flutter from them. Was that a farmer's noon horn which sounded from beyond the woods just now? The hands are coming in to boiled salt beef and cider and Indian bread. Why will men worry themselves so? He that does not eat need not work. I wonder how much they have reaped. Who would live there where a body can never think for the barking of Bose?* And O, the housekeeping! to keep bright the devil's door-knobs, and scour his tubs this bright day! Better not keep a house. Say, some hollow tree; and then for morning calls and dinner-parties! Only a wood-pecker tapping. O, they swarm; the sun is too warm there; they are born too far into life for me. I have water from the spring, and a loaf of brown bread on the shelf.—Hark! I hear a rustling of the leaves. Is it some ill-fed village hound yielding to the instinct of the chase? or the lost pig which is said to be in these woods, whose tracks I saw after the rain? It comes on apace; my sumachs and sweet-briers tremble.—Eh, Mr. Poet, is it you? How do you like the world today?

Poet. See those clouds; how they hang! That's the greatest thing I have seen to-day. There's nothing like it in old paintings, nothing like it in foreign lands,—unless when we were off the coast of Spain. That's a true Mediterranean sky. I thought, as I have my living to get, and have not

eaten to-day, that I might go a-fishing. That's the true
industry for poets. It is the only trade I have learned. Come,
let's along.

Hermit. I cannot resist. My brown bread will soon be gone.
I will go with you gladly soon, but I am just concluding a
serious meditation. I think that I am near the end of it. Leave
me alone, then, for a while. But that we may not be delayed,
you shall be digging the bait meanwhile. Angle-worms are
rarely to be met with in these parts, where the soil was never
fattened with manure; the race is nearly extinct. The sport
of digging the bait is nearly equal to that of catching the fish,
when one's appetite is not too keen; and this you may have
all to yourself to-day. I would advise you to set in the spade
down yonder among the ground-nuts, where you see the
johnswort waving. I think that I may warrant you one worm
to every three sods you turn up, if you look well in among
the roots of the grass, as if you were weeding. Or if you choose
to go farther, it will not be unwise, for I have found the
increase of fair bait to be very nearly as the squares of
the distances.

Hermit alone. Let me see; where was I? Methinks I was
nearly in this frame of mind; the world lay about at this angle.
Shall I go to heaven or a-fishing? If I should soon bring this
meditation to an end, would another so sweet occasion be
likely to offer? I was as near being resolved into the essence of
things as ever I was in my life. I fear my thoughts will not
come back to me. If it would do any good, I would whistle for
them. When they make us an offer, is it wise to say, We will
think of it? My thoughts have left no track, and I cannot find
the path again. What was it that I was thinking of? It was
a very hazy day. I will just try these three sentences of
Con-fut-see;* they may fetch that state about again. I know
not whether it was the dumps or a budding ecstasy. Mem.*
There never is but one opportunity of a kind.

Poet. How now, Hermit, is it too soon? I have got just
thirteen whole ones, beside several which are imperfect or

undersized; but they will do for the smaller fry; they do not cover up the hook so much. Those village worms are quite too large; a shiner may make a meal off one without finding the skewer.

Hermit. Well, then, let's be off. Shall we to the Concord? There's good sport there if the water be not too high.

Why do precisely these objects which we behold make a world? Why has man just these species of animals for his neighbors; as if nothing but a mouse could have filled this crevice? I suspect that Pilpay* & Co. have put animals to their best use, for they are all beasts of burden, in a sense, made to carry some portion of our thoughts.

The mice which haunted my house were not the common ones, which are said to have been introduced into the country, but a wild native kind not found* in the village. I sent one to a distinguished naturalist,* and it interested him much. When I was building, one of these had its nest underneath the house, and before I had laid the second floor, and swept out the shavings, would come out regularly at lunch time and pick up the crums at my feet. It probably had never seen a man before; and it soon became quite familiar, and would run over my shoes and up my clothes.* It could readily ascend the sides of the room by short impulses, like a squirrel, which it resembled in its motions. At length, as I leaned with my elbow on the bench one day, it ran up my clothes, and along my sleeve, and round and round the paper which held my dinner, while I kept the latter close, and dodged and played at bo-peep* with it; and when at last I held still a piece of cheese between my thumb and finger, it came and nibbled it, sitting in my hand, and afterward cleaned its face and paws, like a fly, and walked away.

A phœbe soon built in my shed, and a robin for protection in a pine which grew against the house. In June the partridge, *(Tetrao umbellus,)* which is so shy a bird, led her brood past my windows, from the woods in the rear to the front of my house, clucking and calling to them like a hen, and in all her behavior proving herself the hen of the woods. The young suddenly disperse on your approach, at a signal from the mother, as if a whirlwind had swept them away, and they so exactly resemble the dried leaves and twigs that many a traveller has placed his foot in the midst of a brood, and heard the whir of the old bird as she flew off, and her anxious calls and mewing, or seen her trail her wings to attract his attention, without suspecting their neighborhood. The parent will sometimes roll and spin round before you in such a dishabille,* that you cannot, for a few moments, detect what kind of creature it is. The young squat still and flat, often running their heads under a leaf, and mind only their mother's directions given from a distance, nor will your approach make them run again and betray themselves. You may even tread on them, or have your eyes on them for a minute, without discovering them. I have held them in my open hand at such a time, and still their only care, obedient to their mother and their instinct, was to squat there without fear or trembling. So perfect is this instinct, that once, when I had laid them on the leaves again, and one accidentally fell on its side, it was found with the rest in exactly the same position ten minutes afterward. They are not callow* like the young of most birds, but more perfectly developed and precocious even than chickens. The remarkably adult yet innocent expression of their open and serene eyes is very memorable. All intelligence seems reflected in them. They suggest not merely the purity of infancy, but a wisdom clarified by experience. Such an eye was not born when the bird was, but is coeval with the sky it reflects. The woods do not yield another such gem. The traveller does not often look into such a limpid well. The ignorant or reckless sportsman often shoots the parent at such a time, and leaves these innocents to fall a prey to some prowling beast or bird, or gradually mingle with the decaying leaves which they so much resemble. It is said that when hatched by a hen they will directly disperse on some alarm, and so are lost, for they never hear the mother's call which gathers them again. These were my hens and chickens.

It is remarkable how many creatures live wild and free though secret in the woods, and still sustain themselves in the neighborhood of towns, suspected by hunters only. How retired the otter manages to live here! He grows to be four feet long, as big as a small boy, perhaps without any human being getting a glimpse of him. I formerly saw the raccoon in the woods behind where my house is built, and probably still heard their whinnering at night. Commonly I rested an hour or two in the shade at noon, after planting, and ate my lunch, and read a little by a spring which was the source of a swamp and of a brook, oozing from under Brister's Hill,* half a mile from my field. The approach to this was through a succession of descending grassy hollows, full of young pitch-pines, into a larger wood about the swamp. There, in a very secluded and shaded spot, under a spreading white-pine, there was yet a clean firm sward to sit on. I had dug out the spring and made a well of clear gray water, where I could dip up a pailful without roiling it, and thither I went for this purpose almost every day in midsummer, when the pond was warmest. Thither too the wood-cock led her brood, to probe the mud for worms, flying but a foot above them down the bank, while they ran in a troop beneath; but at last, spying me, she would leave her young and circle round and round me, nearer and nearer till within four or five feet, pretending broken wings and legs, to attract my attention, and get off her young, who would already have taken up their march, with faint wiry peep, single file through the swamp, as she directed. Or I heard the peep of the young when I could not see the parent bird. There too the turtle-doves sat over the spring, or fluttered from bough to bough of the soft white-pines over my head; or the red squirrel, coursing down the nearest bough, was particularly familiar and inquisitive. You only need sit still long enough in some attractive spot in the woods that all its inhabitants may exhibit themselves to you by turns.

I was witness to events of a less peaceful character. One day when I went out to my wood-pile, or rather my pile of stumps, I observed two large ants, the one red, the other much larger, nearly half an inch long, and black, fiercely contending with one another. Having once got hold they never let go, but struggled and wrestled and rolled on the chips incessantly. Looking farther, I was surprised to find that the chips were covered with such combatants, that it was not a *duellum*, but a *bellum*,* a war between two races of ants, the red always pitted against the black, and frequently two red ones to one black. The legions of these Myrmidons* covered all the hills and vales in my wood-yard, and the ground was already strewn with the dead and dying, both red and black. It was the only battle which I have ever witnessed, the only battle-field I ever trod while the battle was raging; internecine war; the red republicans* on the one hand, and the black imperialists* on the other. On every side they were engaged in deadly combat, yet without any noise that I could hear, and human soldiers never fought so resolutely. I watched a couple that were fast locked in each other's embraces, in a little sunny valley amid the chips, now at noon-day prepared to fight till the sun went down, or life went out. The smaller red champion had fastened himself like a vice to his adversary's front, and through all the tumblings on that field never for an instant ceased to gnaw at one of his feelers near the root, having already caused the other to go by the board; while the stronger black one dashed him from side to side, and, as I saw on looking nearer, had already divested him of several of his members. They fought with more pertinacity than bull-dogs. Neither manifested the least disposition to retreat. It was evident that their battle-cry was Conquer or die. In the mean while there came along a single red ant on the hill-side of this valley, evidently full of excitement, who either had despatched his foe, or had not yet taken part in the battle; probably the latter, for he had lost none of his limbs; whose mother had charged him to return with his shield or upon it.* Or perchance he was some

Achilles, who had nourished his wrath apart, and had now come to avenge or rescue his Patroclus.* He saw this unequal combat from afar,—for the blacks were nearly twice the size of the red,—he drew near with rapid pace till he stood on his guard within half an inch of the combatants; then, watching his opportunity, he sprang upon the black warrior, and commenced his operations near the root of his right fore-leg, leaving the foe to select among his own members; and so there were three united for life, as if a new kind of attraction had been invented which put all other locks and cements to shame. I should not have wondered by this time to find that they had their respective musical bands stationed on some eminent chip, and playing their national airs the while, to excite the slow and cheer the dying combatants. I was myself excited somewhat even as if they had been men. The more you think of it, the less the difference. And certainly there is not the fight recorded in Concord history, at least, if in the history of America, that will bear a moment's comparison with this, whether for the numbers engaged in it, or for the patriotism and heroism displayed. For numbers and for carnage it was an Austerlitz or Dresden.* Concord Fight! Two killed on the patriots' side, and Luther Blanchard wounded! Why here every ant was a Buttrick,—"Fire! for God's sake fire!"—and thousands shared the fate of Davis and Hosmer.* There was not one hireling there. I have no doubt that it was a principle they fought for, as much as our ancestors, and not to avoid a three-penny tax on their tea;* and the results of this battle will be as important and memorable to those whom it concerns as those of the battle of Bunker Hill,* at least.

I took up the chip on which the three I have particularly described were struggling, carried it into my house, and placed it under a tumbler on my window-sill, in order to see the issue. Holding a microscope to the first-mentioned red ant, I saw that, though he was assiduously gnawing at the near fore-leg of his enemy, having severed his remaining feeler, his own breast was all torn away, exposing what vitals he had there to the jaws of the black warrior, whose breast-plate was apparently too thick for him to pierce; and the dark carbuncles of the sufferer's eyes shone with ferocity such as war only could excite. They struggled half an hour longer under the tumbler, and when I looked again the black soldier had severed the heads of his foes from their bodies, and the still living heads were hanging on either side of him like ghastly trophies at his saddle-bow, still apparently as firmly fastened as ever, and he was endeavoring with feeble struggles, being without feelers and with only the remnant of a leg, and I know not how many other wounds, to divest himself of them; which at length, after half an hour more, he accomplished. I raised the glass, and he went off over the window-sill in that crippled state. Whether he finally survived that combat, and spent the remainder of his days in some Hotel des Invalides,* I do not know; but I thought that his industry would not be worth much thereafter. I never learned which party was victorious, nor the cause of the war; but I felt for the rest of that day as if I had had my feelings excited and harrowed by witnessing the struggle, the ferocity and carnage, of a human battle before my door.

Kirby and Spence* tell us that the battles of ants have long been celebrated and the date of them recorded, though they say that Huber* is the only modern author who appears to have witnessed them. "Æneas Sylvius," say they, "after giving a very circumstantial account of one contested with great obstinacy by a great and small species on the trunk of a pear tree," adds that "'This action was fought in the pontificate of Eugenius the Fourth, in the presence of Nicholas Pistoriensis, an eminent lawyer, who related the whole history of the battle with the greatest fidelity.' A similar engagement between great and small ants is recorded by Olaus Magnus, in which the small ones, being victorious, are said to have buried the bodies of their own soldiers, but left those of their giant enemies a prey to the birds. This event happened previous to the expulsion of the tyrant Christiern the Second* from Sweden." The battle which I witnessed took place in the Presidency of Polk,* five years before the passage of Webster's Fugitive-Slave Bill.*

Many a village Bose,* fit only to course a mud-turtle in a victualling cellar, sported his heavy quarters in the woods, without the knowledge of his master, and ineffectually smelled at old fox burrows and woodchucks' holes; led perchance by some slight cur which nimbly threaded the wood, and might still inspire a natural terror in its denizens;—now far behind his guide, barking like a canine bull toward some small squirrel which had treed itself for scrutiny, then, cantering off, bending the bushes with his weight, imagining that he is on the track of some stray member of the jerbilla family. Once I was surprised to see a cat walking along the stony shore of the pond, for they rarely wander so far from home. The surprise was mutual. Nevertheless the most domestic cat, which has lain on a rug all her days, appears quite at home in the woods, and, by her sly and stealthy behavior, proves herself more native there than the regular inhabitants. Once, when berrying, I met with a cat with young kittens in the woods, quite wild, and they all, like their mother, had their backs up and were fiercely spitting at me. A few years before I lived in the woods there was what was called a "winged cat" in one of the farm-houses in Lincoln nearest the pond, Mr. Gilian Baker's.* When I called to see her in June, 1842, she was gone a-hunting in the woods, as was her wont, (I am not sure whether it was a male or female, and so use the more common pronoun,) but her mistress told me that she came into the neighborhood a little more than a year before, in April, and was finally taken into their house; that she was of a dark brownish-gray color, with a white spot on her throat, and white feet, and had a large bushy tail like a fox; that in the winter the fur grew thick and flatted out along her sides, forming strips ten or twelve inches long by two and a half wide, and under her chin like a muff, the upper side loose, the under matted like felt, and in the spring these appendages dropped off. They gave me a pair of her "wings," which I keep still. There is no appearance of a membrane about them. Some thought it was part flying-squirrel or some other wild animal, which is not impossible, for, according to naturalists, prolific hybrids have been produced by the union of the marten and domestic cat. This would have been the right kind of cat for me to keep, if I had kept any; for why should not a poet's cat be winged as well as his horse?*

In the fall the loon (*Colymbus glacialis*) came, as usual, to moult and bathe in the pond, making the woods ring with his wild laughter before I had risen. At rumor of his arrival all the Mill-dam* sportsmen are on the alert, in gigs and on foot, two by two and three by three, with patent rifles and conical balls and spy-glasses. They come rustling through the woods like autumn leaves, at least ten men to one loon. Some station themselves on this side of the pond, some on that, for the poor bird cannot be omnipresent; if he dive here he must come up there. But now the kind October wind rises, rustling the leaves and rippling the surface of the water, so that no loon can be heard or seen, though his foes sweep the pond with spy-glasses, and make the woods resound with their discharges. The waves generously rise and dash angrily, taking sides with all waterfowl, and our sportsmen must beat a retreat to town and shop and unfinished jobs. But they were too often successful. When I went to get a pail of water early in the morning I frequently saw this stately bird sailing out of my cove within a few rods. If I endeavored to overtake him in a boat, in order to see how he would manœuvre, he would dive and be completely lost, so that I did not discover him again, sometimes, till the latter part of the day. But I was more than a match for him on the surface. He commonly went off in a rain.

As I was paddling along the north shore one very calm October afternoon, for such days especially they settle on to the lakes, like the milkweed down, having looked in vain over

the pond for a loon, suddenly one, sailing out from the shore
toward the middle a few rods in front of me, set up his wild
laugh and betrayed himself. I pursued with a paddle and
he dived, but when he came up I was nearer than before. He
dived again, but I miscalculated the direction he would take,
and we were fifty rods apart when he came to the surface this
time, for I had helped to widen the interval; and again he
laughed long and loud, and with more reason than before.
He manœuvred so cunningly that I could not get within half a
dozen rods of him. Each time, when he came to the surface,
turning his head this way and that, he coolly surveyed the
water and the land, and apparently chose his course so that he
might come up where there was the widest expanse of water
and at the greatest distance from the boat. It was surprising
how quickly he made up his mind and put his resolve into
execution. He led me at once to the widest part of the pond,
and could not be driven from it. While he was thinking one
thing in his brain, I was endeavoring to divine his thought in
mine. It was a pretty game, played on the smooth surface of
the pond, a man against a loon. Suddenly your adversary's
checker* disappears beneath the board, and the problem is to
place yours nearest to where his will appear again. Sometimes
he would come up unexpectedly on the opposite side of me,
having apparently passed directly under the boat. So long-
winded was he and so unweariable, that when he had swum
farthest he would immediately plunge again, nevertheless;
and then no wit could divine where in the deep pond, beneath
the smooth surface, he might be speeding his way like a fish,
for he had time and ability to visit the bottom of the pond in
its deepest part. It is said that loons have been caught in the
New York lakes eighty feet beneath the surface, with hooks
set for trout,—though Walden is deeper than that. How surprised must the fishes be to see this ungainly visitor from
another sphere speeding his way amid their schools! Yet he
appeared to know his course as surely under water as on the
surface, and swam much faster there. Once or twice I saw a

ripple where he approached the surface, just put his head out to reconnoitre, and instantly dived again. I found that it was as well for me to rest on my oars and wait his reappearing as to endeavor to calculate where he would rise; for again and again, when I was straining my eyes over the surface one way, I would suddenly be startled by his unearthly laugh behind me. But why, after displaying so much cunning, did he invariably betray himself the moment he came up by that loud laugh? Did not his white breast enough betray him? He was indeed a silly loon, I thought. I could commonly hear the plash of the water when he came up, and so also detected him. But after an hour he seemed as fresh as ever, dived as willingly and swam yet farther than at first. It was surprising to see how serenely he sailed off with unruffled breast when he came to the surface, doing all the work with his webbed feet beneath. His usual note was this demoniac laughter, yet somewhat like that of a water-fowl; but occasionally, when he had balked me most successfully and come up a long way off, he uttered a long-drawn unearthly howl, probably more like that of a wolf than any bird; as when a beast puts his muzzle to the ground and deliberately howls. This was his looning,—perhaps the wildest sound that is ever heard here, making the woods ring far and wide. I concluded that he laughed in derision of my efforts, confident of his own resources. Though the sky was by this time overcast, the pond was so smooth that I could see where he broke the surface when I did not hear him. His white breast, the stillness of the air, and the smoothness of the water were all against him. At length, having come up fifty rods off, he uttered one of those prolonged howls, as if calling on the god of loons to aid him, and immediately there came a wind from the east and rippled the surface, and filled the whole air with misty rain, and I was impressed as if it were the prayer of the loon answered, and his god was angry with me; and so I left him disappearing far away on the tumultuous surface.

For hours, in fall days, I watched the ducks cunningly tack

and veer and hold the middle of the pond, far from the sportsman; tricks which they will have less need to practise in Louisiana bayous.* When compelled to rise they would sometimes circle round and round and over the pond at a considerable height, from which they could easily see to other ponds and the river, like black motes in the sky; and, when I thought they had gone off thither long since, they would settle down by a slanting flight of a quarter of a mile on to a distant part which was left free; but what beside safety they got by sailing in the middle of Walden I do not know, unless they love its water for the same reason that I do.

HOUSE-WARMING

In October I went a-graping to the river meadows, and loaded myself with clusters more precious for their beauty and fragrance than for food. There too I admired, though I did not gather, the cranberries, small waxen gems, pendants of the meadow grass, pearly and red, which the farmer plucks with an ugly rake,* leaving the smooth meadow in a snarl, heedlessly measuring them by the bushel and the dollar only, and sells the spoils of the meads to Boston and New York; destined to be *jammed*,* to satisfy the tastes of lovers of Nature there. So butchers rake the tongues of bison out of the prairie grass,* regardless of the torn and drooping plant. The barberry's brilliant fruit was likewise food for my eyes merely; but I collected a small store of wild apples for coddling, which the proprietor and travellers had overlooked. When chestnuts were ripe I laid up half a bushel for winter. It was very exciting at that season to roam the then boundless chestnut woods of Lincoln,—they now sleep their long sleep* under the railroad,—with a bag on my shoulder, and a stick to open burrs with in my hand, for I did not always wait for the frost, amid the rustling of leaves and the loud reproofs of the red-squirrels and the jays, whose half-consumed nuts I sometimes stole, for the burrs which they had selected were sure to contain sound ones. Occasionally I climbed and shook the trees. They grew also behind my house, and one large tree which almost overshadowed it, was, when in flower, a bouquet which scented the whole neighborhood, but the squirrels and the jays got most of its fruit; the last coming in flocks early in the morning and picking the nuts out of the burrs before they fell. I relinquished these trees to them and visited the more distant woods composed wholly of chestnut. These nuts, as far as they went, were a good substitute for bread. Many other substitutes might, perhaps, be found. Digging one day for fish-worms I discovered the ground-nut (*Apios tuberosa*) on its string, the potato of the aborigines, a sort of fabulous fruit, which I had begun to doubt if I had ever dug and eaten in childhood, as I had told, and had not dreamed it. I had often since seen its crimpled red velvety blossom supported by the stems of other plants without knowing it to be the same. Cultivation has well nigh exterminated it. It has a sweetish taste, much like that of a frostbitten potato, and I found it better boiled than roasted. This tuber seemed like a faint promise of Nature to rear her own children and feed them simply here at some future period. In these days of fatted cattle and waving grain-fields, this humble root, which was once the *totem* of an Indian tribe, is quite forgotten, or known only by its flowering vine; but let wild Nature reign here once more, and the tender and luxurious English grains will probably disappear before a myriad of foes, and without the care of man the crow may carry back even the last seed of corn to the great corn-field of the Indian's God in the south-west, whence he is said to have brought it; but the now almost exterminated ground-nut will perhaps revive and flourish in spite of frosts and wildness, prove itself indigenous, and resume its ancient importance and dignity as the diet of the hunter tribe. Some Indian Ceres or Minerva* must have been the inventor and bestower of it; and when the reign of poetry commences here, its leaves and string of nuts may be represented on our works of art.

Already, by the first of September, I had seen two or three small maples turned scarlet across the pond, beneath where the white stems of three aspens diverged, at the point of a promontory, next the water. Ah, many a tale their color told!* And gradually from week to week the character of each tree came out, and it admired itself reflected in the smooth mirror of the lake. Each morning the manager of this gallery substituted some new picture, distinguished by more brilliant or harmonious coloring, for the old upon the walls.

The wasps came by the thousands to my lodge in October, as to winter quarters, and settled on my windows within and on the walls over-head, sometimes deterring visitors from entering. Each morning, when they were numbed with cold, I swept some of them out, but I did not trouble myself much to get rid of them; I even felt complimented by their regarding my house as a desirable shelter. They never molested me seriously, though they bedded with me; and they gradually disappeared, into what crevices I do not know, avoiding winter and unspeakable cold.*

Like the wasps, before I finally went into winter quarters* in November, I used to resort to the north-east side of Walden, which the sun, reflected from the pitch-pine woods and the stony shore, made the fire-side of the pond; it is so much pleasanter and wholesomer to be warmed by the sun while you can be, than by an artificial fire. I thus warmed myself by the still glowing embers which the summer, like a departed hunter, had left.

When I came to build my chimney I studied masonry. My bricks being second-hand ones required to be cleaned with a trowel, so that I learned more than usual of the qualities of bricks and trowels. The mortar on them was fifty years old, and was said to be still growing harder; but this is one of those sayings which men love to repeat whether they are true or not. Such sayings themselves grow harder and adhere more firmly with age, and it would take many blows with a trowel to clean an old wiseacre of them. Many of the villages of Mesopotamia are built of second-hand bricks of a very good quality, obtained from the ruins of Babylon, and the cement on them is older and probably harder still. However that may be, I was struck by the peculiar toughness of the steel which bore so many violent blows without being worn out. As my bricks had been in a chimney before, though I did not read the name of Nebuchadnezzar* on them, I picked out as many fireplace bricks as I could find, to save work and waste, and I filled the spaces between the bricks about the fireplace with stones from the pond shore, and also made my mortar with the white sand from the same place. I lingered most about the fireplace, as the most vital part of the house. Indeed, I worked so deliberately, that though I commenced at the ground in the morning, a course of bricks raised a few inches above the floor served for my pillow at night; yet I did not get a stiff neck for it that I remember; my stiff neck is of older date. I took a poet to board for a fortnight about those times, which caused me to be put to it for room.* He brought his own knife, though I had two, and we used to scour them by thrusting them into the earth. He shared with me the labors of cooking. I was pleased to see my work rising so square and solid by degrees, and reflected, that, if it proceeded slowly, it was calculated to endure a long time. The chimney is to some extent an inde-pendent structure, standing on the ground and rising through the house to the heavens; even after the house is burned it still stands sometimes, and its importance and independence are apparent. This was toward the end of summer. It was now November.

The north wind had already begun to cool the pond, though it took many weeks of steady blowing to accomplish it, it is so deep. When I began to have a fire at evening, before I plas-tered my house, the chimney carried smoke particularly well, because of the numerous chinks between the boards. Yet I passed some cheerful evenings in that cool and airy apart-ment, surrounded by the rough brown boards full of knots, and rafters with the bark on high over-head. My house never pleased my eye so much after it was plastered, though I was obliged to confess that it was more comfortable. Should not every apartment in which man dwells be lofty enough to create some obscurity over-head, where flickering shadows may play at evening about the rafters? These forms are more agreeable to the fancy and imagination than fresco paintings or other the most expensive furniture. I now first began to

inhabit my house, I may say, when I began to use if for
warmth as well as shelter. I had got a couple of old fire-dogs
to keep the wood from the hearth, and it did me good to see
the soot form on the back of the chimney which I had built,
and I poked the fire with more right and more satisfaction
than usual. My dwelling was small, and I could hardly entertain an echo in it; but it seemed larger for being a single
apartment and remote from neighbors. All the attractions of a
house were concentrated in one room; it was kitchen, chamber, parlor, and keeping-room;* and whatever satisfaction
parent or child, master or servant, derive from living in a
house, I enjoyed it all. Cato says, the master of a family
(*patremfamilias*) must have in his rustic villa "cellam
oleariam, vinariam, dolia multa, uti lubeat caritatem
expectare, et rei, et virtuti, et gloriæ erit," that is, "an oil and
wine celler, many casks, so that it may be pleasant to expect
hard times; it will be for his advantage, and virtue, and
glory."* I had in my cellar a firkin* of potatoes, about two
quarts of peas with the weevil in them, and on my shelf a little
rice, a jug of molasses, and of rye and Indian meal a peck
each.

I sometimes dream of a larger and more populous house,
standing in a golden age, of enduring materials, and without
ginger-bread work,* which shall still consist of only one room,
a vast, rude, substantial, primitive hall, without ceiling or
plastering, with bare rafters and purlins supporting a sort of
lower heaven over one's head,—useful to keep off rain and
snow; where the king and queen posts* stand out to receive
your homage, when you have done reverence to the prostrate
Saturn* of an older dynasty on stepping over the sill; a cavernous house, wherein you must reach up a torch upon a pole to
see the roof; where some may live in the fire-place, some in
the recess of a window, and some on settles,* some at one end
of the hall, some at another, and some aloft on rafters with the
spiders, if they choose; a house which you have got into when
you have opened the outside door, and the ceremony is over;

where the weary traveller may wash, and eat, and converse, and sleep, without further journey; such a shelter as you would be glad to reach in a tempestuous night, containing all the essentials of a house, and nothing for house-keeping; where you can see all the treasures of the house at one view, and every thing hangs upon its peg that a man should use; at once kitchen, pantry, parlor, chamber, store-house, and garret; where you can see so necessary a thing as a barrel or a ladder, so convenient a thing as a cupboard, and hear the pot boil, and pay your respects to the fire that cooks your dinner and the oven that bakes your bread, and the necessary furniture and utensils are the chief ornaments; where the washing is not put out, nor the fire, nor the mistress, and perhaps you are sometimes requested to move from off the trap-door, when the cook would descend into the cellar, and so learn whether the ground is solid or hollow beneath you without stamping. A house whose inside is as open and manifest as a bird's nest, and you cannot go in at the front door and out at the back without seeing some of its inhabitants; where to be a guest is to be presented with the freedom of the house, and not to be carefully excluded from seven eighths of it, shut up in a particular cell, and told to make yourself at home there,— in solitary confinement. Nowadays the host does not admit you to *his* hearth, but has got the mason to build one for yourself somewhere in his alley, and hospitality is the art of *keeping** you at the greatest distance. There is as much secrecy about the cooking as if he had a design to poison you. I am aware that I have been on many a man's premises, and might have been legally ordered off, but I am not aware that I have been in many men's houses. I might visit in my old clothes a king and queen who lived simply in such a house as I have described, if I were going their way; but backing out* of a modern palace will be all that I shall desire to learn, if ever I am caught in one.

It would seem as if the very language of our parlors would lose all its nerve and degenerate into *parlaver** wholly, our

lives pass at such remoteness from its symbols, and its meta-
phors and tropes are necessarily so far fetched, through slides
and dumb-waiters,* as it were; in other words, the parlor is so
far from the kitchen and workshop. The dinner even is only
the parable of a dinner, commonly. As if only the savage dwelt
near enough to Nature and Truth to borrow a trope from
them. How can the scholar, who dwells away in the North
West Territory* or the Isle of Man, tell what is parliamentary*
in the kitchen?

However, only one or two of my guests were ever bold
enough to stay and eat a hasty-pudding with me; but when
they saw that crisis approaching they beat a hasty retreat
rather, as if it would shake the house to its foundations.
Nevertheless, it stood through a great many hasty-puddings.*

I did not plaster till it was freezing weather.* I brought over
some whiter and cleaner sand for this purpose from the opposite shore of the pond in a boat, a sort of conveyance which
would have tempted me to go much farther if necessary. My
house had in the mean while been shingled down to the
ground on every side. In lathing I was pleased to be able to
send home each nail with a single blow of the hammer, and it
was my ambition to transfer the plaster from the board to the
wall neatly and rapidly. I remembered the story of a conceited
fellow, who, in fine clothes, was wont to lounge about the
village once, giving advice to workmen. Venturing one day to
substitute deeds for words, he turned up his cuffs, seized a
plasterer's board, and having loaded his trowel without mis-
hap, with a complacent look toward the lathing overhead,
made a bold gesture thitherward; and straightway, to his complete discomfiture, received the whole contents in his ruffled
bosom. I admired anew the economy and convenience of
plastering, which so effectually shuts out the cold and takes a
handsome finish, and I learned the various casualties to which
the plasterer is liable. I was surprised to see how thirsty the
bricks were which drank up all the moisture in my plaster
before I had smoothed it, and how many pailfuls of water it

takes to christen a new hearth. I had the previous winter made a small quantity of lime by burning the shells of the *Unio fluviatilis*, which our river affords, for the sake of the experiment; so that I knew where my materials came from. I might have got good limestone within a mile or two and burned* it myself, if I had cared to do so.

The pond had in the mean while skimmed over in the shadiest and shallowest coves, some days or even weeks before the general freezing. The first ice is especially interesting and perfect, being hard, dark, and transparent, and affords the best opportunity that ever offers for examining the bottom where it is shallow; for you can lie at your length on ice only an inch thick, like a skater insect on the surface of the water, and study the bottom at your leisure, only two or three inches distant, like a picture behind a glass, and the water is necessarily always smooth then. There are many furrows in the sand where some creature has travelled about and doubled on its tracks; and, for wrecks, it is strewn with the cases of cadis worms made of minute grains of white quartz. Perhaps these have creased it, for you find some of their cases in the furrows, though they are deep and broad for them to make. But the ice itself is the object of most interest, though you must improve the earliest opportunity to study it. If you examine it closely the morning after it freezes, you find that the greater part of the bubbles, which at first appeared to be within it, are against its under surface, and that more are continually rising from the bottom; while the ice is as yet comparatively solid and dark, that is, you see the water through it. These bubbles are from an eightieth to an eighth of an inch in diameter, very clear and beautiful, and you see your face reflected in them through the ice. There may be thirty or forty of them to a square inch. There are also already within the ice narrow oblong perpendicular bubbles about half an inch long, sharp cones with the apex upward; or oftener, if the ice is quite fresh, minute spherical bubbles one directly above another, like a string of beads. But these within the ice are not so numerous nor obvious as those beneath. I sometimes used to cast on stones to try the strength of the ice, and those which broke through carried in air with them, which formed very large and conspicuous white bubbles beneath. One day when I came to the same place forty-eight hours afterward, I found that those large bubbles were still perfect, though an inch more of ice had formed, as I could see distinctly by the seam in the edge of a cake. But as the last two days had been very warm, like an Indian summer, the ice was not now transparent, showing the dark green color of the water, and the bottom, but opaque and whitish or gray, and though twice as thick was hardly stronger than before, for the air bubbles had greatly expanded under this heat and run together, and lost their regularity; they were no longer one directly over another, but often like silvery coins poured from a bag, one overlapping another, or in thin flakes, as if occupying slight cleavages. The beauty of the ice was gone, and it was too late to study the bottom. Being curious to know what position my great bubbles occupied with regard to the new ice, I broke out a cake containing a middling sized one, and turned it bottom upward. The new ice had formed around and under the bubble, so that it was included between the two ices. It was wholly in the lower ice, but close against the upper, and was flattish, or perhaps slightly lenticular, with a rounded edge, a quarter of an inch deep by four inches in diameter; and I was surprised to find that directly under the bubble the ice was melted with great regularity in the form of a saucer reversed, to the height of five eighths of an inch in the middle, leaving a thin partition there between the water and the bubble, hardly an eighth of an inch thick; and in many places the small bubbles in this partition had burst out downward, and probably there was no ice at all under the largest bubbles, which were a foot in diameter. I inferred that the infinite number of minute bubbles which I had first seen against the under surface of the ice were now frozen in likewise, and that each, in its degree, had operated like a burning glass on the ice beneath to melt and rot it. These are the little air-guns which contribute to make the ice crack and whoop.

At length the winter set in in good earnest, just as I had finished plastering, and the wind began to howl around the house as if it had not had permission to do so till then. Night after night the geese came lumbering in in the dark with a clangor and a whistling of wings, even after the ground was covered with snow, some to alight in Walden, and some flying low over the woods toward Fair Haven,* bound for Mexico. Several times, when returning from the village at ten or eleven o'clock at night, I heard the tread of a flock of geese, or else ducks, on the dry leaves in the woods by a pond-hole behind my dwelling, where they had come up to feed, and the faint honk or quack of their leader as they hurried off. In 1845 Walden froze entirely over for the first time on the night of the 22d of December, Flints' and other shallower ponds and the river having been frozen ten days or more; in '46, the 16th; in '49, about the 31st; and in '50, about the 27th of December; in '52, the 5th of January; in '53, the 31st of December. The snow had already covered the ground since the 25th of November, and surrounded me suddenly with the scenery of winter. I withdrew yet farther into my shell, and endeavored to keep a bright fire both within my house and within my breast. My employment out of doors now was to collect the dead wood in the forest, bringing it in my hands or on my shoulders, or sometimes trailing a dead pine tree under each arm to my shed. An old forest fence which had seen its best days was a great haul for me. I sacrificed it to Vulcan, for it was past serving the god Terminus.* How much more interesting an event is that man's supper who has just been forth in the snow to hunt, nay, you might say, steal, the fuel to cook it with! His bread and meat are sweet.* There are enough fagots and waste wood of all kinds in the forests of most of our towns to support many fires, but which at present warm none, and, some think, hinder the growth of the young wood. There was also the drift-wood of the pond. In the course of the summer I had discovered a raft of pitch-pine logs with the bark on, pinned together by the Irish when the railroad was built. This I hauled up partly on the shore. After soaking two years and then lying high six months it was perfectly sound, though waterlogged past drying. I amused myself one winter day with sliding this piecemeal across the pond, nearly half a mile, skating behind with one end of a log fifteen feet long on my shoulder, and the other on the ice; or I tied several logs together with a birch withe,* and then, with a longer birch or alder which had a hook at the end, dragged them across. Though completely waterlogged and almost as heavy as lead, they not only burned long, but made a very hot fire; nay, I thought that they burned better for the soaking, as if the pitch, being confined by the water, burned longer as in a lamp.

Gilpin, in his account of the forest borderers of England, says that "the encroachments of trespassers, and the houses and fences thus raised on the borders of the forest," were "considered as great nuisances by the old forest law, and were severely punished under the name of *purprestures*, as tending *ad terrorem ferarum—ad nocumentum forestæ*, &c.," to the frightening of the game and the detriment of the forest.* But I was interested in the preservation of the venison and the vert* more than the hunters or wood-choppers, and as much as though I had been the Lord Warden* himself; and if any part was burned, though I burned it myself by accident,* I grieved with a grief that lasted longer and was more inconsolable than that of the proprietors; nay, I grieved when it was cut down by the proprietors themselves. I would that our farmers when they cut down a forest felt some of that awe which the old Romans did when they came to thin, or let in the light to, a consecrated grove, (*lucum conlucare,*)* that is, would believe that it is sacred to some god. The Roman made an expiatory offering, and prayed, Whatever god or goddess thou art to whom this grove is sacred, be propitious to me, my family, and children, &c.

It is remarkable what a value is still put upon wood even in this age and in this new country, a value more permanent and universal than that of gold. After all our discoveries and inventions no man will go by a pile of wood. It is as precious to us as it was to our Saxon and Norman ancestors. If they made their bows of it, we make our gun-stocks of it. Michaux, more than thirty years ago, says that the price of wood for fuel in New York and Philadelphia "nearly equals, and sometimes exceeds, that of the best wood in Paris, though this immense capital annually requires more than three hundred thousand cords, and is surrounded to the distance of three hundred miles by cultivated plains."* In this town the price of wood rises almost steadily, and the only question is, how much higher it is to be this year than it was the last. Mechanics and tradesmen who come in person to the forest on no other errand, are sure to attend the wood auction, and even pay a high price for the privilege of gleaning after the wood-chopper. It is now many years that men have resorted to the forest for fuel and the materials of the arts; the New Englander and the New Hollander, the Parisian and the Celt, the farmer and Robinhood, Goody Blake and Harry Gill,* in most parts of the world the prince and the peasant, the scholar and the savage, equally require still a few sticks from the forest to warm them and cook their food. Neither could I do without them.

Every man looks at his wood-pile with a kind of affection. I loved to have mine before my window, and the more chips the better to remind me of my pleasing work. I had an old axe which nobody claimed, with which by spells in winter days, on the sunny side of the house, I played about the stumps which I had got out of my bean-field. As my driver prophesied when I was ploughing, they warmed me twice, once while I was splitting them, and again when they were on the fire, so that no fuel could give out more heat. As for the axe, I was advised to get the village blacksmith to "jump" it;* but I jumped him, and, putting a hickory helve from the woods into it, made it do. If it was dull, it was at least hung true.

A few pieces of fat pine were a great treasure. It is interesting to remember how much of this food for fire is still concealed in the bowels of the earth. In previous years I had often gone "prospecting" over some bare hill-side, where a pitch-pine wood had formerly stood, and got out the fat pine roots. They are almost indestructible. Stumps thirty or forty years old, at least, will still be sound at the core, though the sapwood* has all become vegetable mould, as appears by the scales of the thick bark forming a ring level with the earth four or five inches distant from the heart. With axe and shovel you explore this mine, and follow the marrowy store, yellow as beef tallow, or as if you had struck on a vein of gold, deep into the earth. But commonly I kindled my fire with the dry leaves of the forest, which I had stored up in my shed before the snow came. Green hickory finely split makes the wood-chopper's kindlings, when he has a camp in the woods. Once in a while I got a little of this. When the villagers were lighting their fires beyond the horizon, I too gave notice to the various wild inhabitants of Walden vale, by a smoky streamer from my chimney, that I was awake.—

> Light-winged Smoke, Icarian bird,
> Melting thy pinions in thy upward flight,
> Lark without song, and messenger of dawn,
> Circling above the hamlets as thy nest;
> Or else, departing dream, and shadowy form
> Of midnight vision, gathering up thy skirts;
> By night star-veiling, and by day
> Darkening the light and blotting out the sun;
> Go thou my incense upward from this hearth,
> And ask the gods to pardon this clear flame.*

Hard green wood just cut, though I used but little of that, answered my purpose better than any other. I sometimes left

a good fire when I went to take a walk in a winter afternoon;
and when I returned, three or four hours afterward, it would
be still alive and glowing. My house was not empty though I
was gone. It was as if I had left a cheerful housekeeper
behind. It was I and Fire that lived there; and commonly my
housekeeper proved trustworthy. One day, however, as I was
splitting wood, I thought that I would just look in at the
window and see if the house was not on fire; it was the only
time I remember to have been particularly anxious on this
score; so I looked and saw that a spark had caught my bed, and
I went in and extinguished it when it had burned a place as big
as my hand. But my house occupied so sunny and sheltered a
position, and its roof was so low, that I could afford to let the
fire go out in the middle of almost any winter day.

The moles nested in my cellar, nibbling every third potato,
and making a snug bed even there of some hair* left after
plastering and of brown paper; for even the wildest animals
love comfort and warmth as well as man, and they survive the
winter only because they are so careful to secure them. Some
of my friends spoke as if I was coming to the woods on
purpose to freeze myself. The animal merely makes a bed,
which he warms with his body in a sheltered place; but man,
having discovered fire, boxes up some air in a spacious apartment, and warms that, instead of robbing himself, makes that
his bed, in which he can move about divested of more cum-
brous clothing, maintain a kind of summer in the midst of
winter, and by means of windows even admit the light, and
with a lamp lengthen out the day. Thus he goes a step or two
beyond instinct, and saves a little time for the fine arts.
Though, when I had been exposed to the rudest blasts a long
time, my whole body began to grow torpid, when I reached
the genial atmosphere of my house I soon recovered my
faculties and prolonged my life. But the most luxuriously
housed has little to boast of in this respect, nor need we
trouble ourselves to speculate how the human race may be at
last destroyed. It would be easy to cut their threads* any time

with a little sharper blast from the north. We go on dating from Cold Fridays* and Great Snows;* but a little colder Friday, or greater snow, would put a period to man's existence on the globe.

The next winter I used a small cooking-stove for economy, since I did not own the forest; but it did not keep a fire so well as the open fire-place. Cooking was then, for the most part, no longer a poetic, but merely a chemic process. It will soon be forgotten, in these days of stoves, that we used to roast potatoes in the ashes, after the Indian fashion. The stove not only took up room and scented the house, but it concealed the fire, and I felt as if I had lost a companion. You can always see a face in the fire. The laborer, looking into it at evening, purifies his thoughts of the dross and earthiness which they have accumulated during the day. But I could no longer sit and look into the fire, and the pertinent words of a poet recurred to me with new force.—

"Never, bright flame, may be denied to me
Thy dear, life imaging, close sympathy.
What but my hopes shot upward e'er so bright?
What but my fortunes sunk so low in night?

Why art thou banished from our hearth and hall,
Thou who art welcomed and beloved by all?
Was thy existence then too fanciful
For our life's common light, who are so dull?
Did thy bright gleam mysterious converse hold
With our congenial souls? secrets too bold?
Well, we are safe and strong, for now we sit
Beside a hearth where no dim shadows flit,
Where nothing cheers nor saddens, but a fire
Warms feet and hands—nor does to more aspire;
By whose compact utilitarian heap
The present may sit down and go to sleep,
Nor fear the ghosts who from the dim past walked,
And with us by the unequal light of the old wood fire talked."*

FORMER INHABITANTS;
AND WINTER VISITORS

I WEATHERED some merry snow storms, and spent some cheerful winter evenings by my fire-side, while the snow whirled wildly without, and even the hooting of the owl was hushed. For many weeks I met no one in my walks but those who came occasionally to cut wood and sled it to the village. The elements, however, abetted me in making a path through the deepest snow in the woods, for when I had once gone through the wind blew the oak leaves into my tracks, where they lodged, and by absorbing the rays of the sun melted the snow, and so not only made a dry bed for my feet, but in the night their dark line was my guide. For human society I was obliged to conjure up the former occupants of these woods. Within the memory of many of my townsmen the road near which my house stands resounded with the laugh and gossip of inhabitants, and the woods which border it were notched and dotted here and there with their little gardens and dwellings, though it was then much more shut in by the forest than now. In some places, within my own remembrance, the pines would scrape both sides of a chaise at once, and women and children who were compelled to go this way to Lincoln* alone and on foot did it with fear, and often ran a good part of the distance. Though mainly but a humble route to neighboring villages, or for the woodman's team, it once amused the traveller more than now by its variety, and lingered longer in his memory. Where now firm open fields stretch from the village to the woods, it then ran through a maple swamp on a foundation of logs,* the remnants of which, doubtless, still underlie the present dusty highway, from the Stratten,* now the Alms House, Farm, to Brister's Hill.*

East of my bean-field, across the road, lived Cato
Ingraham, slave of Duncan Ingraham,* Esquire, gentleman
of Concord village; who built his slave a house, and gave him
permission to live in Walden Woods;—Cato, not Uticensis,*
but Concordiensis. Some say that he was a Guinea Negro.
There are a few who remember his little patch among the
walnuts, which he let grow up till he should be old and need
them; but a younger and whiter speculator* got them at last.
He too, however, occupies an equally narrow house at
present. Cato's half-obliterated cellar hole* still remains,
though known to few, being concealed from the traveller by a
fringe of pines. It is now filled with the smooth sumach, (*Rhus
glabra*,) and one of the earliest species of golden-rod
(*Solidago stricta*) grows there luxuriantly.

Here, by the very corner of my field, still nearer to town,
Zilpha, a colored woman, had her little house, where she spun
linen for the townsfolk, making the Walden Woods ring with
her shrill singing, for she had a loud and notable voice. At
length, in the war of 1812, her dwelling was set on fire by
English soldiers, prisoners on parole,* when she was away,
and her cat and dog and hens were all burned up together.
She led a hard life, and somewhat inhumane. One old frequenter of these woods remembers, that as he passed her
house one noon he heard her muttering to herself over her
gurgling pot,—"Ye are all bones, bones!" I have seen bricks
among the oak copse there.

Down the road, on the right hand, on Brister's Hill, lived
Brister Freeman, "a handy Negro," slave of Squire
Cummings* once,—there where grow still the apple-trees
which Brister planted and tended; large old trees now, but
their fruit still wild and ciderish to my taste. Not long since I
read his epitaph* in the old Lincoln burying-ground, a little
on one side, near the unmarked graves of some British grena-
diers who fell in the retreat from Concord,—where he is
styled "Sippio Brister,"—Scipio Africanus he had some title to
be called,—"a man of color," as if he were discolored. It also

told me, with staring emphasis, when he died; which was but an indirect way of informing me that he ever lived. With him dwelt Fenda, his hospitable wife, who told fortunes, yet pleasantly,—large, round, and black, blacker than any of the children of night, such a dusky orb as never rose on Concord before or since.

Farther down the hill, on the left, on the old road in the woods, are marks of some homestead of the Stratten* family; whose orchard once covered all the slope of Brister's Hill, but was long since killed out by pitch-pines, excepting a few stumps, whose old roots furnish still the wild stocks of many a thrifty village tree.

Nearer yet to town, you come to Breed's location,* on the other side of the way, just on the edge of the wood; ground famous for the pranks of a demon not distinctly named in old mythology, who has acted a prominent and astounding part in our New England life, and deserves, as much as any mythological character, to have his biography written one day; who first comes in the guise of a friend or hired man, and then robs and murders the whole family,—New-England Rum. But history must not yet tell the tragedies enacted here; let time intervene in some measure to assuage and lend an azure tint to them. Here the most indistinct and dubious tradition says that once a tavern stood; the well the same, which tempered the traveller's beverage and refreshed his steed. Here then men saluted one another, and heard and told the news, and went their ways again.

Breed's hut was standing only a dozen years ago, though it had long been unoccupied. It was about the size of mine. It was set on fire by mischievous boys, one Election night,* if I do not mistake. I lived on the edge of the village then, and had just lost myself over Davenant's Gondibert,* that winter that I labored with a lethargy,—which, by the way, I never knew whether to regard as a family complaint, having an uncle* who goes to sleep shaving himself, and is obliged to sprout potatoes in a cellar Sundays, in order to keep awake and keep

the Sabbath, or as the consequence of my attempt to read
Chalmers' collection* of English poetry without skipping. It
fairly overcame my Nervii.* I had just sunk my head on this
when the bells rung fire, and in hot haste the engines rolled
that way, led by a straggling troop of men and boys, and I
among the foremost, for I had leaped the brook.* We thought
it was far south over the woods,—we who had run to fires
before,—barn, shop, or dwelling-house, or all together. "It's
Baker's barn," cried one. "It is the Codman Place,"* affirmed
another. And then fresh sparks went up above the wood, as if
the roof fell in, and we all shouted "Concord to the rescue!"
Wagons shot past with furious speed and crushing loads, bearing, perchance, among the rest, the agent of the Insurance
Company,* who was bound to go however far; and ever and
anon the engine bell tinkled behind, more slow and sure, and
rearmost of all, as it was afterward whispered, came they who
set the fire and gave the alarm. Thus we kept on like true
idealists, rejecting the evidence of our senses, until at a turn in
the road we heard the crackling and actually felt the heat of
the fire from over the wall, and realized, alas! that we were
there. The very nearness of the fire but cooled our ardor. At
first we thought to throw a frog-pond on to it; but concluded
to let it burn, it was so far gone and so worthless. So we stood
round our engine, jostled one another, expressed our sentiments through speaking trumpets, or in lower tone referred to
the great conflagrations which the world had witnessed, including Bascom's shop, and, between ourselves, we thought
that, were we there in season with our "tub,"* and a full frog-
pond by, we could turn that threatened last and universal one
into another flood. We finally retreated without doing any
mischief,—returned to sleep and Gondibert. But as for
Gondibert, I would except that passage in the preface about
wit being the soul's powder,*—"but most of mankind are
strangers to wit, as Indians are to powder."

It chanced that I walked that way across the fields the
following night, about the same hour, and hearing a low

moaning at this spot, I drew near in the dark, and discovered the only survivor of the family that I know, the heir of both its virtues and its vices, who alone was interested in this burning, lying on his stomach and looking over the cellar wall at the still smouldering cinders beneath, muttering to himself, as is his wont. He had been working far off in the river meadows all day, and had improved the first moments that he could call his own to visit the home of his fathers and his youth. He gazed into the cellar from all sides and points of view by turns, always lying down to it, as if there was some treasure, which he remembered, concealed between the stones, where there was absolutely nothing but a heap of bricks and ashes. The house being gone, he looked at what there was left. He was soothed by the sympathy which my mere presence implied, and showed me, as well as the darkness permitted, where the well was covered up; which, thank Heaven, could never be burned; and he groped long about the wall to find the well-sweep* which his father had cut and mounted, feeling for the iron hook or staple by which a burden had been fastened to the heavy end,—all that he could now cling to,—to convince me that it was no common "rider."* I felt it, and still remark it almost daily in my walks, for by it hangs the history of a family.

Once more, on the left, where are seen the well and lilac bushes by the wall, in the now open field, lived Nutting and Le Grosse. But to return toward Lincoln.

Farther in the woods than any of these, where the road approaches nearest to the pond, Wyman the potter squatted, and furnished his townsmen with earthen ware, and left descendants to succeed him. Neither were they rich in worldly goods, holding the land by sufferance while they lived; and there often the sheriff came in vain to collect the taxes, and "attached a chip,"* for form's sake, as I have read in his accounts, there being nothing else that he could lay his hands on. One day in midsummer, when I was hoeing, a man who was carrying a load of pottery to market stopped his horse against my field and inquired concerning Wyman the younger. He had long ago bought a potter's wheel of him, and wished to know what had become of him. I had read of the potter's clay and wheel* in Scripture,* but it had never occurred to me that the pots we use were not such as had come down unbroken from those days, or grown on trees like gourds somewhere, and I was pleased to hear that so fictile an art was ever practised in my neighborhood.

The last inhabitant of these woods before me was an Irishman, Hugh Quoil,* (if I have spelt his name with coil enough,) who occupied Wyman's tenement,—Col. Quoil, he was called. Rumor said that he had been a soldier at Waterloo. If he had lived I should have made him fight his battles over again. His trade here was that of a ditcher. Napoleon went to St. Helena;* Quoil came to Walden Woods. All I know of him is tragic. He was a man of manners, like one who had seen the world, and was capable of more civil speech than you could well attend to. He wore a great coat in mid-summer, being affected with the trembling delirium, and his face was the color of carmine. He died in the road at the foot of Brister's Hill shortly after I came to the woods, so that I have not remembered him as a neighbor. Before his house was pulled down, when his comrades avoided it as "an unlucky castle," I visited it. There lay his old clothes curled up by use, as if they were himself, upon his raised plank bed. His pipe lay broken on the hearth, instead of a bowl broken at the fountain.* The last could never have been the symbol of his death, for he confessed to me that, though he had heard of Brister's Spring,* he had never seen it; and soiled cards, kings of diamonds spades and hearts, were scattered over the floor. One black chicken which the administrator* could not catch, black as night and as silent, not even croaking, awaiting Reynard,* still went to roost in the next apartment. In the rear there was the dim outline of a garden, which had been planted but had never received its first hoeing, owing to those terrible shaking fits, though it was now harvest time. It was over-run with Roman wormwood and beggar-ticks, which last stuck to my clothes for all fruit. The skin of a woodchuck was freshly stretched upon the back of the house, a trophy of his last Waterloo; but no warm cap or mittens would he want more.

Now only a dent in the earth marks the site of these dwellings, with buried cellar stones, and strawberries, raspberries, thimble-berries, hazel-bushes, and sumachs growing in the sunny sward there; some pitch-pine or gnarled oak occupies what was the chimney nook, and a sweet-scented black-birch, perhaps, waves where the door-stone was. Sometimes the well dent is visible, where once a spring oozed; now dry and tearless grass; or it was covered deep,—not to be discovered till some late day,—with a flat stone under the sod, when the last of the race departed. What a sorrowful act that must be,— the covering up of wells! coincident with the opening of wells of tears. These cellar dents, like deserted fox burrows, old holes, are all that is left where once were the stir and bustle of human life, and "fate, free-will, foreknowledge absolute,"* in some form and dialect or other were by turns discussed. But all I can learn of their conclusions amounts to just this, that "Cato and Brister pulled wool;"* which is about as edifying as the history of more famous schools of philosophy.

Still grows the vivacious lilac a generation after the door and lintel and the sill are gone, unfolding its sweet-scented flowers each spring, to be plucked by the musing traveller; planted and tended once by children's hands, in front-yard plots,—now standing by wall-sides in retired pastures, giving place to new-rising forests;—the last of that stirp,* sole survivor of that family. Little did the dusky children think that the puny slip with its two eyes only, which they stuck in the ground in the shadow of the house and daily watered, would root itself so, and outlive them, and house itself in the rear that shaded it, and grown man's garden and orchard, and tell their story faintly to the lone wanderer a half century after they had grown up and died,—blossoming as fair, and smelling as sweet, as in that first spring. I mark its still tender, civil, cheerful, lilac colors.

But this small village, germ of something more, why did it fail while Concord kept its ground?* Were there no natural advantages,—no water privileges,* forsooth? Ay, the deep Walden Pond and cool Brister's Spring,—privilege to drink long and healthy draughts at these, all unimproved by these men but to dilute their glass. They were universally a thirsty race. Might not the basket, stable-broom, mat-making, corn-parching,* linen-spinning, and pottery business have thrived here, making the wilderness to blossom* like the rose, and a numerous posterity have inherited the land of their fathers? The sterile soil would at least have been proof against a low-land degeneracy. Alas! how little does the memory of these human inhabitants enhance the beauty of the land-scape! Again, perhaps, Nature will try, with me for a first settler, and my house raised last spring to be the oldest in the hamlet.

I am not aware that any man has ever built on the spot which I occupy. Deliver me from a city built on the site of a more ancient city, whose materials are ruins, whose gardens cemeteries. The soil is blanched and accursed there, and before that becomes necessary the earth itself will be de-stroyed. With such reminiscences I repeopled the woods and lulled myself asleep.

At this season I seldom had a visitor. When the snow lay deepest no wanderer ventured near my house for a week or fortnight at a time, but there I lived as snug as a meadow mouse, or as cattle and poultry which are said to have survived for a long time buried in drifts, even without food; or like that early settler's family in the town of Sutton,* in this state, whose cottage was completely covered by the great snow* of 1717 when he was absent, and an Indian found it only by the hole which the chimney's breath made in the drift, and so relieved the family. But no friendly Indian concerned himself about me; nor needed he, for the master of the house was at home. The Great Snow! How cheerful it is to hear of! When the farmers could not get to the woods and swamps with their teams, and were obliged to cut down the shade trees before their houses, and when the crust was harder cut off the trees in the swamps ten feet from the ground, as it appeared the next spring.

In the deepest snows, the path which I used from the highway to my house, about half a mile long, might have been represented by a meandering dotted line, with wide intervals between the dots. For a week of even weather I took exactly the same number of steps, and of the same length, coming and going, stepping deliberately and with the precision of a pair of dividers in my own deep tracks,—to such routine the winter reduces us,—yet often they were filled with heaven's own blue. But no weather interfered fatally with my walks, or rather my going abroad, for I frequently tramped eight or ten miles through the deepest snow to keep an appointment with a beech-tree, or a yellow-birch, or an old acquaintance among the pines; when the ice and snow causing their limbs to droop, and so sharpening their tops, had changed the pines into fir-trees; wading to the tops of the highest hills when the snow was nearly two feet deep on a level, and shaking down another snow-storm on my head at every step; or sometimes creeping and floundering thither on my hands and knees, when the hunters had gone into winter quarters. One afternoon I amused myself by watching a barred owl (*Strix nebulosa*) sitting on one of the lower dead limbs of a white-pine, close to the trunk, in broad daylight, I standing within a rod of him. He could hear me when I moved and cronched the snow with my feet, but could not plainly see me. When I made most noise he would stretch out his neck, and erect his neck feathers, and open his eyes wide; but their lids soon fell again, and he began to nod. I too felt a slumberous influence after watching him half an hour, as he sat thus with his eyes half open, like a cat, winged brother of the cat. There was only a narrow slit left between their lids, by which he preserved a peninsular relation to me; thus, with half-shut eyes, looking out from the land of dreams, and endeavoring to realize me, vague object or mote that interrupted his visions. At length, on some louder noise or my nearer approach, he would grow uneasy and sluggishly turn about on his perch, as if impatient at having his dreams disturbed; and when he launched himself off and flapped through the pines, spreading his wings to unexpected breadth, I could not hear the slightest sound from them. Thus, guided amid the pine boughs rather by a delicate sense of their neighborhood than by sight, feeling his twilight way as it were with his sensitive pinions, he found a new perch, where he might in peace await the dawning of his day.

As I walked over the long causeway* made for the railroad through the meadows, I encountered many a blustering and nipping wind, for nowhere has it freer play; and when the frost had smitten me on one cheek, heathen as I was, I turned* to it the other also. Nor was it much better by the carriage road from Brister's Hill. For I came to town still, like a friendly Indian, when the contents of the broad open fields were all piled up between the walls of the Walden road, and half an hour sufficed to obliterate the tracks of the last traveller. And when I returned new drifts would have formed, through which I floundered, where the busy north-west wind had been depositing the powdery snow round a sharp angle in the road, and not a rabbit's track, nor even the fine print, the small type, of a meadow mouse* was to be seen. Yet I rarely failed to find, even in mid-winter, some warm and springy swamp where the grass and the skunk-cabbage still put forth with perennial verdure, and some hardier bird occasionally awaited the return of spring.

Sometimes, notwithstanding the snow, when I returned from my walk at evening I crossed the deep tracks of a woodchopper* leading from my door, and found his pile of whittlings on the hearth, and my house filled with the odor of his pipe. Or on a Sunday afternoon, if I chanced to be at home, I heard the cronching of the snow made by the step of a long-headed farmer,* who from far through the woods sought my house, to have a social "crack;"* one of the few of his vocation who are "men on their farms;" who donned a frock instead of a professor's gown, and is as ready to extract the moral out of church or state as to haul a load of manure from his barn-yard. We talked of rude and simple times, when men sat about large fires in cold bracing weather, with clear heads; and when other dessert failed, we tried our teeth on many a nut which wise squirrels have long since abandoned, for those which have the thickest shells are commonly empty.

The one who came from farthest to my lodge, through deepest snows and most dismal tempests, was a poet.* A farmer, a hunter, a soldier, a reporter, even a philosopher, may be daunted; but nothing can deter a poet, for he is actuated by pure love. Who can predict his comings and goings? His business calls him out at all hours, even when doctors sleep. We made that small house ring with boisterous mirth and resound with the murmur of much sober talk, making amends then to Walden vale* for the long silences. Broadway* was still and deserted in comparison. At suitable intervals there were regular salutes of laughter, which might have been referred indifferently to the last uttered or the forth-coming jest. We made many a "bran new"* theory of life over a thin dish of gruel, which combined the advantages of conviviality with the clear-headedness which philosophy requires.

I should not forget that during my last winter at the pond there was another welcome visitor,* who at one time came through the village, through snow and rain and darkness, till he saw my lamp through the trees, and shared with me some long winter evenings. One of the last of the philosophers,— Connecticut gave him to the world,—he peddled first her wares, afterwards, as he declares, his brains. These he peddles still, prompting God and disgracing man, bearing for fruit his brain only, like the nut its kernel. I think that he must be the man of the most faith of any alive. His words and attitude always suppose a better state of things than other men are acquainted with, and he will be the last man to be disappointed as the ages revolve. He has no venture in the present. But though comparatively disregarded now, when his day comes, laws unsuspected by most will take effect, and masters of families and rulers will come to him for advice.—

"How blind that cannot see serenity!"*

A true friend of man; almost the only friend of human progress. An Old Mortality,* say rather an Immortality, with unwearied patience and faith making plain the image engraven in men's bodies, the God of whom they are but defaced and leaning monuments. With his hospitable intellect he embraces children, beggars, insane, and scholars, and entertains the thought of all, adding to it commonly some breadth and elegance. I think that he should keep a caravansary on the world's highway, where philosophers of all nations might put up, and on his sign should be printed, "Entertainment for man, but not for his beast. Enter ye that have leisure and a quiet mind, who earnestly seek the right road." He is perhaps the sanest man and has the fewest crotchets of any I chance to know; the same yesterday and to-morrow. Of yore we had sauntered and talked, and effectually put the world behind us; for he was pledged to no institution in it, freeborn, *ingenuus*. Whichever way we turned, it seemed that the heavens and the earth had met together, since he enhanced the beauty of the landscape. A blue-robed man, whose fittest roof is the overarching sky which reflects his serenity. I do not see how he can ever die; Nature cannot spare him.

Having each some shingles of thought well dried, we sat and whittled them, trying our knives, and admiring the clear yellowish grain of the pumpkin pine. We waded so gently and reverently, or we pulled together so smoothly, that the fishes of thought were not scared from the stream, nor feared any angler on the bank, but came and went grandly, like the clouds which float through the western sky, and the mother-o'-pearl flocks which sometimes form and dissolve there. There we worked, revising mythology, rounding a fable* here and there, and building castles in the air for which earth offered no worthy foundation. Great Looker! Great Expecter! to converse with whom was a New England Night's Entertainment.* Ah! such discourse we had, hermit and philosopher, and the old settler* I have spoken of,—we three,—it expanded and racked my little house; I should not dare to say how many pounds' weight there was above the atmospheric pressure on every circular inch; it opened its seams so that they had to be calked with much dulness thereafter to stop the consequent leak;—but I had enough of that kind of oakum* already picked.

There was one other* with whom I had "solid seasons," long to be remembered, at his house in the village, and who looked in upon me from time to time; but I had no more for society there.

There too, as every where, I sometimes expected the Visitor who never comes. The Vishnu Purana* says, "The householder is to remain at eventide in his court-yard as long as it takes to milk a cow, or longer if he pleases, to await the arrival of a guest." I often performed this duty of hospitality, waited long enough to milk a whole herd of cows, but did not see the man approaching from the town.*

WINTER ANIMALS

WHEN the ponds were firmly frozen, they afforded not only new and shorter routes to many points, but new views from their surfaces of the familiar landscape around them. When I crossed Flints' Pond,* after it was covered with snow, though I had often paddled about and skated over it, it was so unexpectedly wide and so strange that I could think of nothing but Baffin's Bay.* The Lincoln hills* rose up around me at the extremity of a snowy plain, in which I did not remember to have stood before; and the fishermen, at an indeterminable distance over the ice, moving slowly about with their wolfish dogs, passed for sealers or Esquimaux, or in misty weather loomed like fabulous creatures, and I did not know whether they were giants or pygmies. I took this course when I went to lecture in Lincoln in the evening, travelling in no road and passing no house between my own hut and the lecture room. In Goose Pond, which lay in my way,* a colony of muskrats dwelt, and raised their cabins high above the ice, though none could be seen abroad when I crossed it. Walden, being like the rest usually bare of snow, or with only shallow and interrupted drifts on it, was my yard, where I could walk freely when the snow was nearly two feet deep on a level elsewhere and the villagers were confined to their streets. There, far from the village street, and except at very long intervals, from the jingle of sleigh-bells, I slid and skated, as in a vast moose-yard* well trodden, overhung by oak woods and solemn pines bent down with snow or bristling with icicles.

For sounds in winter nights, and often in winter days, I heard the forlorn but melodious note of a hooting owl indefinitely far; such a sound as the frozen earth would yield if struck with a suitable plectrum, the very *lingua vernacula** of Walden Wood, and quite familiar to me at last, though I never saw the bird while it was making it. I seldom opened my door in a winter evening without hearing it; *Hoo hoo hoo, hoorer hoo*, sounded sonorously, and the first three syllables accented somewhat like *how der do*; or sometimes *hoo hoo* only. One night in the beginning of winter, before the pond froze over, about nine o'clock, I was startled by the loud honking of a goose, and, stepping to the door, heard the sound of their wings like a tempest in the woods as they flew low over my house. They passed over the pond toward Fair Haven, seemingly deterred from settling by my light, their commodore honking all the while with a regular beat. Suddenly an unmistakable cat-owl from very near me, with the most harsh and tremendous voice I ever heard from any inhabitant of the woods, responded at regular intervals to the goose, as if determined to expose and disgrace this intruder from Hudson's Bay* by exhibiting a greater compass and volume of voice in a native, and *boo-hoo* him out of Concord horizon. What do you mean by alarming the citadel* at this time of night consecrated to me? Do you think I am ever caught napping at such an hour, and that I have not got lungs and a larynx as well as yourself? *Boo-hoo, boo-hoo, boo-hoo!* It was one of the most thrilling discords I ever heard. And yet, if you had a discriminating ear, there were in it the elements of a concord such as these plains never saw nor heard.

I also heard the whooping* of the ice in the pond, my great bed-fellow in that part of Concord, as if it were restless in its bed and would fain turn over, were troubled with flatulency and bad dreams; or I was waked by the cracking of the ground by the frost, as if some one had driven a team against my door, and in the morning would find a crack in the earth a quarter of a mile long and a third of an inch wide.

Sometimes I heard the foxes as they ranged over the snow crust, in moonlight nights, in search of a partridge or other game, barking raggedly and demoniacally like forest dogs, as if laboring with some anxiety, or seeking expression, struggling for light and to be dogs outright and run freely in the streets;

for if we take the ages into our account, may there not be a civilization going on among brutes as well as men? They seemed to me to be rudimental, burrowing men, still standing on their defence, awaiting their transformation. Sometimes one came near to my window, attracted by my light, barked a vulpine curse at me, and then retreated.

Usually the red squirrel (*Sciurus Hudsonius*) waked me in the dawn, coursing over the roof and up and down the sides of the house, as if sent out of the woods for this purpose. In the course of the winter I threw out half a bushel of ears of sweet-corn, which had not got ripe, on to the snow crust by my door, and was amused by watching the motions of the various animals which were baited by it. In the twilight and the night the rabbits came regularly and made a hearty meal. All day long the red squirrels came and went, and afforded me much entertainment by their manœuvres. One would approach at first warily through the shrub-oaks, running over the snow crust by fits and starts like a leaf blown by the wind, now a few paces this way, with wonderful speed and waste of energy, making inconceivable haste with his "trotters," as if it were for a wager, and now as many paces that way, but never getting on more than half a rod at a time; and then suddenly pausing with a ludicrous expression and a gratuitous somerset,* as if all the eyes in the universe were fixed on him,—for all the motions of a squirrel, even in the most solitary recesses of the forest, imply spectators as much as those of a dancing girl,—wasting more time in delay and circumspection than would have sufficed to walk the whole distance,—I never saw one walk,—and then suddenly, before you could say Jack Robinson, he would be in the top of a young pitch-pine, winding up his clock and chiding all imaginary spectators, soliloquizing and talking to all the universe at the same time,—for no reason that I could ever detect, or he himself was aware of, I suspect. At length he would reach the corn, and selecting a suitable ear, brisk* about in the same uncertain trigonometrical* way to the top-most stick of my wood-pile, before my window, where he looked me in the face, and there sit for hours, supplying himself with a new ear from time to time, nibbling at first voraciously and throwing the half-naked cobs about; till at length he grew more dainty still and played with his food, tasting only the inside of the kernel, and the ear, which was held balanced over the stick by one paw, slipped from his careless grasp and fell to the ground, when he would look over at it with a ludicrous expression of uncertainty, as if suspecting that it had life, with a mind not made up whether to get it again, or a new one, or be off; now thinking of corn, then listening to hear what was in the wind. So the little impudent fellow would waste many an ear in a fore-noon; till at last, seizing some longer and plumper one, considerably bigger than himself, and skilfully balancing it, he would set out with it to the woods, like a tiger with a buffalo, by the same zig-zag course and frequent pauses, scratching along with it as if it were too heavy for him and falling all the while, making its fall a diagonal between a perpendicular and horizontal, being determined to put it through at any rate;—a singularly frivolous and whimsical fellow;—and so he would get off with it to where he lived, perhaps carry it to the top of a pine tree forty or fifty rods distant, and I would afterwards find the cobs strewn about the woods in various directions.

At length the jays arrive, whose discordant screams were heard long before, as they were warily making their approach an eighth of a mile off, and in a stealthy and sneaking manner they flit from tree to tree, nearer and nearer, and pick up the kernels which the squirrels have dropped. Then, sitting on a pitch-pine bough, they attempt to swallow in their haste a kernel which is too big for their throats and chokes them; and after great labor they disgorge it, and spend an hour in the endeavor to crack it by repeated blows with their bills. They were manifestly thieves, and I had not much respect for them; but the squirrels, though at first shy, went to work as if they were taking what was their own.

Meanwhile also came the chicadees in flocks, which,

picking up the crums the squirrels had dropped, flew to the nearest twig, and, placing them under their claws, hammered away at them with their little bills, as if it were an insect in the bark, till they were sufficiently reduced for their slender throats. A little flock of these tit-mice came daily to pick a dinner out of my wood-pile, or the crums at my door, with faint flitting lisping notes, like the tinkling of icicles in the grass, or else with sprightly *day day day*, or more rarely, in spring-like days, a wiry summery *phe-be* from the wood-side. They were so familiar that at length one alighted on an armful of wood which I was carrying in, and pecked at the sticks without fear. I once had a sparrow alight upon my shoulder for a moment while I was hoeing in a village garden, and I felt that I was more distinguished by that circumstance than I should have been by any epaulet I could have worn. The squirrels also grew at last to be quite familiar, and occasionally stepped upon my shoe, when that was the nearest way.

When the ground was not yet quite covered, and again near the end of the winter, when the snow was melted on my south hill-side and about my wood-pile, the partridges came out of the woods morning and evening to feed there. Whichever side you walk in the woods the partridge bursts away on whirring wings, jarring the snow from the dry leaves and twigs on high, which comes sifting down in the sun-beams like golden dust; for this brave bird is not to be scared by winter. It is frequently covered up by drifts, and, it is said, "sometimes plunges from on wing into the soft snow, where it remains concealed for a day or two."* I used to start them in the open land also, where they had come out of the woods at sunset to "bud" the wild apple-trees. They will come regularly every evening to particular trees, where the cunning sportsman lies in wait for them, and the distant orchards next the woods suffer thus not a little. I am glad that the partridge gets fed, at any rate. It is Nature's own bird which lives on buds and diet-drink.*

In dark winter mornings, or in short winter afternoons, I sometimes heard a pack of hounds threading all the woods with hounding cry and yelp, unable to resist the instinct of the chase, and the note of the hunting horn at intervals, proving that man was in the rear. The woods ring again, and yet no fox bursts forth on to the open level of the pond, nor following pack pursuing their Actæon.* And perhaps at evening I see the hunters returning with a single brush trailing from their sleigh for a trophy, seeking their inn. They tell me that if the fox would remain in the bosom of the frozen earth he would be safe, or if he would run in a straight line away no fox-hound could overtake him; but, having left his pursuers far behind, he stops to rest and listen till they come up, and when he runs he circles round to his old haunts, where the hunters await him. Sometimes, however, he will run upon a wall many rods, and then leap off far to one side, and he appears to know that water will not retain his scent. A hunter told me that he once saw a fox pursued by hounds burst out on to Walden when the ice was covered with shallow puddles, run part way across, and then return to the same shore. Ere long the hounds arrived, but here they lost the scent. Sometimes a pack hunting by themselves would pass my door, and circle round my house, and yelp and hound without regarding me, as if afflicted by a species of madness, so that nothing could divert them from the pursuit. Thus they circle until they fall upon the recent trail of a fox, for a wise hound will forsake every thing else for this. One day a man came to my hut from Lexington* to inquire after his hound that made a large track, and had been hunting for a week by himself. But I fear that he was not the wiser for all I told him, for every time I attempted to answer his questions he interrupted me by asking, "What do you do here?" He had lost a dog, but found a man.

One old hunter who has a dry tongue, who used to come to bathe in Walden once every year when the water was warmest, and at such times looked in upon me, told me, that many years ago he took his gun one afternoon and went out for a cruise in Walden Wood; and as he walked the Wayland* road he heard the cry of hounds approaching, and ere long a fox leaped the wall into the road, and as quick as thought leaped the other wall out of the road, and his swift bullet had not touched him. Some way behind came an old hound and her three pups in full pursuit, hunting on their own account, and disappeared again in the woods. Late in the afternoon, as he was resting in the thick woods south of Walden, he heard the voice of the hounds far over toward Fair Haven still pursuing the fox; and on they came, their hounding cry which made all the woods ring sounding nearer and nearer, now from Well-Meadow,* now from the Baker Farm. For a long time he stood still and listened to their music, so sweet to a hunter's ear, when suddenly the fox appeared, threading the solemn aisles with an easy coursing pace, whose sound was concealed by a sympathetic rustle of the leaves, swift and still, keeping the ground, leaving his pursuers far behind; and, leaping upon a rock amid the woods, he sat erect and listening, with his back to the hunter. For a moment compassion restrained the latter's arm; but that was a short-lived mood, and as quick as thought can follow thought his piece was levelled, and *whang!*—the fox rolling over the rock lay dead on the ground. The hunter still kept his place and listened to the hounds. Still on they came, and now the near woods resounded through all their aisles with their demoniac cry. At length the old hound burst into view with muzzle to the ground, and snapping the air as if possessed, and ran directly to the rock; but spying the dead fox she suddenly ceased her hounding, as if struck dumb with amazement, and walked round and round him in silence; and one by one her pups arrived, and, like their mother, were sobered into silence by the mystery. Then the hunter came forward and stood in their midst, and the mystery was solved. They waited in silence while he skinned the fox, then followed the brush a while, and at length turned off into the woods again. That evening a Weston Squire came to the Concord hunter's cottage to inquire for his hounds, and told how for a week they had been hunting on their own account from Weston* woods. The Concord hunter told him what he knew and offered him the skin; but the other declined it and departed. He did not find his hounds that night, but the next day learned that they had crossed the river and put up at a farmhouse for the night, whence, having been well fed, they took their departure early in the morning.

The hunter who told me this could remember one Sam Nutting,* who used to hunt bears on Fair Haven Ledges, and exchange their skins for rum in Concord village; who told him, even, that he had seen a moose there. Nutting had a famous fox-hound named Burgoyne,*—he pronounced it Bugine,— which my informant used to borrow. In the "Wast Book"* of an old trader of this town, who was also a captain, town-clerk, and representative, I find the following entry. Jan. 18th, 1742– 3, "John Melven Cr. by 1 Grey Fox 0—2—3;" they are not now found here; and in his leger, Feb. 7th, 1743, Hezekiah Stratton has credit "by ½ a Catt skin 0—1—4½;" of course, a wild-cat, for Stratton was a sergeant in the old French war, and would not have got credit for hunting less noble game. Credit is given for deer skins also, and they were daily sold. One man still preserves the horns of the last deer that was killed in this vicinity, and another has told me the particulars of the hunt in which his uncle was engaged. The hunters were formerly a numerous and merry crew here. I remember well one gaunt Nimrod* who would catch up a leaf by the roadside and play a strain on it wilder and more melodious, if my memory serves me, than any hunting horn.

At midnight, when there was a moon, I sometimes met with hounds in my path prowling about the woods, which would skulk out of my way, as if afraid, and stand silent amid the bushes till I had passed.

Squirrels and wild mice disputed for my store of nuts. There were scores of pitch-pines around my house, from one to four inches in diameter, which had been gnawed by mice the previous winter,—a Norwegian winter for them, for the snow lay long and deep, and they were obliged to mix a large proportion of pine bark with their other diet. These trees were alive and apparently flourishing at mid-summer, and many of them had grown a foot, though completely girdled; but after another winter such were without exception dead. It is remarkable that a single mouse should thus be allowed a whole pine tree for its dinner, gnawing round instead of up and down it; but perhaps it is necessary in order to thin these trees, which are wont to grow up densely.

The hares (*Lepus Americanus*) were very familiar. One had her form under my house all winter, separated from me only by the flooring, and she startled me each morning by her hasty departure when I began to stir,—thump, thump, thump, striking her head against the floor timbers in her hurry. They used to come round my door at dusk to nibble the potato parings which I had thrown out, and were so nearly the color of the ground that they could hardly be distinguished when still. Sometimes in the twilight I alternately lost and recovered sight of one sitting motionless under my window. When I opened my door in the evening, off they would go with a squeak and a bounce. Near at hand they only excited my pity. One evening one sat by my door two paces from me, at first trembling with fear, yet unwilling to move; a poor wee thing, lean and bony, with ragged ears and sharp nose, scant tail and slender paws. It looked as if Nature no longer contained the breed of nobler bloods, but stood on her last toes. Its large eyes appeared young and unhealthy, almost dropsical.* I took a step, and lo, away it scud with an elastic spring over the snow crust, straightening its body and its limbs into graceful length, and soon put the forest between me and itself,—the wild free venison,* asserting its vigor and the dignity of Nature. Not without reason was its slenderness. Such then was its nature. (*Lepus, levipes*, light-foot, some think.*)

What is a country without rabbits and partridges? They are among the most simple and indigenous animal products; ancient and venerable families known to antiquity as to modern times; of the very hue and substance of Nature, nearest allied to leaves and to the ground,—and to one another; it is either winged or it is legged. It is hardly as if you had seen a wild creature when a rabbit or a partridge bursts away, only a natural one, as much to be expected as rustling leaves. The partridge and the rabbit are still sure to thrive, like true natives of the soil, whatever revolutions occur. If the forest is cut off, the sprouts and bushes which spring up afford them concealment, and they become more numerous than ever. That must be a poor country indeed that does not support a hare. Our woods teem with them both, and around every swamp may be seen the partridge or rabbit walk, beset with twiggy fences* and horse-hair snares, which some cow-boy* tends.

THE POND IN WINTER

AFTER a still winter night I awoke with the impression that some question had been put to me, which I had been endeavoring in vain to answer in my sleep, as what—how—when—where? But there was dawning Nature, in whom all creatures live, looking in at my broad windows with serene and satisfied face, and no question on *her* lips. I awoke to an answered question, to Nature and daylight. The snow lying deep on the earth dotted with young pines, and the very slope of the hill on which my house is placed, seemed to say, Forward! Nature puts no question and answers none which we mortals ask. She has long ago taken her resolution. "O Prince, our eyes contemplate with admiration and transmit to the soul the wonderful and varied spectacle of this universe. The night veils without a doubt a part of this glorious creation; but day comes to reveal to us this great work, which extends from earth even into the plains of the ether."*

Then to my morning work. First I take an axe and pail and go in search of water, if that be not a dream. After a cold and snowy night it needed a divining rod to find it. Every winter the liquid and trembling surface of the pond, which was so sensitive to every breath, and reflected every light and shadow, becomes solid to the depth of a foot or a foot and a half, so that it will support the heaviest teams, and perchance the snow covers it to an equal depth, and it is not to be distinguished from any level field. Like the marmots in the surrounding hills, it closes its eye-lids and becomes dormant for three months or more. Standing on the snow-covered plain, as if in a pasture amid the hills, I cut my way first through a foot of snow, and then a foot of ice, and open a window under my feet, where, kneeling to drink, I look down into the quiet parlor of the fishes, pervaded by a softened light as through a window of ground glass, with its bright sanded floor the same as in summer; there a perennial wave-less serenity reigns as in the amber twilight sky, corresponding to the cool and even temperament of the inhabitants. Heaven is under our feet as well as over our heads.

Early in the morning, while all things are crisp with frost, men come with fishing reels and slender lunch, and let down their fine lines through the snowy field to take pickerel and perch; wild men, who instinctively follow other fashions and trust other authorities than their townsmen, and by their goings and comings stitch towns together in parts where else they would be ripped. They sit and eat their luncheon in stout fear-naughts* on the dry oak leaves on the shore, as wise in natural lore as the citizen is in artificial. They never con-sulted with books, and know and can tell much less than they have done. The things which they practise are said not yet to be known. Here is one fishing for pickerel with grown perch for bait. You look into his pail with wonder as into a summer pond, as if he kept summer locked up at home, or knew where she had retreated. How, pray, did he get these in mid-winter? O, he got worms out of rotten logs since the ground froze, and so he caught them. His life itself passes deeper in Nature than the studies of the naturalist penetrate; himself a subject for the naturalist. The latter raises the moss and bark gently with his knife in search of insects; the former lays open logs to their core with his axe, and moss and bark fly far and wide. He gets his living by barking trees. Such a man has some right to fish, and I love to see Nature carried out in him. The perch swal-lows the grub-worm, the pickerel swallows the perch, and the fisherman swallows the pickerel; and so all the chinks in the scale of being* are filled.

When I strolled around the pond in misty weather I was sometimes amused by the primitive mode which some ruder fisherman had adopted. He would perhaps have placed alder branches over the narrow holes in the ice, which were four or five rods apart and an equal distance from the shore, and having fastened the end of the line to a stick to prevent its being pulled through, have passed the slack line over a twig of the alder, a foot or more above the ice, and tied a dry oak leaf to it, which, being pulled down, would show when he had a bite. These alders loomed through the mist at regular intervals as you walked half way around the pond.

Ah, the pickerel of Walden! when I see them lying on the ice, or in the well which the fisherman cuts in the ice, making a little hole to admit the water, I am always surprised by their rare beauty, as if they were fabulous fishes, they are so foreign to the streets, even to the woods, foreign as Arabia to our Concord life. They possess a quite dazzling and transcendent beauty which separates them by a wide interval from the cadaverous cod and haddock whose fame is trumpeted* in our streets. They are not green like the pines, nor gray like the stones, nor blue like the sky; but they have, to my eyes, if possible, yet rarer colors, like flowers and precious stones, as if they were the pearls, the animalized *nuclei* or crystals of the Walden water. They, of course, are Walden all over and all through; are themselves small Waldens in the animal kingdom, Waldenses.* It is surprising that they are caught here,— that in this deep and capacious spring, far beneath the rattling teams and chaises and tinkling sleighs that travel the Walden road, this great gold and emerald fish swims. I never chanced to see its kind in any market; it would be the cynosure of all eyes there. Easily, with a few convulsive quirks, they give up their watery ghosts, like a mortal translated before his time to the thin air of heaven.

As I was desirous to recover the long lost bottom of Walden Pond, I surveyed it carefully, before the ice broke up, early in '46, with compass and chain* and sounding line. There have been many stories told about the bottom, or rather no bottom, of this pond, which certainly had no foundation for themselves. It is remarkable how long men will believe in the bottomlessness of a pond without taking the trouble to sound

I have visited two such Bottomless Ponds,* in one way in this neighborhood. Many have believed that Walden reached quite through to the other side of the globe. Some who have lain flat on the ice for a long time, looking down through the illusive medium, perchance with watery eyes into the bargain, and driven to hasty conclusions by the fear of catching cold in their breasts, have seen vast holes "into which a load of hay might be driven," if there were any body to drive it, the undoubted source of the Styx and entrance to the Infernal Regions from these parts. Others have gone down from the village with a "fifty-six" and a wagon load of inch rope, but yet have failed to find any bottom; for while the "fifty-six" was resting by the way, they were paying out the rope in the vain attempt to fathom their truly immeasurable capacity for marvellousness. But I can assure my readers that Walden has a reasonably tight bottom at a not unreasonable, though at an unusual, depth. I fathomed it easily with a cod-line and a stone weighing about a pound and a half, and could tell accurately when the stone left the bottom, by having to pull so much harder before the water got underneath* to help me. The greatest depth was exactly one hundred and two feet; to which may be added the five feet which it has risen since,* making one hundred and seven. This is a remarkable depth for so small an area; yet not an inch of it can be spared by the imagination. What if all ponds were shallow? Would it not react on the minds of men? I am thankful that this pond was made deep and pure for a symbol. While men believe in the infinite some ponds will be thought to be bottomless.

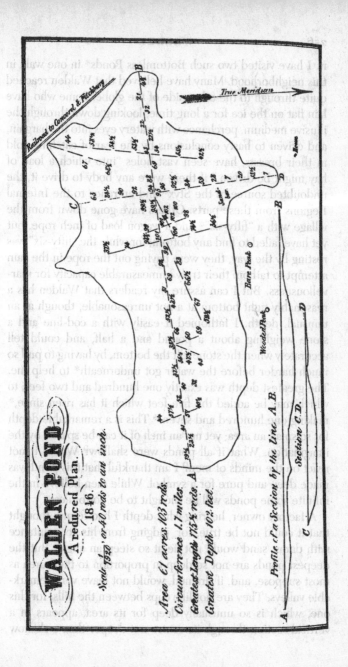

WALDEN POND.
A reduced Plan.
1846.
Scale 1 to 40 rods to an inch.

Area 61 acres 103 rods.
Circumference 1.7 miles.
Greatest Length 175½ rods. A.B.
Greatest Depth 102 feet.

Profile of a Section by the line A.B.

Section C.D.

True Meridian

Railroad to Concord & Fitchburg

Bare Peak ⊙   Wooded Peak ⊙   Arrow

Sandy Bar

A factory-owner, hearing what depth I had found, thought that it could not be true, for, judging from his acquaintance with dams, sand would not lie at so steep an angle. But the deepest ponds are not so deep in proportion to their area as most suppose, and, if laid bare, would not leave a very remarkable valley. They are not like cups between the hills; for this one, which is so unusually deep for its area, appears in a vertical section

it. I have visited two such Bottomless Ponds* in one walk in
this neighborhood. Many have believed that Walden reached
quite through to the other side of the globe. Some who have
lain flat on the ice for a long time, looking down through the
illusive medium, perchance with watery eyes into the bargain,
and driven to hasty conclusions by the fear of catching cold
in their breasts, have seen vast holes "into which a load of
hay might be driven," if there were any body to drive it, the
undoubted source of the Styx* and entrance to the Infernal
Regions from these parts. Others have gone down from the
village with a "fifty-six"* and a wagon load of inch rope, but
yet have failed to find any bottom; for while the "fifty-six" was
resting by the way, they were paying out the rope in the vain
attempt to fathom their truly immeasurable capacity for marvellousness. But I can assure my readers that Walden has a
reasonably tight bottom at a not unreasonable, though at an
unusual, depth. I fathomed it easily with a cod-line and a
stone weighing about a pound and a half, and could tell
accurately when the stone left the bottom, by having to pull so
much harder before the water got underneath* to help me.
The greatest depth was exactly one hundred and two feet; to
which may be added the five feet which it has risen since,*
making one hundred and seven.* This is a remarkable depth
for so small an area; yet not an inch of it can be spared by the
imagination. What if all ponds were shallow? Would it not
react on the minds of men? I am thankful that this pond was
made deep and pure for a symbol. While men believe in the
infinite some ponds will be thought to be bottomless.

A factory owner, hearing what depth I had found, thought
that it could not be true, for, judging from his acquaintance
with dams, sand would not lie at so steep an angle. But the
deepest ponds are not so deep in proportion to their area as
most suppose, and, if drained, would not leave very remarkable valleys. They are not like cups between the hills; for this
one, which is so unusually deep for its area, appears in a
vertical section through its centre not deeper than a shallow

plate. Most ponds, emptied, would leave a meadow no more hollow than we frequently see. William Gilpin, who is so admirable in all that relates to landscapes, and usually so correct, standing at the head of Loch Fyne,* in Scotland, which he describes as "a bay of salt water, sixty or seventy fathoms deep, four miles in breadth," and about fifty miles long, surrounded by mountains, observes, "If we could have seen it immediately after the diluvian crash, or whatever convulsion of Nature occasioned it, before the waters gushed in, what a horrid chasm it must have appeared!

> So high as heaved the tumid hills, so low
> Down sunk a hollow bottom, broad, and deep,
> Capacious bed of waters—."*

But if, using the shortest diameter of Loch Fyne, we apply these proportions to Walden, which, as we have seen, appears already in a vertical section only like a shallow plate, it will appear four times as shallow. So much for the *increased* horrors of the chasm of Loch Fyne when emptied. No doubt many a smiling valley with its stretching cornfields occupies exactly such a "horrid chasm," from which the waters have receded, though it requires the insight and the far sight of the geologist to convince the unsuspecting inhabitants of this fact. Often an inquisitive eye may detect the shores of a primitive lake in the low horizon hills, and no subsequent elevation of the plain have been* necessary to conceal their history. But it is easiest, as they who work on the highways know, to find the hollows by the puddles after a shower. The amount of it is, the imagination, give it the least license, dives deeper and soars higher than Nature goes. So, probably, the depth of the ocean will be found to be very inconsiderable compared with its breadth.

As I sounded through the ice I could determine the shape of the bottom with greater accuracy than is possible in surveying harbors which do not freeze over, and I was surprised at its general regularity. In the deepest part there are several acres

more level than almost any field which is exposed to the sun
wind and plough. In one instance, on a line arbitrarily chosen,
the depth did not vary more than one foot in thirty rods; and
generally, near the middle, I could calculate the variation for
each one hundred feet in any direction beforehand within
three or four inches. Some are accustomed to speak of deep
and dangerous holes even in quiet sandy ponds like this, but
the effect of water under these circumstances is to level all
inequalities. The regularity of the bottom and its conformity
to the shores and the range of the neighboring hills were
so perfect that a distant promontory betrayed itself in the
soundings quite across the pond, and its direction could be
determined by observing the opposite shore. Cape becomes
bar, and plain shoal, and valley and gorge deep water and
channel.

When I had mapped the pond by the scale of ten rods* to
an inch, and put down the soundings, more than a hundred in
all, I observed this remarkable coincidence. Having noticed
that the number indicating the greatest depth was apparently
in the centre of the map, I laid a rule on the map lengthwise,
and then breadthwise, and found, to my surprise, that the line
of greatest length intersected the line of greatest breadth
exactly at the point of greatest depth, notwithstanding that the
middle is so nearly level, the outline of the pond far from
regular, and the extreme length and breadth were got by
measuring into the coves; and I said to myself, Who knows but
this hint would conduct to the deepest part of the ocean as
well as of a pond or puddle? Is not this the rule also for the
height of mountains, regarded as the opposite of valleys? We
know that a hill is not highest at its narrowest part.

Of five coves, three, or all which had been sounded, were
observed to have a bar quite across their mouths and deeper
water within, so that the bay tended to be an expansion of
water within the land not only horizontally but vertically, and
to form a basin or independent pond, the direction of the
two capes showing the course of the bar. Every harbor on

the sea-coast, also, has its bar* at its entrance. In proportion as the mouth of the cove was wider compared with its length, the water over the bar was deeper compared with that in the basin. Given, then, the length and breadth of the cove, and the character of the surrounding shore, and you have almost elements enough to make out a formula for all cases.

In order to see how nearly I could guess, with this experience, at the deepest point in a pond, by observing the outlines of its surface and the character of its shores alone, I made a plan of White Pond, which contains about forty-one acres, and, like this, has no island in it, nor any visible inlet or outlet; and as the line of greatest breadth fell very near the line of least breadth, where two opposite capes approached each other and two opposite bays receded, I ventured to mark a point a short distance from the latter line, but still on the line of greatest length, as the deepest. The deepest part was found to be within one hundred feet of this, still farther in the direction to which I had inclined, and was only one foot deeper, namely, sixty feet. Of course, a stream running through, or an island in the pond, would make the problem much more complicated.

If we knew all the laws of Nature, we should need only one fact, or the description of one actual phenomenon, to infer all the particular results at that point. Now we know only a few laws, and our result is vitiated, not, of course, by any confusion or irregularity in Nature, but by our ignorance of essential elements in the calculation. Our notions of law and harmony are commonly confined to those instances which we detect; but the harmony which results from a far greater number of seemingly conflicting, but really concurring, laws, which we have not detected, is still more wonderful. The particular laws are as our points of view, as, to the traveller, a mountain outline varies with every step, and it has an infinite number of profiles, though absolutely but one form. Even when cleft or bored through it is not comprehended in its entireness.

What I have observed of the pond is no less true in ethics. It is the law of average. Such a rule of the two diameters not only guides us toward the sun in the system and the heart in man, but draws lines through the length and breadth of the aggregate of a man's particular daily behaviors and waves of life into his coves and inlets, and where they intersect will be the height or depth of his character. Perhaps we need only to know how his shores trend* and his adjacent country or circumstances, to infer his depth and concealed bottom. If he is surrounded by mountainous circumstances, an Achillean shore,* whose peaks overshadow and are reflected in his bosom, they suggest a corresponding depth in him. But a low and smooth shore proves him shallow on that side. In our bodies, a bold projecting brow falls off to and indicates a corresponding depth of thought. Also there is a bar across the entrance of our every cove, or particular inclination; each is our harbor for a season, in which we are detained and partially land-locked. These inclinations are not whimsical usually, but their form, size, and direction are determined by the promontories of the shore, the ancient axes of elevation. When this bar is gradually increased by storms, tides, or currents, or there is a subsidence of the waters, so that it reaches to the surface, that which was at first but an inclination in the shore in which a thought was harbored becomes an individual lake, cut off from the ocean, wherein the thought secures its own conditions, changes, perhaps, from salt to fresh, becomes a sweet sea, dead sea, or a marsh. At the advent of each individual into this life, may we not suppose that such a bar has risen to the surface somewhere? It is true, we are such poor navigators that our thoughts, for the most part, stand off and on upon a harborless coast, are conversant only with the bights of the bays of poesy, or steer for the public ports of entry, and go into the dry docks of science, where they merely refit for this world, and no natural currents concur to individualize them.

As for the inlet or outlet of Walden, I have not discovered any but rain or snow and evaporation, though perhaps, with a thermometer and a line, such places may be found, for where the water flows into the pond it will probably be coldest in summer and warmest in winter. When the ice-men were at work here in '46–7,* the cakes sent to the shore were one day rejected by those who were stacking them up there, not being thick enough to lie side by side with the rest; and the cutters thus discovered that the ice over a small space was two or three inches thinner than elsewhere, which made them think that there was an inlet there. They also showed me in another place what they thought was a "leach hole,"* through which the pond leaked out under a hill into a neighboring meadow, pushing me out on a cake of ice to see it. It was a small cavity under ten feet of water; but I think that I can warrant the pond not to need soldering till they find a worse leak than that. One has suggested, that if such a "leach hole" should be found, its connection with the meadow, if any existed, might be proved by conveying some colored powder or sawdust to the mouth of the hole, and then putting a strainer over the spring in the meadow, which would catch some of the particles carried through by the current.

While I was surveying, the ice, which was sixteen inches thick, undulated under a slight wind like water. It is well known that a level cannot be used on ice.* At one rod from shore its greatest fluctuation, when observed by means of a level on land directed toward a graduated staff on the ice, was three quarters of an inch, though the ice appeared firmly attached to the shore. It was probably greater in the middle. Who knows but if our instruments were delicate enough we might detect an undulation in the crust of the earth? When two legs of my level were on the shore and the third on the ice, and the sights were directed over the latter, a rise or fall of the ice of an almost infinitesimal amount made a difference of several feet on a tree across the pond. When I began to cut holes for sounding, there were three or four inches of water on the ice under a deep snow which had sunk it thus far; but the water began immediately to run into these holes, and continued to run for two days in deep streams, which wore away the ice on every side, and contributed essentially, if not mainly, to dry the surface of the pond; for, as the water ran in, it raised and floated the ice. This was somewhat like cutting a hole in the bottom of a ship to let the water out. When such holes freeze, and a rain succeeds, and finally a new freezing forms a fresh smooth ice over all, it is beautifully mottled internally by dark figures, shaped somewhat like a spider's web, what you may call ice rosettes, produced by the channels worn by the water flowing from all sides to a centre. Sometimes, also, when the ice was covered with shallow puddles, I saw a double shadow of myself, one standing on the head of the other, one on the ice, the other on the trees or hill-side.

While yet it is cold January, and snow and ice are thick and solid, the prudent landlord comes from the village to get ice to cool his summer drink; impressively, even pathetically wise, to foresee the heat and thirst of July now in January,—wearing a thick coat and mittens! when so many things are not provided for. It may be that he lays up no treasures* in this world which will cool his summer drink in the next. He cuts and saws the solid pond, unroofs the house of fishes, and carts off their very element and air, held fast by chains and stakes like corded wood, through the favoring winter air, to wintry cellars, to underlie the summer there. It looks like solidified azure, as, far off, it is drawn through the streets. These ice-cutters are a merry race, full of jest and sport, and when I went among them they were wont to invite me to saw pit-fashion* with them, I standing underneath.

In the winter of '46-7 there came a hundred men of Hyperborean* extraction swoop down on to our pond one morning, with many car-loads of ungainly-looking farming tools, sleds, ploughs, drill-barrows,* turf-knives,* spades, saws, rakes, and each man was armed with a double-pointed pike-staff, such as is not described in the New-England Farmer or the Cultivator.* I did not know whether they had come to sow a crop of winter rye, or some other kind of grain recently introduced from Iceland. As I saw no manure, I judged that they meant to skim the land, as I had done, thinking the soil was deep and had lain fallow long enough. They said that a gentleman farmer,* who was behind the scenes, wanted to double his money, which, as I understood, amounted to half a million already; but in order to cover each one of his dollars with another, he took off the only coat, ay, the skin itself, of Walden Pond in the midst of a hard winter. They went to work at once, ploughing, harrowing, rolling,* furrowing, in admirable order, as if they were bent on making this a model farm;* but when I was looking sharp to see what kind of seed they dropped into the furrow, a gang of fellows by my side suddenly began to hook up the virgin mould itself, with a peculiar jerk, clean down to the sand, or rather the water,—for it was a very springy soil,—indeed all the *terra firma** there was,—and haul it away on sleds, and then I guessed that they must be cutting peat in a bog.* So they came and went every day, with a peculiar shriek from the locomotive, from and to some point of the polar regions, as it seemed to me, like a flock of arctic snow-birds. But sometimes Squaw Walden had her revenge, and a hired man, walking behind his team, slipped through a crack in the ground down toward Tartarus,* and he who was so brave before suddenly became but the ninth part of a man,* almost gave up his animal heat, and was glad to take refuge in my house, and acknowledged that there was some virtue in a stove; or sometimes the frozen soil took a piece of steel out of a ploughshare, or a plough got set in the furrow and had to be cut out.

To speak literally, a hundred Irishmen, with Yankee overseers, came from Cambridge* every day to get out the ice. They divided it into cakes by methods too well known* to require description, and these, being sledded to the shore, were rapidly hauled off on to an ice platform, and raised by grappling irons and block and tackle, worked by horses, on to a stack, as surely as so many barrels of flour, and there placed evenly side by side, and row upon row, as if they formed the solid base of an obelisk designed to pierce the clouds. They told me that in a good day they could get out a thousand tons, which was the yield of about one acre. Deep ruts and "cradle holes" were worn in the ice, as on *terra firma*, by the passage of the sleds over the same track, and the horses invariably ate their oats out of cakes of ice hollowed out like buckets. They stacked up the cakes thus in the open air in a pile thirty-five feet high on one side and six or seven rods square, putting hay between the outside layers to exclude the air; for when the wind, though never so cold, finds a passage through, it will wear large cavities, leaving slight supports or studs only here and there, and finally topple it down. At first it looked like a vast blue fort or Valhalla;* but when they began to tuck the coarse meadow hay into the crevices, and this became covered with rime and icicles, it looked like a venerable moss-grown and hoary ruin, built of azure-tinted marble, the abode of Winter, that old man* we see in the almanac,—his shanty, as if he had a design to estivate* with us. They calculated that not a twenty-five per cent of this would reach its destination, and that two or three per cent would be wasted in the cars.* However, a still greater part of this heap had a different destiny from what was intended; for, either because the ice was found not to keep so well as was expected, containing more air than usual, or for some other reason, it never got to market. This heap, made in the winter of '46-7 and estimated to contain ten thousand tons, was finally covered with hay and boards; and though it was unroofed the following July, and a part of it carried off, the rest remaining exposed to the sun, it stood over that summer and the next winter, and was not quite melted till September 1848. Thus the pond recovered the greater part.

Like the water, the Walden ice, seen near at hand, has a green tint, but at a distance is beautifully blue, and you can easily tell it from the white ice of the river, or the merely greenish ice of some ponds, a quarter of a mile off. Sometimes one of those great cakes slips from the ice-man's sled into the village street, and lies there for a week like a great emerald, an object of interest to all passers. I have noticed that a portion of Walden which in the state of water was green will often, when frozen, appear from the same point of view blue. So the hollows about this pond will, sometimes, in the winter, be filled with a greenish water somewhat like its own, but the next day will have frozen blue. Perhaps the blue color of water and ice is due to the light and air they contain, and the most transparent is the bluest. Ice is an interesting subject for contemplation. They told me that they had some in the ice-houses at Fresh Pond* five years old which was as good as ever. Why is it that a bucket of water soon becomes putrid, but frozen remains sweet forever? It is commonly said that this is the difference between the affections and the intellect.

Thus for sixteen days I saw from my window a hundred men at work like busy husbandmen, with teams and horses and apparently all the implements of farming, such a picture as we see on the first page of the almanac; and as often as I looked out I was reminded of the fable of the lark* and the reapers, or the parable of the sower,* and the like; and now they are all gone, and in thirty days more, probably, I shall look from the same window on the pure sea-green Walden water there, reflecting the clouds and the trees, and sending up its evaporations in solitude, and no traces will appear that a man has ever stood there. Perhaps I shall hear a solitary loon laugh as he dives and plumes himself, or shall see a lonely fisher in his boat, like a floating leaf, beholding his form reflected in the waves, where lately a hundred men securely labored.

Thus it appears that the sweltering inhabitants of Charleston and New Orleans, of Madras and Bombay and Calcutta,* drink at my well. In the morning I bathe my intellect in the stupendous and cosmogonal philosophy of the Bhagvat

Geeta,* since whose composition years of the gods have elapsed, and in comparison with which our modern world and its literature seem puny and trivial; and I doubt if that philosophy is not to be referred to a previous state of existence, so remote is its sublimity from our conceptions. I lay down the book and go to my well for water, and lo! there I meet the servant of the Bramin, priest of Brahma and Vishnu and Indra,* who still sits in his temple on the Ganges* reading the Vedas,* or dwells at the root of a tree with his crust and water jug. I meet his servant come to draw water for his master, and our buckets as it were grate together in the same well. The pure Walden water is mingled with the sacred water of the Ganges. With favoring winds it is wafted past the site of the fabulous islands of Atlantis* and the Hesperides,* makes the periplus of Hanno,* and, floating by Ternate and Tidore* and the mouth of the Persian Gulf, melts in the tropic gales of the Indian seas, and is landed in ports of which Alexander* only heard the names.

SPRING

THE opening of large tracts by the ice-cutters commonly causes a pond to break up earlier; for the water, agitated by the wind, even in cold weather, wears away the surrounding ice. But such was not the effect on Walden that year, for she had soon got a thick new garment to take the place of the old. This pond never breaks up so soon as the others in this neighborhood, on account both of its greater depth and its having no stream passing through it to melt or wear away the ice. I never knew it to open in the course of a winter, not excepting that of '52–3, which gave the ponds so severe a trial. It commonly opens about the first of April, a week or ten days later than Flints' Pond and Fair-Haven, beginning to melt on the north side and in the shallower parts where it began to freeze. It indicates better than any water hereabouts the absolute progress of the season, being least affected by transient changes of temperature. A severe cold of a few days' duration in March may very much retard the opening of the former ponds, while the temperature of Walden increases almost uninterruptedly. A thermometer thrust into the middle of Walden on the 6th of March, 1847, stood at 32°, or freezing point; near the shore at 33°; in the middle of Flints' Pond, the same day, at $32\frac{1}{2}°$; at a dozen rods from the shore, in shallow water, under ice a foot thick, at 36°. This difference of three and a half degrees between the temperature of the deep water and the shallow in the latter pond, and the fact that a great proportion of it is comparatively shallow, show why it should break up so much sooner than Walden. The ice in the shallowest part was at this time several inches thinner than in the middle. In mid-winter the middle had been the warmest and the ice thinnest there. So, also, every one who has waded about the shores of a pond in summer must have perceived how much warmer the water is close to the shore, where only three or four inches deep, than a little distance out, and on the surface where it is deep, than near the bottom. In spring the sun not only exerts an influence through the increased temperature of the air and earth, but its heat passes through ice a foot or more thick, and is reflected from the bottom in shallow water, and so also warms the water and melts the under side of the ice, at the same time that it is melting more directly above, making it uneven, and causing the air bubbles which it contains to extend themselves upward and downward until it is completely honey-combed, and at last disappears suddenly in a single spring rain. Ice has its grain as well as wood, and when a cake begins to rot or "comb," that is, assume the appearance of a honey-comb, whatever may be its position, the air cells are at right angles with what was the water surface. Where there is a rock or a log rising near to the surface the ice over it is much thinner, and is frequently quite dissolved by this reflected heat;* and I have been told that in the experiment at Cambridge* to freeze water in a shallow wooden pond, though the cold air circulated underneath, and so had access to both sides, the reflection of the sun from the bottom more than counterbalanced this advantage. When a warm rain in the middle of the winter melts off the snow-ice from Walden, and leaves a hard dark or transparent ice on the middle, there will be a strip of rotten though thicker white ice, a rod or more wide, about the shores, created by this reflected heat. Also, as I have said, the bubbles themselves within the ice operate as burning glasses to melt the ice beneath.

The phenomena of the year take place every day in a pond on a small scale. Every morning, generally speaking, the shallow water is being warmed more rapidly than the deep, though it may not be made so warm after all, and every evening it is being cooled more rapidly until the morning. The day is an epitome of the year. The night is the winter, the morning and evening are the spring and fall, and the noon is the summer. The cracking and booming of the ice indicate a change of temperature. One pleasant morning after a cold night, February 24th, 1850, having gone to Flints' Pond to spend the day, I noticed with surprise, that when I struck the ice with the head of my axe, it resounded like a gong for many rods around, or as if I had struck on a tight drum-head. The pond began to boom about an hour after sunrise, when it felt the influence of the sun's rays slanted upon it from over the hills; it stretched itself and yawned like a waking man with a gradually increasing tumult, which was kept up three or four hours. It took a short siesta at noon, and boomed once more toward night, as the sun was withdrawing his influence. In the right stage of the weather a pond fires its evening gun with great regularity. But in the middle of the day, being full of cracks,* and the air also being less elastic,* it had completely lost its resonance, and probably fishes and muskrats could not then have been stunned by a blow on it. The fishermen say that the "thundering of the pond" scares the fishes and prevents their biting. The pond does not thunder every evening, and I cannot tell surely when to expect its thundering; but though I may perceive no difference in the weather, it does. Who would have suspected so large and cold and thick-skinned a thing to be so sensitive? Yet it has its law to which it thunders obedience when it should as surely as the buds expand in the spring. The earth is all alive and covered with papillæ.* The largest pond is as sensitive to atmospheric changes as the globule of mercury in its tube.

One attraction in coming to the woods to live was that I should have leisure and opportunity to see the spring come in. The ice in the pond at length begins to be honey-combed, and I can set my heel in it as I walk. Fogs and rains and warmer suns are gradually melting the snow; the days have grown sensibly longer; and I see how I shall get through the winter without adding to my wood-pile, for large fires are no longer necessary. I am on the alert for the first signs of spring, to hear the chance note of some arriving bird, or the striped squirrel's

chirp, for his stores must be now nearly exhausted, or see the
woodchuck venture out of his winter quarters. On the 13th of
March, after I had heard the bluebird, song-sparrow, and red-
wing, the ice was still nearly a foot thick. As the weather grew
warmer, it was not sensibly worn away by the water, nor
broken up and floated off as in rivers, but, though it was
completely melted for half a rod in width about the shore, the
middle was merely honey-combed and saturated with water,
so that you could put your foot through it when six inches
thick; but by the next day evening, perhaps, after a warm rain
followed by fog, it would have wholly disappeared, all gone off
with the fog, spirited away. One year I went across the middle
only five days before it disappeared entirely. In 1845 Walden
was first completely open on the 1st of April; in '46, the 25th
of March; in '47, the 8th of April; in '51, the 28th of March; in
'52, the 18th of April; in '53, the 23d of March; in '54, about
the 7th of April.

Every incident connected with the breaking up of the rivers
and ponds and the settling of the weather is particularly interesting to us who live in a climate of so great extremes. When
the warmer days come, they who dwell near the river hear the
ice crack at night with a startling whoop as loud as artillery, as
if its icy fetters were rent from end to end, and within a few
days see it rapidly going out. So the alligator comes out of the
mud with quakings of the earth. One old man, who has been
a close observer of Nature, and seems as thoroughly wise in
regard to all her operations as if she had been put upon the
stocks* when he was a boy, and he had helped to lay her
keel,—who has come to his growth, and can hardly acquire
more of natural lore if he should live to the age of
Methuselah,*—told me, and I was surprised to hear him
express wonder at any of Nature's operations, for I thought
that there were no secrets between them, that one spring day
he took his gun and boat, and thought that he would have a
little sport with the ducks. There was ice still on the meadows,
but it was all gone out of the river, and he dropped down

without obstruction from Sudbury, where he lived, to Fair-Haven Pond,* which he found, unexpectedly, covered for the most part with a firm field of ice. It was a warm day, and he was surprised to see so great a body of ice remaining. Not seeing any ducks, he hid his boat on the north or back side of an island in the pond, and then concealed himself in the bushes on the south side, to await them. The ice was melted for three or four rods from the shore, and there was a smooth and warm sheet of water, with a muddy bottom, such as the ducks love, within, and he thought it likely that some would be along pretty soon. After he had lain still there about an hour he heard a low and seemingly very distant sound, but singularly grand and impressive, unlike any thing he had ever heard, gradually swelling and increasing as if it would have a universal and memorable ending, a sullen rush and roar, which seemed to him all at once like the sound of a vast body of fowl coming in to settle there, and, seizing his gun, he started up in haste and excited; but he found, to his surprise, that the whole body of ice had started while he lay there, and drifted in to the shore, and the sound he had heard was made by its edge grating on the shore,—at first gently nibbled and crumbled off, but at length heaving up and scattering its wrecks along the island to a considerable height before it came to a stand still.

At length the sun's rays have attained the right angle, and warm winds blow up mist and rain and melt the snow banks, and the sun dispersing the mist smiles on a checkered landscape of russet and white smoking with incense, through which the traveller picks his way from islet to islet, cheered by the music of a thousand tinkling rills and rivulets whose veins are filled with the blood of winter which they are bearing off.

Few phenomena* gave me more delight than to observe the forms which thawing sand and clay assume in flowing down the sides of a deep cut on the railroad through which I passed on my way to the village,* a phenomenon not very

common on so large a scale, though the number of freshly
exposed banks of the right material must have been greatly
multiplied since railroads were invented. The material was
sand of every degree of fineness and of various rich colors,
commonly mixed with a little clay. When the frost comes out
in the spring, and even in a thawing day in the winter, the sand
begins to flow down the slopes like lava, sometimes bursting
out through the snow and overflowing it where no sand was to
be seen before. Innumerable little streams overlap and inter-
lace one with another, exhibiting a sort of hybrid product,
which obeys half way the laws of currents, and half way that of
vegetation. As it flows it takes the forms of sappy leaves or
vines, making heaps of pulpy sprays a foot or more in depth,
and resembling, as you look down on them, the laciniated*
lobed and imbricated* thalluses* of some lichens; or you are
reminded of coral, of leopards' paws or birds' feet, of brains or
lungs or bowels, and excrements* of all kinds. It is a truly
*grotesque** vegetation, whose forms and color we see imitated
in bronze, a sort of architectural foliage more ancient and
typical* than acanthus,* chiccory, ivy, vine, or any vegetable
leaves; destined perhaps, under some circumstances, to become a puzzle to future geologists. The whole cut impressed
me as if it were a cave with its stalactites laid open to the light.
The various shades of the sand are singularly rich and agreeable, embracing the different iron colors, brown, gray, yellowish, and reddish. When the flowing mass reaches the drain at
the foot of the bank it spreads out flatter into *strands*, the
separate streams losing their semi-cylindrical form and gradually becoming more flat and broad, running together as they
are more moist, till they form an almost flat *sand*, still variously and beautifully shaded, but in which you can trace the
original forms of vegetation; till at length, in the water itself,
they are converted into *banks*, like those formed off the
mouths of rivers, and the forms of vegetation are lost in the
ripple marks on the bottom.

The whole bank, which is from twenty to forty feet high, is

sometimes overlaid with a mass of this kind of foliage, or sandy rupture, for a quarter of a mile on one or both sides, the produce of one spring day. What makes this sand foliage remarkable is its springing into existence thus suddenly. When I see on the one side the inert bank,—for the sun acts on one side first,—and on the other this luxuriant foliage, the creation of an hour, I am affected as if in a peculiar sense I stood in the laboratory* of the Artist who made the world and me,—had come to where he was still at work, sporting on this bank, and with excess of energy strewing his fresh designs about. I feel as if I were nearer to the vitals of the globe, for this sandy overflow is something such a foliaceous* mass as the vitals of the animal body. You find thus in the very sands an anticipation of the vegetable leaf. No wonder that the earth expresses itself outwardly in leaves, it so labors* with the idea inwardly. The atoms have already learned this law, and are pregnant by it. The overhanging leaf sees here its prototype. *Internally*, whether in the globe or animal body, it is a moist thick *lobe*,* a word especially applicable to the liver and lungs and the *leaves* of fat, (λειβω, *labor*, *lapsus*, to flow or slip downward, a lapsing; λοβος, *globus*, lobe, globe; also lap, flap, and many other words,) *externally* a dry thin *leaf*, even as the *f* and *v* are a pressed and dried *b*. The radicals of lobe are *lb*, the soft mass of the *b* (single lobed, or B, double lobed,) with a liquid *l* behind it pressing it forward. In globe, *glb*, the guttural *g* adds to the meaning the capacity of the throat. The feathers and wings of birds are still drier and thinner leaves. Thus, also, you pass from the lumpish grub in the earth to the airy and fluttering butterfly. The very globe continually transcends and translates itself, and becomes winged in its orbit. Even ice begins with delicate crystal leaves, as if it had flowed into moulds which the fronds of water plants have impressed on the watery mirror. The whole tree itself is but one leaf, and rivers are still vaster leaves whose pulp is inter-vening earth, and towns and cities are the ova of insects in their axils.*

When the sun withdraws the sand ceases to flow, but in the morning the streams will start once more and branch and branch again into a myriad of others. You here see perchance how blood vessels are formed.* If you look closely you observe that first there pushes forward from the thawing mass a stream of softened sand with a drop-like point, like the ball of the finger, feeling its way slowly and blindly downward, until at last with more heat and moisture, as the sun gets higher, the most fluid portion, in its effort to obey the law to which the most inert also yields, separates from the latter and forms for itself a meandering channel or artery within that, in which is seen a little silvery stream glancing like lightning from one stage of pulpy leaves or branches to another, and ever and anon swallowed up in the sand. It is wonderful how rapidly yet perfectly the sand organizes itself as it flows, using the best material its mass affords to form the sharp edges of its channel. Such are the sources of rivers. In the silicious* matter which the water deposits is perhaps the bony system, and in the still finer soil and organic matter the fleshy fibre or cellular tissue. What is man but a mass of thawing clay? The ball of the human finger is but a drop congealed. The fingers and toes flow to their extent from the thawing mass of the body. Who knows what the human body would expand and flow out to under a more genial heaven? Is not the hand a spreading *palm* leaf with its lobes and veins? The ear may be regarded, fancifully, as a lichen, *umbilicaria*,* on the side of the head, with its lobe or drop. The lip*—*labium*, from *labor* (?)—laps or lapses from the sides of the cavernous mouth. The nose is a manifest congealed drop or stalactite. The chin is a still larger drop, the confluent dripping of the face. The cheeks are a slide from the brows into the valley of the face, opposed and diffused by the cheek bones. Each rounded lobe of the vegetable leaf, too, is a thick and now loitering drop, larger or smaller; the lobes are the fingers of the leaf; and as many lobes as it has, in so many directions it tends to flow, and more heat or other genial influences would have caused it to flow yet farther.

Thus it seemed that this one hillside illustrated the principle of all the operations of Nature. The Maker of this earth but patented a leaf. What Champollion* will decipher this hieroglyphic* for us, that we may turn over a new leaf at last? This phenomenon is more exhilarating to me than the luxuriance and fertility of vineyards. True, it is somewhat excrementitious in its character, and there is no end to the heaps of liver lights and bowels, as if the globe were turned wrong side outward; but this suggests at least that Nature has some bowels,* and there again is mother of humanity. This is the frost coming out of the ground; this is Spring. It precedes the green and flowery spring, as mythology precedes regular poetry. I know of nothing more purgative of winter fumes and indigestions. It convinces me that Earth is still in her swaddling clothes, and stretches forth baby fingers on every side. Fresh curls spring from the baldest brow. There is nothing inorganic. These foliaceous heaps lie along the bank like the slag of a furnace, showing that Nature is "in full blast"* within. The earth is not a mere fragment of dead history, stratum upon stratum like the leaves of a book, to be studied by geologists and antiquaries chiefly, but living poetry like the leaves of a tree, which precede flowers and fruit,—not a fossil earth, but a living earth; compared with whose great central life all animal and vegetable life is merely parasitic. Its throes will heave our exuviæ* from their graves. You may melt your metals and cast them into the most beautiful moulds you can; they will never excite me like the forms which this molten earth flows out into. And not only it, but the institutions upon it, are plastic like clay in the hands of the potter.*

Ere long, not only on these banks, but on every hill and plain and in every hollow, the frost comes out of the ground like a dormant quadruped from its burrow, and seeks the sea with music, or migrates to other climes in clouds. Thaw with his gentle persuasion is more powerful than Thor* with his hammer. The one melts, the other but breaks in pieces.

When the ground was partially bare of snow, and a few warm days had dried its surface somewhat, it was pleasant to compare the first tender signs of the infant year just peeping forth with the stately beauty of the withered vegetation which had withstood the winter,—life-everlasting, golden-rods, pinweeds, and graceful wild grasses, more obvious and interesting frequently than in summer even, as if their beauty was not ripe till then; even cotton-grass, cat-tails, mulleins, johnswort, hard-hack, meadow-sweet, and other strong stemmed plants, those unexhausted granaries which entertain the earliest birds,—decent weeds,* at least, which widowed Nature wears. I am particularly attracted by the arching and sheaf-like top of the wool-grass; it brings back the summer to our winter memories, and is among the forms which art loves to copy, and which, in the vegetable kingdom, have the same relation to types* already in the mind of man that astronomy has. It is an antique style older than Greek or Egyptian. Many of the phenomena of Winter are suggestive of an inexpressible tenderness and fragile delicacy. We are accustomed to hear this king described as a rude and boisterous tyrant; but with the gentleness of a lover he adorns the tresses of Summer.

At the approach of spring the red-squirrels got under my house, two at a time, directly under my feet as I sat reading or writing, and kept up the queerest chuckling and chirruping and vocal pirouetting and gurgling sounds that ever were heard; and when I stamped they only chirruped the louder, as if past all fear and respect in their mad pranks, defying humanity to stop them. No you don't—chickaree—chickaree. They were wholly deaf to my arguments, or failed to perceive their force, and fell into a strain of invective that was irresistible.

The first sparrow of spring! The year beginning with younger hope than ever! The faint silvery warblings heard over the partially bare and moist fields from the blue-bird, the song-sparrow, and the red-wing, as if the last flakes of winter

tinkled as they fell! What at such a time are histories, chrono-
logies, traditions, and all written revelations? The brooks sing
carols and glees to the spring. The marsh-hawk sailing low
over the meadow is already seeking the first slimy life that
awakes. The sinking sound of melting snow is heard in all
dells, and the ice dissolves apace in the ponds. The grass
flames up on the hillsides like a spring fire,—"et primitus
oritur herba imbribus primoribus evocata,"*—as if the earth
sent forth an inward heat to greet the returning sun; not
yellow but green is the color of its flame;—the symbol of
perpetual youth, the grass-blade, like a long green ribbon,
streams from the sod into the summer, checked indeed by the
frost, but anon pushing on again, lifting its spear of last year's
hay with the fresh life below. It grows as steadily as the rill
oozes out of the ground. It is almost identical with that, for in
the growing days of June, when the rills are dry, the grass
blades are their channels, and from year to year the herds
drink at this perennial green stream, and the mower draws
from it betimes their winter supply. So our human life but
dies down to its root, and still puts forth its green blade to
eternity.

Walden is melting apace. There is a canal two rods wide
along the northerly and westerly sides, and wider still at the
east end. A great field of ice has cracked off from the main
body. I hear a song-sparrow singing from the bushes on the
shore,—*olit, olit, olit,—chip, chip, chip, che char,—che wiss,
wiss, wiss.* He too is helping to crack it. How handsome the
great sweeping curves in the edge of the ice, answering somewhat to those of the shore, but more regular! It is unusually
hard, owing to the recent severe but transient cold, and all
watered or waved like a palace floor.* But the wind slides
eastward over its opaque surface in vain, till it reaches the
living surface beyond. It is glorious to behold this ribbon of
water sparkling in the sun, the bare face of the pond full of
glee and youth, as if it spoke the joy of the fishes within it, and
of the sands on its shore,—a silvery sheen as from the scales of

a *leuciscus*, as it were all one active fish. Such is the contrast between winter and spring. Walden was dead and is alive again.* But this spring it broke up more steadily, as I have said.

The change from storm and winter to serene and mild weather, from dark and sluggish hours to bright and elastic ones, is a memorable crisis which all things proclaim. It is seemingly instantaneous at last. Suddenly an influx of light filled my house, though the evening was at hand, and the clouds of winter still overhung it, and the eaves were dripping with sleety rain. I looked out the window, and lo! where yesterday was cold gray ice there lay the transparent pond already calm and full of hope as in a summer evening, reflecting a summer evening sky in its bosom, though none was visible overhead, as if it had intelligence with some remote horizon. I heard a robin in the distance, the first I had heard for many a thousand years, me thought, whose note I shall not forget for many a thousand more,—the same sweet and powerful song as of yore. O the evening robin, at the end of a New England summer day! If I could ever find the twig he sits upon! I mean *he*; I mean *the twig*. This at least is not the *Turdus migratorius*. The pitch-pines and shrub-oaks about my house, which had so long drooped, suddenly resumed their several characters, looked brighter, greener, and more erect and alive, as if effectually cleansed and restored by the rain. I knew that it would not rain any more. You may tell by looking at any twig of the forest, ay, at your very wood-pile, whether its winter is past or not. As it grew darker, I was startled by the *honking* of geese flying low over the woods, like weary travellers getting in late from southern lakes, and indulging at last in unrestrained complaint and mutual consolation. Standing at my door, I could hear the rush of their wings; when, driving toward my house, they suddenly spied my light, and with hushed clamor wheeled and settled in the pond. So I came in, and shut the door, and passed my first spring night in the woods.

In the morning I watched the geese from the door through the mist, sailing in the middle of the pond, fifty rods off, so large and tumultuous that Walden appeared like an artificial pond for their amusement. But when I stood on the shore they at once rose up with a great flapping of wings at the signal of their commander, and when they had got into rank circled about over my head, twenty-nine of them, and then steered straight to Canada, with a regular *honk* from the leader at intervals, trusting to break their fast in muddier pools. A "plump"* of ducks rose at the same time and took the route to the north in the wake of their noisier cousins.

For a week I heard the circling groping clangor of some solitary goose in the foggy mornings, seeking its companion, and still peopling the woods with the sound of a larger life than they could sustain. In April the pigeons were seen again flying express in small flocks, and in due time I heard the martins twittering over my clearing, though it had not seemed that the township contained so many that it could afford me any,* and I fancied that they were peculiarly of the ancient race that dwelt in hollow trees ere white men came. In almost all climes the tortoise and the frog are among the precursors and heralds of this season, and birds fly with song and glancing plumage, and plants spring and bloom, and winds blow, to correct this slight oscillation of the poles and preserve the equilibrium of Nature.

As every season seems best to us in its turn, so the coming in of spring is like the creation of Cosmos out of Chaos and the realization of the Golden Age.*—

> "Eurus ad Auroram, Nabathacaque* regna recessit,
> Persidaque, et radiis juga subdita matutinis."

"The East-Wind withdrew to Aurora* and the Nabathæan kingdom,*
And the Persian, and the ridges* placed under the morning rays.

    °   °   °   °

Man was born. Whether that Artificer of things,
The origin of a better world, made him from the divine seed;

Or the earth being recent and lately sundered from the high
Ether, retained some seeds of cognate heaven."

A single gentle rain makes the grass many shades greener.
So our prospects brighten on the influx of better thoughts. We
should be blessed if we lived in the present always, and took
advantage of every accident that befell us, like the grass which
confesses the influence of the slightest dew that falls on it; and
did not spend our time in atoning for the neglect of past
opportunities, which we call doing our duty. We loiter in
winter while it is already spring. In a pleasant spring morning
all men's sins are forgiven. Such a day is a truce to vice. While
such a sun holds out to burn, the vilest sinner may return.*
Through our own recovered innocence we discern the innocence of our neighbors. You may have known your neighbor
yesterday for a thief, a drunkard, or a sensualist, and merely
pitied or despised him, and despaired of the world; but the
sun shines bright and warm this first spring morning, re-
creating the world, and you meet him at some serene work,
and see how his exhausted and debauched veins expand with
still joy and bless the new day, feel the spring influence with
the innocence of infancy, and all his faults are forgotten.
There is not only an atmosphere of good will about him, but
even a savor of holiness groping for expression, blindly and
ineffectually perhaps, like a new-born instinct, and for a short
hour the south hill-side echoes to no vulgar jest. You see some
innocent fair shoots preparing to burst from his gnarled rind
and try another year's life, tender and fresh as the youngest
plant. Even he has entered into the joy of his Lord.* Why the
jailer does not leave open his prison doors,—why the judge
does not dismiss his case,—why the preacher does not dismiss
his congregation! It is because they do not obey the hint which
God gives them, nor accept the pardon which he freely offers
to all.

"A return to goodness produced each day in the tranquil
and beneficent breath of the morning, causes that in respect

to the love of virtue and the hatred of vice, one approaches a little the primitive nature of man, as the sprouts of the forest which has been felled. In like manner the evil which one does in the interval of a day prevents the germs of virtues which began to spring up again from developing themselves and destroys them.

"After the germs of virtue have thus been prevented many times from developing themselves, then the beneficent breath of evening does not suffice to preserve them. As soon as the breath of evening does not suffice longer to preserve them, then the nature of man does not differ much from that of the brute. Men seeing the nature of this man like that of the brute, think that he has never possessed the innate faculty of reason. Are those the true and natural sentiments of man?"*

"The Golden Age was first created, which without any avenger
Spontaneously without law cherished fidelity and rectitude.
Punishment and fear were not; nor were threatening words read
On suspended brass; nor did the suppliant crowd fear
The words of their judge; but were safe without an avenger.
Not yet the pine felled on its mountains had descended
To the liquid waves that it might see a foreign world,
And mortals knew no shores but their own.

There was eternal spring, and placid zephyrs with warm
Blasts soothed the flowers born without seed."*

On the 29th of April, as I was fishing from the bank of the river near the Nine-Acre-Corner bridge,* standing on the quaking grass and willow roots, where the muskrats lurk, I heard a singular rattling sound, somewhat like that of the sticks which boys play with their fingers,* when, looking up, I observed a very slight and graceful hawk, like a night-hawk, alternately soaring like a ripple and tumbling a rod or two over and over, showing the underside of its wings, which gleamed like a satin ribbon in the sun, or like the pearly inside of a shell. This sight reminded me of falconry and what nobleness and poetry are associated with that sport. The Merlin it seemed to me it might be called: but I care not for its name. It was the most ethereal flight I had ever witnessed. It did not simply flutter like a butterfly, nor soar like the larger hawks, but it sported with proud reliance in the fields of air; mounting again and again with its strange chuckle, it repeated its free and beautiful fall, turning over and over like a kite, and then recovering from its lofty tumbling, as if it had never set its foot on *terra firma*. It appeared to have no companion in the universe,—sporting there alone,—and to need none but the morning and the ether with which it played. It was not lonely, but made all the earth lonely beneath it. Where was the parent which hatched it, its kindred, and its father in the heavens? The tenant of the air, it seemed related to the earth but by an egg hatched some time in the crevice of a crag;—or was its native nest made in the angle of a cloud, woven of the rainbow's trimmings and the sunset sky, and lined with some soft midsummer haze caught up from earth? Its eyry now some cliffy cloud.

Beside this I got a rare mess of golden and silver and bright cupreous* fishes, which looked like a string of jewels. Ah! I have penetrated to those meadows on the morning of many a first spring day, jumping from hummock to hummock, from willow root to willow root, when the wild river valley and the woods were bathed in so pure and bright a light as would have waked the dead, if they had been slumbering in their graves, as some suppose. There needs no stronger proof of immortality. All things must live in such a light. O Death, where was thy sting? O Grave, where was thy victory, then?*

Our village life would stagnate if it were not for the unexplored forests and meadows which surround it. We need the tonic of wildness,—to wade sometimes in marshes where the bittern and the meadow-hen lurk, and hear the booming of the snipe; to smell the whispering sedge where only some wilder and more solitary fowl builds her nest, and the mink crawls with its belly close to the ground. At the same time that we are earnest to explore and learn all things, we require that all things be mysterious and unexplorable, that land and sea be infinitely wild, unsurveyed and unfathomed by us because unfathomable. We can never have enough of Nature. We must be refreshed by the sight of inexhaustible vigor, vast and Titanic features, the sea-coast with its wrecks, the wilderness with its living and its decaying trees, the thunder cloud, and the rain which lasts three weeks and produces freshets. We need to witness our own limits transgressed, and some life pasturing freely where we never wander. We are cheered when we observe the vulture feeding on the carrion which disgusts and disheartens us and deriving health and strength from the repast. There was a dead horse in the hollow by the path to my house, which compelled me sometimes to go out of my way, especially in the night when the air was heavy, but the assurance it gave me of the strong appetite and inviolable health of Nature was my compensation for this. I love to see that Nature is so rife with life that myriads can be afforded to be sacrificed and suffered to prey on one another; that tender organizations can be so serenely squashed out of existence like pulp,—tadpoles which herons gobble up, and tortoises and toads run over in the road; and that sometimes it has rained flesh and blood! With the liability to accident, we must see how little account is to be made of it. The impression made on a wise man is that of universal innocence. Poison is not poisonous after all, nor are any wounds fatal. Compassion is a very untenable ground. It must be expeditious. Its pleadings will not bear to be stereotyped.

Early in May, the oaks, hickories, maples, and other trees, just putting out amidst the pine woods around the pond, imparted a brightness like sunshine to the landscape, especially in cloudy days, as if the sun were breaking through mists and shining faintly on the hill-sides here and there. On the third or fourth of May I saw a loon in the pond, and during the first week of the month I heard the whippoorwill, the brown-thrasher, the veery, the wood-pewee, the chewink, and other birds. I had heard the wood-thrush long before. The phoebe had already come once more and looked in at my door and window, to see if my house was cavern-like enough for her, sustaining herself on humming wings with clinched talons, as if she held by the air, while she surveyed the premises. The sulphur-like pollen of the pitch-pine soon covered the pond and the stones and rotten wood along the shore, so that you could have collected a barrel-ful. This is the "sulphur showers" we hear of. Even in Calidas' drama of Sacontala,* we read of "rills dyed yellow with the golden dust of the lotus." And so the seasons went rolling on into summer, as one rambles into higher and higher grass.

Thus was my first year's life in the woods completed; and the second year was similar to it. I finally left Walden September 6th, 1847.

CONCLUSION

————————

To the sick the doctors wisely recommend a change of air and scenery. Thank Heaven, here is not all the world. The buck-eye does not grow in New England, and the mocking-bird is rarely heard here. The wild-goose is more of a cosmopolite than we; he breaks his fast in Canada, takes a luncheon in the Ohio, and plumes himself for the night in a southern bayou. Even the bison, to some extent, keeps pace with the seasons, cropping the pastures of the Colorado only till a greener and sweeter grass awaits him by the Yellowstone. Yet we think that if rail-fences are pulled down, and stone-walls* piled up on our farms, bounds are henceforth set to our lives and our fates decided. If you are chosen town-clerk,* forsooth, you cannot go to Tierra del Fuego* this summer: but you may go to the land of infernal fire nevertheless. The universe is wider than our views of it.

Yet we should oftener look over the tafferel* of our craft, like curious passengers, and not make the voyage like stupid sailors picking oakum.* The other side of the globe is but the home of our correspondent. Our voyaging is only great-circle sailing,* and the doctors prescribe for diseases of the skin merely. One hastens to Southern Africa to chase the giraffe; but surely that is not the game he would be after. How long, pray, would a man hunt giraffes if he could? Snipes and woodcocks also may afford rare sport; but I trust it would be nobler game to shoot one's self.——

> "Direct your eye right inward, and you'll find
> A thousand regions in your mind
> Yet undiscovered. Travel them, and be
> Expert in home-cosmography."*

What does Africa,—what does the West stand for? Is not our own interior white on the chart?* black though it may prove, like the coast, when discovered. Is it the source of the Nile, or the Niger, or the Mississippi, or a North-West Passage* around this continent, that we would find? Are these the problems which most concern mankind? Is Franklin* the only man who is lost, that his wife should be so earnest to find him? Does Mr. Grinnell* know where he himself is? Be rather the Mungo Park,* the Lewis and Clarke* and Frobisher,* of your own streams and oceans; explore your own higher latitudes,— with shiploads of preserved meats* to support you, if they be necessary; and pile the empty cans sky-high for a sign. Were preserved meats invented to preserve meat merely? Nay, be a Columbus to whole new continents and worlds within you, opening new channels, not of trade, but of thought. Every man is the lord of a realm beside which the earthly empire of the Czar is but a petty state, a hummock* left by the ice. Yet some can be patriotic who have no *self*-respect, and sacrifice the greater to the less. They love the soil which makes their graves, but have no sympathy with the spirit which may still animate their clay. Patriotism is a maggot in their heads. What was the meaning of that South-Sea Exploring Expedition,* with all its parade and expense, but an indirect recognition of the fact, that there are continents and seas in the moral world, to which every man is an isthmus or an inlet, yet unexplored by him, but that it is easier to sail many thousand miles through cold and storm and cannibals, in a government ship, with five hundred men and boys to assist one, than it is to explore the private sea, the Atlantic and Pacific Ocean of one's being alone.—

> "Erret, et extremos alter scrutetur Iberos.
> Plus habet hic vitæ, plus habet ille viae."*

Let them wander and scrutinize the outlandish Australians.
I have more of God, they more of the road.

It is not worth the while to go round the world to count the cats in Zanzibar.* Yet do this even till you can do better, and you may perhaps find some "Symmes' Hole"* by which to get at the inside at last. England and France, Spain and Portugal, Gold Coast and Slave Coast, all front on this private sea; but no bark* from them has ventured out of sight of land, though it is without doubt the direct way to India. If you would learn to speak all tongues and conform to the customs of all nations, if you would travel farther than all travellers, be naturalized in all climes, and cause the Sphinx to dash her head against a stone,* even obey the precept of the old philosopher, and Explore thyself.* Herein are demanded the eye and the nerve. Only the defeated and deserters go to the wars, cowards that run away and enlist. Start now on that farthest western way,* which does not pause at the Mississippi or the Pacific, nor conduct toward a worn-out China or Japan, but leads on direct a tangent to this sphere, summer and winter, day and night, sun down, moon down, and at last earth down too.

It is said that Mirabeau* took to highway robbery "to ascertain what degree of resolution was necessary in order to place one's self in formal opposition to the most sacred laws of society." He declared that a "soldier who fights in the ranks does not require half so much courage as a foot-pad,"*—"that honor and religion have never stood in the way of a well-considered and a firm resolve." This was manly, as the world goes; and yet it was idle, if not desperate. A saner man would have found himself often enough "in formal opposition" to what are deemed "the most sacred laws of society," through obedience to yet more sacred laws, and so have tested his resolution without going out of his way. It is not for a man to put himself in such an attitude to society, but to maintain himself in whatever attitude he find himself through obedience to the laws of his being, which will never be one of opposition to a just government, if he should chance to meet with such.

I left the woods for as good a reason* as I went there. Perhaps it seemed to me that I had several more lives to live, and could not spare any more time for that one. It is remarkable how easily and insensibly we fall into a particular route, and make a beaten track for ourselves. I had not lived there a week before my feet wore a path from my door to the pond-side; and though it is five or six years since I trod it, it is still quite distinct. It is true, I fear that others may have fallen into it, and so helped to keep it open. The surface of the earth is soft and impressible by the feet of men; and so with the paths which the mind travels. How worn and dusty, then, must be the highways of the world, how deep the ruts of tradition and conformity! I did not wish to take a cabin passage, but rather to go before the mast and on the deck of the world, for there I could best see the moonlight amid the mountains.* I do not wish to go below now.*

I learned this, at least, by my experiment; that if one advances confidently in the direction of his dreams, and endeavors to live the life which he has imagined, he will meet with a success unexpected in common hours. He will put some things behind, will pass an invisible boundary; new, universal, and more liberal laws will begin to establish themselves around and within him; or the old laws be expanded, and interpreted in his favor in a more liberal sense, and he will live with the license of a higher order of beings. In proportion as he simplifies his life, the laws of the universe will appear less complex, and solitude will not be solitude, nor poverty poverty, nor weakness weakness. If you have built castles in the air, your work need not be lost; that is where they should be. Now put the foundations under them.

It is a ridiculous demand which England and America make, that you shall speak so that they can understand you. Neither men nor toad-stools grow so. As if that were important, and there were not enough to understand you without them. As if Nature could support but one order of understandings, could not sustain birds as well as quadrupeds, flying as well as creeping things, and *hush* and *who*, which Bright* can understand, were the best English. As if there were safety in stupidity alone. I fear chiefly lest my expression may not be *extra-vagant** enough, may not wander far enough beyond the narrow limits of my daily experience, so as to be adequate to the truth of which I have been convinced. *Extra vagance!* it depends on how you are yarded.* The migrating buffalo, which seeks new pastures in another latitude, is not extravagant like the cow which kicks over the pail, leaps the cow-yard fence, and runs after her calf, in milking time. I desire to speak somewhere *without* bounds; like a man in a waking moment, to men in their waking moments; for I am convinced that I cannot exaggerate enough even to lay the foundation of a true expression. Who that has heard a strain of music feared then lest he should speak extravagantly any more forever? In view of the future or possible, we should live quite laxly and undefined in front, our outlines dim and misty on that side; as our shadows reveal an insensible perspiration toward the sun. The volatile truth of our words should continually betray the inadequacy of the residual statement. Their truth is instantly *translated*;* its literal monument alone remains. The words which express our faith and piety are not definite; yet they are significant and fragrant like frankincense to superior natures.

Why level downward to our dullest perception always, and praise that as common sense? The commonest sense is the sense of men asleep, which they express by snoring. Sometimes we are inclined to class those who are once-and-a-half witted with the half-witted, because we appreciate only a third part of their wit. Some would find fault with the morning-red, if they ever got up early enough. "They pretend," as I hear, "that the verses of Kabir have four different senses; illusion, spirit, intellect, and the exoteric doctrine of the Vedas;"* but in this part of the world it is considered a ground for complaint if a man's writings admit of more than one interpretation. While England endeavors to cure the potato-rot,* will not any endeavor to cure the brain-rot, which prevails so much more widely and fatally?

I do not suppose that I have attained to obscurity, but I

should be proud if no more fatal fault were found with my pages on this score than was found with the Walden ice. Southern customers objected to its blue color, which is the evidence of its purity, as if it were muddy, and preferred the Cambridge ice,* which is white, but tastes of weeds. The purity men love is like the mists which envelop the earth, and not like the azure ether beyond.

Some are dinning in our ears that we Americans, and moderns generally, are intellectual dwarfs compared with the ancients, or even the Elizabethan men. But what is that to the purpose? A living dog is better than a dead lion.* Shall a man go and hang himself because he belongs to the race of pygmies, and not be the biggest pygmy that he can? Let every one mind his own business, and endeavor to be what he was made.

Why should we be in such desperate haste to succeed, and in such desperate enterprises? If a man does not keep pace with his companions, perhaps it is because he hears a different drummer. Let him step to the music which he hears, however measured or far away. It is not important that he should mature as soon as an apple-tree or an oak. Shall he turn his spring into summer? If the condition of things which we were made for is not yet, what were any reality which we can substitute? We will not be shipwrecked on a vain reality. Shall we with pains erect a heaven of blue glass over ourselves, though when it is done we shall be sure to gaze still at the true ethereal heaven far above, as if the former were not?

There was an artist in the city of Kouroo* who was disposed to strive after perfection. One day it came into his mind to make a staff. Having considered that in an imperfect work time is an ingredient, but into a perfect work time does not enter, he said to himself, It shall be perfect in all respects, though I should do nothing else in my life. He proceeded instantly to the forest for wood, being resolved that it should not be made of unsuitable material; and as he searched for and rejected stick after stick, his friends gradually deserted him, for they grew old in their works and died, but he grew not older by a moment. His singleness of purpose and resolution, and his elevated piety, endowed him, without his knowledge, with perennial youth. As he made no compromise with Time, Time kept out of his way, and only sighed at a distance because he could not overcome him. Before he had found a stock in all respects suitable the city of Kouroo was a hoary ruin, and he sat on one of its mounds to peel the stick. Before he had given it the proper shape the dynasty of the Candahars was at an end, and with the point of the stick he wrote the name of the last of that race in the sand, and then resumed his work. By the time he had smoothed and polished the staff Kalpa* was no longer the pole-star; and ere he had put on the ferule and the head adorned with precious stones, Brahma had awoke* and slumbered many times. But why do I stay to mention these things? When the finishing stroke was put to his work, it suddenly expanded before the eyes of the astonished artist into the fairest of all the creations of Brahma. He had made a new system in making a staff, a world with full and fair proportions; in which, though the old cities and dynasties had passed away, fairer and more glorious ones had taken their places. And now he saw by the heap of shavings still fresh at his feet, that, for him and his work, the former lapse of time had been an illusion, and that no more time had elapsed than is required for a single scintillation from the brain of Brahma to fall on and inflame the tinder of a mortal brain. The material was pure, and his art was pure; how could the result be other than wonderful?

No face which we can give to a matter will stead us so well at last as the truth. This alone wears well. For the most part, we are not where we are, but in a false position. Through an infirmity of our natures, we suppose a case, and put ourselves into it, and hence are in two cases at the same time, and it is doubly difficult to get out. In sane moments we regard only the facts, the case that is. Say what you have to say, not what you ought. Any truth is better than make-believe. Tom

Hyde,* the tinker, standing on the gallows, was asked if he had any thing to say. "Tell the tailors," said he, "to remember to make a knot in their thread before they take the first stitch." His companion's prayer is forgotten.

However mean your life is, meet it and live it; do not shun it and call it hard names. It is not so bad as you are. It looks poorest when you are richest. The fault-finder will find faults even in paradise. Love your life, poor as it is. You may perhaps have some pleasant, thrilling, glorious hours, even in a poorhouse. The setting sun is reflected from the windows of the alms-house as brightly as from the rich man's abode; the snow melts before its door as early in the spring. I do not see but a quiet mind may live as contentedly there, and have as cheering thoughts, as in a palace. The town's poor seem to me often to live the most independent lives of any. May be they are simply great enough to receive without misgiving. Most think that they are above being supported by the town; but it oftener happens that they are not above supporting themselves by dishonest means, which should be more disreputable. Cultivate poverty like a garden herb, like sage. Do not trouble yourself much to get new things, whether clothes or friends. Turn the old; return to them. Things do not change; we change. Sell your clothes and keep your thoughts. God will see that you do not want society. If I were confined to a corner of a garret all my days, like a spider, the world would be just as large to me while I had my thoughts about me. The philosopher said: "From an army of three divisions one can take away its general, and put it in disorder; from the man the most abject and vulgar one cannot take away his thought."* Do not seek so anxiously to be developed, to subject yourself to many influences to be played on; it is all dissipation. Humility like darkness reveals the heavenly lights. The shadows of poverty and meanness gather around us, "and lo! creation widens to our view."* We are often reminded that if there were bestowed on us the wealth of Croesus,* our aims must still be the same, and our means essentially the same. Moreover, if you are restricted in your range by poverty, if you cannot buy books and newspapers, for instance, you are but confined to the most significant and vital experiences; you are compelled to deal with the material which yields the most sugar and the most starch. It is life near the bone where it is sweetest. You are defended from being a trifler. No man loses ever on a lower level by magnanimity on a higher. Superfluous wealth can buy superfluities only. Money is not required to buy one necessary of the soul.

I live in the angle of a leaden wall, into whose composition was poured a little alloy of bell metal. Often, in the repose of my mid-day, there reaches my ears a confused *tintin-nabulum** from without. It is the noise of my contemporaries. My neighbors tell me of their adventures with famous gentlemen and ladies, what notabilities they met at the dinner-table; but I am no more interested in such things than in the contents of the Daily Times. The interest and the conversation are about costume and manners chiefly; but a goose is a goose still, dress it as you will. They tell me of California and Texas, of England and the Indies, of the Hon. Mr. —— of Georgia or of Massachusetts,* all transient and fleeting phenomena, till I am ready to leap from their court-yard like the Mameluke bey.* I delight to come to my bearings,—not walk in procession with pomp and parade, in a conspicuous place, but to walk even with the Builder of the universe, if I may,—not to live in this restless, nervous, bustling, trivial Nineteenth Century, but stand or sit thoughtfully while it goes by. What are men celebrating? They are all on a committee of arrangements, and hourly expect a speech from somebody. God is only the president of the day, and Webster* is his orator. I love to weigh, to settle, to gravitate toward that which most strongly and rightfully attracts me;—not hang by the beam of the scale and try to weigh less,—not suppose a case, but take the case that is; to travel the only path I can, and that on which no power can resist me. It affords me no satisfaction to commence to spring an arch before I have got a solid foundation.

Let us not play at kittlybenders.* There is a solid bottom every where. We read that the traveller asked the boy if the swamp* before him had a solid bottom. The boy replied that it had. But presently the traveller's horse sank in up to the girths, and he observed to the boy, "I thought you said that this bog had a hard bottom." "So it has," answered the latter, "but you have not got half way to it yet." So it is with the bogs and the quicksands of society; but he is an old boy that knows it. Only what is thought said or done at a certain rare coincidence is good. I would not be one of those who will foolishly drive a nail into mere lath and plastering; such a deed would keep me awake nights. Give me a hammer, and let me feel for the furrowing.* Do not depend on the putty. Drive a nail home and clinch it so faithfully that you can wake up in the night and think of your work with satisfaction,—a work at which you would not be ashamed to invoke the Muse.* So will help you God, and so only. Every nail driven should be as another rivet in the machine of the universe, you carrying on the work.

Rather than love, than money, than fame, give me truth. I sat at a table where were rich food and wine in abundance, and obsequious attendance, but sincerity and truth were not; and I went away hungry from the inhospitable board. The hospitality was as cold as the ices. I thought that there was no need of ice to freeze them. They talked to me of the age of the wine and the fame of the vintage; but I thought of an older, a newer, and purer wine, of a more glorious vintage, which they had not got, and could not buy. The style, the house and grounds and "entertainment" pass for nothing with me. I called on the king, but he made me wait in his hall, and conducted like a man incapacitated for hospitality. There was a man in my neighborhood who lived in a hollow tree.* His manners were truly regal. I should have done better had I called on him.

How long shall we sit in our porticoes practising idle and musty virtues, which any work would make impertinent? As if

one were to begin the day with long-suffering, and hire a man
to hoe his potatoes; and in the afternoon go forth to practise
Christian meekness and charity with goodness aforethought!
Consider the China pride and stagnant self-complacency of
mankind. This generation reclines a little to congratulate itself
on being the last of an illustrious line; and in Boston and
London and Paris and Rome, thinking of its long descent, it
speaks of its progress in art and science and literature with
satisfaction. There are the Records of the Philosophical Societies, and the public Eulogies of *Great Men!* It is the good
Adam contemplating his own virtue. "Yes, we have done great
deeds, and sung divine songs, which shall never die,"—that is,
as long as *we* can remember them. The learned societies and
great men of Assyria,—where are they? What youthful philosophers and experimentalists we are! There is not one of my
readers who has yet lived a whole human life. These may be
but the spring months in the life of the race. If we have had
the seven-years' itch,* we have not seen the seventeen-year
locust yet in Concord. We are acquainted with a mere pellicle* of the globe on which we live. Most have not delved six
feet beneath the surface, nor leaped as many above it. We
know not where we are. Beside, we are sound asleep nearly
half our time. Yet we esteem ourselves wise, and have an
established order on the surface. Truly, we are deep thinkers,
we are ambitious spirits! As I stand over the insect crawling
amid the pine needles on the forest floor, and endeavoring to
conceal itself from my sight, and ask myself why it will cherish
those humble thoughts, and hide its head from me who might,
perhaps, be its benefactor, and impart to its race some cheering information, I am reminded of the greater Benefactor and
Intelligence that stands over me the human insect.

There is an incessant influx of novelty into the world, and
yet we tolerate incredible dulness. I need only suggest what
kind of sermons are still listened to in the most enlightened
countries. There are such words as joy and sorrow, but they
are only the burden of a psalm, sung with a nasal twang, while

we believe in the ordinary and mean. We think that we can change our clothes only. It is said that the British Empire is very large and respectable, and that the United States are a first-rate power. We do not believe that a tide rises and falls behind every man which can float the British Empire like a chip, if he should ever harbor it in his mind. Who knows what sort of seventeen-year locust will next come out of the ground? The government of the world I live in was not framed, like that of Britain, in after-dinner conversations over the wine.

The life in us is like the water in the river. It may rise this year higher than man has ever known it, and flood the parched uplands; even this may be the eventful year, which will drown out all our muskrats.* It was not always dry land where we dwell. I see far inland the banks which the stream anciently washed, before science began to record its freshets. Every one has heard the story which has gone the rounds of New England, of a strong and beautiful bug which came out of the dry leaf of an old table* of apple-tree wood, which had stood in a farmer's kitchen for sixty years, first in Connecticut, and afterward in Massachusetts,—from an egg deposited in the living tree many years earlier still, as appeared by counting the annual layers beyond it; which was heard gnawing out for several weeks, hatched perchance by the heat of an urn. Who does not feel his faith in a resurrection and immortality strengthened by hearing of this? Who knows what beautiful and winged life, whose egg has been buried for ages under many concentric layers of woodenness in the dead dry life of society, deposited at first in the alburnum* of the green and living tree, which has been gradually converted into the semblance of its well-seasoned tomb,—heard perchance gnawing out now for years by the astonished family of man, as they sat round the festive board,—may unexpectedly come forth from amidst society's most trivial and handselled* furniture, to enjoy its perfect summer life at last!

I do not say that John or Jonathan* will realize all this; but such is the character of that morrow which mere lapse of time can never make to dawn. The light which puts out our eyes is darkness to us. Only that day dawns to which we are awake. There is more day to dawn. The sun is but a morning star.

THE END

APPENDIX

WALDEN'S ANIMALS AND VEGETABLES

The purpose of this appendix is to help the reader to identify as many as possible of the plants and animals to be found in *Walden*. The terms have been alphabetized; otherwise they appear as in the book, followed by their scientific (Latin) names, their common (English) names, and, where relevant, a brief description. The Latin is in the order of *Genus* and *species*—not as T himself would have known them (since over a century of botanical and zoological research has now produced different designations for most of the genera and many of the species he noted), but as they might be found now in a good popular field guide to American flora and fauna, such as the series published by Harper & Row, or the National Audubon Society.

The reader may wish to consult one or more of the following books, from which most of the information here has been taken.

American Fisheries Society, *A List of Common and Scientific Names of Fishes from the United States and Canada* (Ann Arbor, 1959).

Ray Angelo, *Botanical Index to the Journal of Henry David Thoreau*, companion volume to the 14-vol. edition of Thoreau's *Journal*, ed. Bradford Torrey and Francis J. Allen (Salt Lake City: Gibbs M. Smith, Inc., Peregrine Smith Books, 1984).

Helen Cruickshank [Cruick.], *Thoreau on Birds: Selections from his Writing, Compiled and with Commentary* (New York: McGraw Hill, 1964).

Richard Jefferson Eaton [Eaton], *A Flora of Concord: An Account of the Flowering Plants, Ferns and Fern-Allies Known to Have Occurred Without Cultivation in Concord, Massachusetts from Thoreau's Time to the Present Day* (Cambridge, Mass.: Special Publication No. 4, The Museum of Comparative Zoology, Harvard University, 1974).

[Harper & Row], *Complete Field Guide to North American Wildlife: Eastern Edition* (New York: Harper & Row, 1981).

Richard Howard and Alick Moore (eds.), *A Complete Checklist of Birds of the World*, 2nd edn. (London/New York: Academic Press/ Harcourt Brace Jovanovich, 1991).

National Audubon Society, *Field Guide to North American Fishes, Whales & Dolphins* (New York: Knopf, 1995).

—— *Field Guide to North American Mammals* (New York: Knopf, 1995).

Alwyne Wheeler, *Fishes of the World: An Illustrated Dictionary* (London: Ferndale Editions, 1975).

Index of Flora and Fauna Mentioned in *Walden*

Alder (*Alnus* spp.): possibly *A. serrulata*, the common alder, found on pond margins and riverbanks.

Alder-berry (probably the berry of *Ilex verticillata*): the black alder, or common winterberry.

Angle-worms: American usage for earth worms; any of the round, segmented, burrowing worms, of which the most common sp. is *Lumbricus terrestris*; so-called because anglers use them for bait.

Ant . . . black; Ant . . . red: there are something over 580 spp. of ant in North America; conflict of the sort described by T on pp. 206–9 can take place between two spp. of the same genus, as—for illustration only, not identification—*Lasius niger*, the large black garden ant and *Lasius flavus*, the yellow meadow ant, both of which conform to T's description.

Barberry (*Berberis vulgaris*): the European barberry, the English pipperidge bush.

Barred owl (*Strix varia*): about 18 in. long; has a dark round head without 'horns'; not more than a single pair nest in the Concord area today (Cruick.).

Bass (*Tilia americana*): the American basswood, a linden, or lime tree. T distinguished between this, 'bass' or 'basswood', and the European basswood (*Tilia europaea*), which he called 'linden'.

Bear (*Ursus americanus*): the American black bear; found formerly throughout the eastern half of North America except northern Canada; vanished from Concord area by T's time.

Beech (*Fagus grandifolia*): the American beech.

Beggar-ticks (*Bidens* spp.): the prickly one-seeded fruits of the bur marigold.

Birch (*Betula* spp.): see also black-birch; yellow-birch.

Bison (*Bison bison*): the American bison, or (in the vernacular, buffalo); by then chiefly found on the Great Plains.

Bittern (*Botaurus lentiginosus*): the American bittern, a nocturnal heron, at home in fresh or salt bogs and marshes.

Black ash (*Fraxinus nigra*): quite common.

Black-birch (*Betula lenta*).

Black-spruce (*Picea mariana*): or double spruce; see also Single spruce.

Blackberry (*Rubus* spp.): common bramble.

Blue flag (*Iris versicolor*): a water plant.

Blue-bird (*Sialus sialus*): the Eastern bluebird, an early spring migrant.

Blue-jay (*Cyanocitta cristata*): a cry as distinctive as its appearance; still quite common in the Concord area.

Blueberry (*Vaccinium* spp.): (probably V. *corymbosum* or V. *atrococcum*), common; likes bogs, swamps, shores, or even dry uplands.

Brake (*Pteridium aquilinum*): bracken.

Bream (*Lepomis* spp.): one of the American freshwater sunfish, possibly L. *macrochirus*, the bluegill.

Brown-thrasher (*Toxotoma rufum*): also called red mavis; an abundant nesting bird in T's day, much depleted since the late nineteenth century.

Buck eye (*Aesculus* spp.): the American horse chestnut; commonly grows further south and west than New England.

Bullfrog (*Rana catesbeiana*): a large sp. of frog found in the US; dusky brown colour mixed with yellowish green and spotted with black.

Bulrush (*Scirpus* spp.).

Buttonwoods (*Platanus occidentalis*): the American sycamore.

Cadis worm: or codworm, the larva of the caddis fly (order *Trichoptera*, 40 families, 650 genera, 7,000 spp.) that lives in a silken case covered with bits of debris.

Caribou (*Rangifer tarandus*): the American reindeer; now inhabits arctic tundra and surrounding coniferous forest; in T's time more widely distributed in North America.

Cat (*Felis catus*): the domestic cat; descended from the wild cat of North Africa (F. *sylvestris libyca*); readily reverts to the wild.

Cat-owl (*Bubo virginianus*): more often referred to as the great horned owl.

Cat-tail (*Typha latifolia*): popular name for a strong, tall weed growing in or near water.

Cedar (*Juniperus virginiana*): the eastern red cedar.

Chestnut (*Castanea dentata*): the American chestnut.

Chewink (*Pipilo erythrophtalmus*): named for the sound of its call; a common eastern finch now known as the rufous-sided towhee. T also referred to it as that, and also as the 'ground robin'.

Chicadee (*Parus atricapillus*): the black-capped chicadee; very sociable; a species of titmouse. T often referred to them as titmice.

Chivin (*Semotilus corporalis*), local name for a chub.

Cinquefoil (*Potentilla* spp.): a common weed with compound leaves of five leaflets.

Common sweet flag (*Acorus calamus*): or calamus, a water plant.

Corn (*Zea mays*): maize, or Indian corn.

Cotton-grass (*Eriophorum* spp.).

Cranberry (*Vaccinium* spp.): probably either the large cranberry (*V. macrocarpon*) or the small cranberry (*V. oxycoccus*).

Creeping juniper (*Juniperus communis*): the ground juniper.

Dandelion (*Taraxacum officinale*): the common dandelion.

Deer (*Odocoileus* spp.): T is probably referring to *O. virginianus*, the white-tailed deer.

Deer mouse (*Peromyscus maniculatus*): has a body 3–4 in. in length, and a tail of from 2 to 5 in.

Dogwood (probably *Cornus florida*): the flowering dogwood.

Double spruce (*Picea mariana*): the native spruce in the Concord area.

Duck (*Anas* spp.): the context suggests one of several migrating ducks, such as *A. platyrhnchos*, the mallard, or *A. rubripes*, the North American black duck.

Eagle: either (*Haliaeetus leucocephalus*): the American bald eagle; now rare, and found mainly in central and southern US; or (more likely) (*Aquila chrysaetos*): the golden eagle which still breeds from northern Canada south to New York, and also in the Appalachians.

Eel (*Anguilla rostrata*): the American eel, found in both coastal seawater and freshwater; can grow up to 5 ft long.

False elm (*Celtis occidentalis*): the common hackberry.

Field sparrow (*Spizella pusilla*): a nesting bird in the Concord area; now less common than in T's day.

Fir-tree (probably *Abies balsama*): the balsam fir.

Fishhawk: no listing in the standard reference books; almost certainly one of the many *Buteo* spp.; hawk names vary widely according to local usage.

Fish-worms: i.e. a segmented worm of the phylum *Annelida*; see also Angle-worms.

Flag (probably one of the *Typha* spp.): cat-tail, or common cat-tail, qv.

Flying squirrel (*Glaucomys sabrinus*): the northern flying squirrel; flying squirrels do not fly, but glide on the flattened loose skin between extended fore and rear legs.

Fox (*Vulpes* spp.): probably *V. vulpes*, the red fox.

Frog: if one of the true frogs (*Rana* spp.), then possibly *R. clamitans*, the green frog, or *R. pipiens*, the northern leopard frog.

Gerbille family: strictly speaking, the *Gerbillinae* are a sub-family of the family *Muridae*, or old-world rats and mice: a family of small rodents with 15 genera.

Golden-rod (*Solidago* spp.): very common; there are 22 species in Middlesex County alone; on p. 230 T mentions the wand-like goldenrod (*S. stricta*).

Ground-nut (*Apios americana*): the American groundnut; described on p. 215, where it is misidentified as *A. tuberosa*; not the peanut (*Arachis hupogoea*), but a leguminous, twining plant with roots pleasant to the taste.

Grub-worm: i.e. the larva of an insect.

Hard-hack (*Spiraea tomentosa*): a low shrub common in New England.

Hare (*Lepus americanus*): the snowshoe hare, or (less commonly) the snowshoe rabbit; 'very familiar' T says.

Hawk (*Buteo* spp.): three spp. are common in New England; see Fishhawk, Hen-hawk.

Hazel (*Corylus* spp.): or hazelnut.

Heart-leaves: (*Nymphoides cordata*): floating-heart, a pond plant.

Hemlock (*Tsuga canadensis* or *Abies canadensis*), the eastern hemlock, a fir tree, not the poisonous umbelliferous plant (*Conium maculatum*) used to execute Socrates.

Hen-harrier (*Circus cyaneus*): the marsh-hawk, or more commonly now, the northern harrier; also called by T elsewhere the frog-hawk, because in the spring it feeds on peepers, or baby frogs;

according to H (295 n.) the local farmers referred to most of the larger hawks as 'hen-hawks' or 'hen-harriers'.

Hen-hawk (*Buteo jamaicencis*): not the hen-harrier (qv) which the locals called the larger hawks indiscriminately, but the red-tailed hawk, the largest of the three spp. of *Buteo* common in New England.

Heron: either (*Ardea herodias*): the great blue heron, or (*Egretta caerulea*): the little blue heron.

Hickory (*Carya* spp.): five species in the Concord area, of which *C. ovata* commonest; now rare.

High-blueberry (*Vaccinium corymbosum*): common highbush blueberry, common in the Concord area.

Hooting owl: if actually hooting, then either (*Bubo virginianus*), the great horned owl, or the screech owl (qv).

Hornbeam (*Carpinis caroliniana*): the American hornbeam.

Horned pout (*Ictalurus nebulosus*): or brown bullhead, a species of catfish, up to 19 in. long; native to eastern US.

Horsefly (*Tabanus* spp.): possibly *T. atratus*, a large, swift moving black fly, the females of which are voracious blood suckers.

Huckleberry (*Gaylusaccia* spp.): red or black huckleberry; North American equivalent to the whortleberry, similar to blueberry; abundant, wild.

Humble-bee: mid-century American usage for bumblebee (*Bombus* spp.).

Jay: see Blue jay.

Jerbilla family (*Gerbillinae*): in his own copy of W, T corrected 'jerbilla' to 'gerbille'; see Gerbille family.

Johnswort (*Hypericum* spp.): or St. John's-wort; T probably has *H. majus* in mind, common around W.

Kingfisher: probably (*Ceryle alcyon*): the belted kingfisher.

Lamprey (*Lampetra appendix*): the American brook lamprey; feeds on organic matter, bottoms of ponds; not to be confused with the seawater parasite.

Lark: the eastern meadowlark (*Sturnella magna*); frequents prairies, hayfields, and pastures.

Laughing loon: see Loon.

Leuciscus (spp.): old notation for a genus of freshwater fish; T mentions it variously under the names of chivin, dace, roach, 'cousin-trout', and shiner.

Life-everlasting (*Gnaphalium* spp.): or cudweed; common.

Lily, . . . white (*Nymphaea odorata*): sweet-scented water lily.

Lily, yellow (*Nuphar* spp.): bullhead lily, or small yellow pond lily.

Loon: (*Gavia immer*): American term for the great northern diver; emits wild, ringing calls like whoops of laughter; usually only a single pair nest on a lake.

Maple (*Acer* spp.).

Marmot: see woodchuck.

Marsh-hawk: see Hen-harrier.

Martins: the context (p. 279) suggests one of the *Hirundo* spp., probably *H. rustica*, the barn swallow.

Meadow grass (could be *Poa* spp.): but T uses it in a generalized sense.

Meadow-hen (*Rallus limicola*): like marsh-hen, another name for the Virginia rail, a chicken-like bird living in dense vegetation in marshes.

Meadow mouse (*Zapus hudsonius*): a jumping mouse. In his own copy of *W*, T inserted 'deer' between 'meadow' and 'mouse'. See Deer mouse.

Meadow-sweet (*Spiraea latifolia*): the common meadowsweet.

Merlin: the reference on p. 282 is not to a specific bird, but the nearest American merlin would be *Falco columbarius*.

Millet grass (*Setaria glauca*): not millet (*Panicum miliaceum*), but a local name for foxtail, an abundant weed on cultivated ground, roadsides, waste places.

Mink (*Mustela vison*): in T's time, common in Massachusetts, and in ponds and streams throughout America.

Mocking-bird (*Mimus polyglottos*): when T was writing, much commoner in the southern US than in New England, but has since spread further north.

Moose (*Alces alces*): as implied on p. 249, no longer found in Concord area by T's time.

Mouse . . . not the common: in his own copy of *W*, T added 'Mus leucopus'; so he was probably thinking of the white-footed mouse, now classified as *Peromyscus leucopus*; it has a body of between $3\frac{1}{2}$ and 4 in. long, and a tail of from 2 to 4 in.

Mud-turtle (*Kinosternon subrubrum*): the eastern mud turtle; small (up to 4 in. long), dwells on bottoms of freshwater ponds and brooks.

Mullein (*Verbascum thapsus*): the common mullein; grows in roads and neglected fields.

Muscle [Mussel] (*Elliptio complanata* or *Las Migona complanata*): one of the New World freshwater clams, formerly classified as of the genus *Unio*.

Muskrat (*Ondatra zibethicus*): the large American aquatic rodent, valued for its fur.

Musquash: see Muskrat.

Nettle-tree *Celtis occidentalis* (common hackberry): damp to dry slopes, borders of woods; rare.

Night-hawk (*Chordeiles minor*): also called the booming night-hawk; in the courting display the male plunges towards the earth, only to zoom upwards with a booming sound just before touching down.

Oak (*Quercus* spp.).

Oriole (*Icterus* spp.): possibly *I. spurius*, the northern oriole; nests even in shade-trees in towns.

Otter (*Lutra canadensis*): the river otter; feeds mainly on fish.

Owl (*Otus* spp.): most likely genus of so-called 'owl' in North America.

Partridge (*Bonasa umbellus*): the American ruffled grouse; common in varied habitats around Concord.

Peetweet: (*Actitis macularia*): local name, after its call, for spotted sandpiper, the most widely distributed nesting sandpiper in the US.

Perch: (*Perca* spp.): probably *P. flavescens*, the yellow perch; well distributed in North American lakes and slow-flowing rivers; still popular with anglers.

Pewee (*Sayornis phoebe*): the eastern phoebe; often called pewee or bridge pewee; the first flycatcher to arrive in Concord in spring, usually in late March.

Phoebe: see Pewee.

Pickerel (*Esox* spp.): W, pp. 166–7, mentions at least three different 'kinds', but there are only five spp. of *Esox* in the whole of North America. If T really observed three kinds, they might have comprised, in order of likeliness, the redfin pickerel (*E. americanus americanus*); the chain pickerel (*E. niger*); the northern pike (*E. lucius*). A more distant possibility is the muskellunge (*E. masquinongy*). As for the different colours observed by T, all the pikes are predators whose diets change according to age;

so the various fish seen by T might have exhibited different camouflage according to their age and/or the part of the pond in which they hid to take their prey.

Pickerel-weed (*Pontederia cordata*), an aquatic plant.

Pigweed (*Chenopodium lanceolatum*): or goosefoot, found on waste or cultivated ground; naturalized from Europe.

Pinweed: (*Lechea* spp.): any of the herbs of the rock-rose family; four spp. common in the Concord area.

Piper grass (*Agropyron repens*): or quack grass, a common weed.

Pipewort (*Eriocaulon septangulare*), the seven-angled pipewort, described on p. 176.

Pitch-pine (*Pinus rigida*): one of two species of pine that occurred naturally in the Concord area; common second growth in T's time; now much rarer than e.g. (*Pinus strobus*), the eastern white pine.

Potomogeton (*Potomogeton* spp.): pondweed.

Pout: see Horned pout.

Puff ball (*Lycoperdon* spp.): a genus of fungi so-called because the mature plant bursts with a loud report to scatter its spores.

Pumpkin pine (*Pinus strobus*): first-growth white pine (qv), so-called because of the yellowish colour of its wood (H).

Purslane (*Portulaca oleracea*): common purslane; an annual plant with fleshy leaves, often used as a pot-herb, in salads and for garnishing and pickling (Web).

Rabbit (*Sylvilagus* spp.): probably one of the cottontails common in the eastern states: the eastern cottontail (*S. floridanus*) or the New England cottontail (*S. transitiolanis*); see also Hare.

Raccoon (*Procyon lotor*): the American carniverous mammal, somewhat larger than a fox, with a sharp snout and dark rings around its tail.

Raspberry (*Rubus* spp.): used interchangeably for the raspberry and blackberry.

Red huckleberry (*Gaylussacia baccata*, forma *leucocarpa*): common, often abundant.

Red mavis: see Brown-thrasher.

Red pine (*Pinus resinosa*): mostly in northern states of America.

Red squirrel (*Tamiasciurus hudsonicus*): the North American red squirrel.

Red-wing (*Agelaius phoenicus*): T's term for the red-winged blackbird; migrants arrive in the Concord area in early spring.

Reed-bird (*Dolichonyx oryzivorus*): probably the bobolink.

Roach (*Lepomis* spp.): like the robin (see below), one of those American animals named nostalgically, but wrongly, for a European genus it resembles; the American roach is one of the freshwater sunfish, not the member of the carp family, *Rutilus rutilus*.

Robin (*Turdus migratorius*): like the roach, a nostalgic misnomer; though it has a red breast, the American robin is a thrush, much larger than the English robin—around 10 in. long as against 5 in.

Roman wormwood (*Ambrosia artemisiifolia*): also (like wormwood, qv) called common ragweed, but a different genus; abundant 'pernicious weed' (Eaton, 183), the pollen of which is a common allergen.

Rush (either *Scirpus* spp.): bullrush (or *Juncus* spp.): rush.

Sage (*Salvia officinalis*).

Sand-cherry (*Prunus susquehanae*): likes scrub-oak woods; sterile, sandy, or other open, acid habitats; uncommon.

Scarlet Tanager (*Piranga olivacea*): 'gleam and shine like satin' (Cruick.); in the autumn males lose their bright red plumage, which is replaced with green.

Screech-owl (*Otus asio*): the eastern screech owl; fairly common in the Concord area.

Sedge (*Carex* spp.): or sedge grass; there were over seventy species in the Concord area.

Seventeen-year locust (*Cicada septemdecim*): the commonest American cicada, found chiefly in the middle and southern Atlantic states; it lives from 13 to 17 years underground in the larval condition, from which it emerges as an adult to survive for only a few weeks.

Shiner (*Notemigonus crysoleucas*): or golden shiner, or New York shiner; an American minnow of the carp family; reaches lengths of up to $6^{1}/_{2}$ in.; in T's day generally taken for pickerel bait.

Shingle tree (possibly *Quercus imbricaria*): the shingle oak, common in North America.

Shrub-oak (*Quercus prinoides* or *Q. ilicifolia*): or scrub oak; *Q. prinoides* is more common; both like dry, sterile woods and gravel plains.

Single spruce (*Picea glauca*): or white spruce; grows further north than the Concord area—in Maine, e.g.; T applied it in error to

Picea mariana, the double, or black spruce, and corrected 'single spruce' to 'double spruce' in his own copy of *W*.

Skater insect: popular American usage for the *Gerris* spp., or water striders, but in his own copy of *W*, T wrote 'Hydrometer' here. The *Hydrometra* spp., or water measurers, also travel on the surface tension of pond and river water, but are much longer and thinner than the *Gerridae*.

Skunk-cabbage (*Symplocarpus foetidus*): or skunkweed; very common in low, wet, or swampy woods (Eaton).

Snipe: (*Gallinago gallinago*): migrants of the common snipe arrive in April; they 'boom' because when flushed, a migrant flock makes a loud, rasping note.

Song-sparrow (*Melaspiza medolia*): widely distributed in North America; abundant in the Concord area during migration periods.

Sorrel (*Rumex acetosella*): or sheep sorrel; ubiquitous weed, naturalized from Europe; the leaves, which grow in clumps, have an acid taste when chewed.

Sparrow (*Spizella* spp.): see Field-sparrow; Song-sparrow.

Spotted salamander (*Ambystoma maculatum*): around $9\frac{1}{2}$ in. long; found in most of the eastern US.

Spruce (*Picea* spp.): see also Double spruce; Single spruce.

Squirrel (*Sciurus* spp.): possibly *S. carolinensis*, the grey squirrel; see also Red-squirrel; Striped squirrel.

Strawberry (*Fragaria virginiae*): the common or wild strawberry; abundant in fields or on banks, moist or dry.

Striped squirrel (*Tamias striatus*): the eastern chipmunk.

Sucker (*Catostomus* spp.): either *C. catostomus*, the longnose sucker, or *C. commersoni*, the white sucker; i.e. not the saltwater lump-sucker, but a freshwater fish related to the carp.

Sumach (*Rhus glabra*): the smooth sumac; common volunteer growth; described closely on pp. 104–5.

Swamp-pink (not clear: possibly *Silene caroliniana*): the common wild pink.

Sweet-briar (*Rosa eglanteria*): or eglantine; infrequent; naturalized from Europe.

Sweet-fern (*Comptonia peregrina*): often dominant.

Thimble-berry (*Rubus occidentalis*): another name for the black raspberry.

Toad: if a true toad, then (*Bufo* spp.): probably *B. americanus*, the American toad, or *B. woodhousei*, Woodhouse's toad.

Toad-stool (possibly *Amanita muscaria*): the fly agaric; deadly poisonous; or one of the other *Amanita* spp.

Tortoise: this cannot be right; *Gopheris polyphemus*, the gopher tortoise, is strictly terrestrial and is distributed mainly in the southeastern US; T may mean one of the freshwater box turtles, e.g. *Clemmys insculpta*, the wood turtle, or *Clemmys guttata*, the spotted turtle.

Trout: either (*Salvelinus fontinalis*), the brook trout; or (*S. namayacush*), the lake trout. Unlike the rainbow and brown trout of the western US and Canada, not of the salmon genus., but spp. of charr; widely distributed in fresh water from Canada to New England.

Turdus migratorius: see Robin.

Turtle-dove (*Streptopelia turtur*): used illustratively; not a New England species.

Unio fluviatilus: a species of freshwater clam, now classified *Elliptio complanata* or *Las migona complanata*.

Usnea lichen (*Usnea florida*): the commonest lichen in the Concord region, especially rich on maples in swampy areas.

Veery (*Cathurus Hylocichla fuscescens*): a North American thrush, common in the eastern US.

Vulture: very general; could be any of the New World vultures of the family *Cathartidae*.

Wasp: the context (p. 216) suggests one of the *Vespa* spp., or social wasps.

Water bug (probably *Gerris remigis*): the large water strider.

Water target (*Brasenia schreberi*): or water shield; an aquatic plant with underparts coated with a jelly-like slime.

Waxwork (*Celastrus scandens*): or climbing bittersweet.

Whippoorwill (*Caprimulgus vociferus*): a goatsucker, related to the nighthawk and the nightjar; usually heard rather than seen; not uncommon.

White grape (*Vitus labrusca*, forma *alba*): or fox grape; likes wet and dry thickets, wooded riverbanks, roadsides; the less common variety of the the extremely prolific common wild grape.

White oak (*Quercus alba*): common in the Concord area.

White-bellied swallow (*Iridoprocne bicolor*): T's name for the tree swallow.

White-pine: see pitch-pine.

Wild apple (*Pyrus malus*): the common apple; T does not mean a separate species by 'wild'.

Wild bee: (*Apis mellifera*): the domestic honeybee; by 'wild bee' mid-century Americans meant the domestic honeybee gone wild.

Wild goose (*Branta canadensis*): the Canadian goose; now (in England as in New England) often nests locally; in T's time genuinely migratory, moving down from Canada to the Gulf Coast in the autumn.

Wild mouse: see Mouse . . . not the common.

Wild pigeon (*Ectopistes migratorius*): the passenger pigeon (see note to p. 106); already scarce in T's day, now extinct.

Wild-cat: could be a house cat (*Felis catus*), gone feral; see Cat.

Wild-holly (*Nemopanthus mucronata*): the mountain holly, as distinct from the American holly (*Ilex opaca*).

Willow (*Salix* spp.): frequently mentioned in J.

Wood-cock: see Woodcock.

Wood-pewee (*Cantopus virens*): the eastern wood-pewee, a flycatcher; much less common in the Concord area than in T's day.

Wood-thrush (*Hylocichla mustelina*): according to H, T often confused the wood thrush with the hermit thrush (*Catharus guttatus*).

Woodchuck (*Marmota monax*): or ground hog; common American species of marmot; hibernates in winter; its name is folk etymology for Ojibwa *otching* or Cree *otcheck*: 'the fisher'.

Woodcock (*Scolopax minor*): the American woodcock; allied to the snipe, but with a more robust bill.

Wool-grass (*Scirpus* spp.).

Wormwood (*Artemisia artemisiifolia*): common ragweed; not the bitter herb used to flavour vermouth.

Yellow birch (*Betula lutea*).

Yellow pine: see Pitch pine.

Yellow violet (*Viola pensylvanica* or *V. pubescens*): smooth yellow, or downy yellow violet; likes rich, usually deciduous woods; uncommon.

EXPLANATORY NOTES

ABBREVIATIONS USED IN THE NOTES

AS	Anglo-Saxon
Barber	John Warner Barber, *Historical Collections . . . relating to the history and antiquities of every town in Massachusetts . . .* (Worcester: Dorr, Howland & Co., 1839).
Chalmers	Alexander Chalmers (ed.), *The Works of the English Poets, from Chaucer to Cowper . . .* (21 vols.; London, 1810); many of T's citations from English poetry and drama come from Chalmers.
DAE	*Dictionary of American English*
Du	Dutch
fl.	flourished
Fr	French
Gk	Greek
H	Walter Harding (ed.), *The Variorum* Walden (New York: Twayne, 1962).
It	Italian
J (NS)	(*Journal*, New Style), Henry D. Thoreau, *Journal* (Princeton: Princeton University Press, 1981–).
J (OS)	(*Journal*, Old Style), *The Journal of Henry David Thoreau* (14 vols.; Salt Lake City: Peregrine Smith Books, 1984). (Both editions of the Journal are cited as 2.142, for vol. 2, p. 142.)
L	Latin
LA	[The Library of America edition of] Henry David Thoreau, *A Week on the Concord and Merrimack Rivers; Walden; or, Life in the Woods; The Maine Woods; Cape Cod*, ed. Robert F. Sayre (New York: The Library of America, 1985).
MS	manuscript of *Walden*, Huntington Library, HM 924.
OED	*Oxford English Dictionary*
p. proofs	page proofs of *Walden*, Huntington Library, HM 925.
Pauthier	Jean-Pierre-Guillaume Pauthier, *Confucius et Mencius ou les quatres livres de philosophie moral et politique de la Chine* (Paris, 1841). Most of T's quotations from Confucius are his translations from Pauthier's French.
Prince	[The Princeton edition of] Henry D. Thoreau, *Walden*,

ed. J. Lyndon Shanley (Princeton: Princeton University Press, 1971).

Rom   Roman

Scriptores   *Scriptores rei Rusticae* . . . [*Writers on country matters*: Cato, Varro, Columnella, Palladius] (Heidelberg, 1595).

Shakespeare Most of T's citations from Shakespeare come from the 2-vol. edition of *The dramatic works* (Burlington, NJ, 1774).

T's copy   Thoreau's own copy of *Walden*

T   Thoreau

VDS   Philip Van Doren Stern (ed.), *The Annotated Walden* ([New York]: Clarkson N. Potter Inc., 1970).

W   Henry David Thoreau, *Walden* (Boston: Ticknor and Fields, 1854).

W   Walden Pond

W7NWD   *Webster's Seventh New World Dictionary*

Web   Webster's *Dictionary of the English Language* (1854).

Week   Henry David Thoreau, *A Week on the Concord and Merrimack Rivers* (Boston: James Munroe and Company, 1849).

1 [*title*]: The first edition of *W* was titled *Walden; or, Life in the Woods*, but T instructed his publishers to drop the half-title for the second printing of 1862, presumably to distinguish his book from other contemporary productions with similar titles, such as J. T. Headley's *The Adirondack; or, Life in the Woods* (1849) and Charles Lane's essay in *The Dial*, 4 (1844), 415, 'Life in the Woods'. Headley's book is a sort of Hudson-River-School landscape in print. Expressed as letters to friends from 'The Backwoods' of Upstate New York, it is full of 'glorious prospects' of the sublime and the beautiful, as well as stories of camping out, hunting, and fishing. Lane seriously reviews the desirability and possibility of living in the real woods, actually removed from civilized settlement, concluding that the practical demands of housing, clothing, and the gathering and preparation of food would preclude any expansion of the intellect or sensibilities.

5 *I do not . . . ode to dejection*: from the second chapter of *W*; published (as here) as the epigraph to the first edition, but omitted from the title-page of many modern editions. Cf. Samuel Taylor Coleridge, 'Dejection: An Ode' (1802).

chanticleer: traditional literary periphrasis for rooster, from the name of the cock in the fable of Reynard the Fox.

When I wrote . . . pages: T lived at Walden Pond from July 1845 to September 1847, during which he wrote the equivalent of about half the length of the finished text of *Walden*.

sojourner: to sojourn is, literally, to spend the day (Fr); so the point of the joke is that since living on the Pond, T feels his residence back in 'civilized life' to be at best temporary and provisional.

lonesome: American usage for lonely (Web), though not always implying unhappiness in the state.

6 *Sandwich Islanders*: inhabitants of the Hawaiian Islands; still in his Lyceum mode at this stage of *Walden*, T is hitting off against a popular topic for public lectures, accounts of travel to exotic places.

Bramins: or Brahmins, upper-caste Hindus, who underwent various penances as acts of devotion (H, who could not find the source of the quote).

twelve labors: Hercules, the hero of Gk legend, was set twelve labours to perform. They included confronting the nine-headed hydra, which he killed with the help of his servant Iolus, and cleansing the stables of Augeas, where 3,000 oxen had been kept for 30 years.

7 *suckled by a wolf*: as Romulus and Remus were, the twin foundlings who, according to legend, grew up to found Rome.

peck of dirt: cf. the English proverb 'We must eat up a peck of dirt before we die' (H). Equal to eight quarts, or one-quarter of a bushel, the peck was still a commonly used measure in T's day, as contemporary grocers' bills make clear.

old book: the Bible; cf. Matthew 6: 19: 'Lay not up for yourselves treasures upon earth, where moth and rust doth corrupt, and where thieves break through and steal.'

Deucalion and Pyrrha: in the Gk version of the biblical story of the flood, Deucalion, son of Prometheus, and his wife Pyrrha were the only mortals to be saved when Zeus decided to exterminate the degenerate race of men. After the holocaust they veiled their faces and cast stones over their shoulders which grew into men and women and thus repopulated the earth. The story is told in Ovid, *Metamorphoses* (Philadelphia, 1923), book I.

[*verse*]: Ovid, *Metamorphoses*, i. 414–15, the end of the Deucalion and Pyrrha story, cited by Walter Ralegh in his *The Historie of the World* (1614), in *The Works of Sir Walter Ralegh . . .* (8 vols.; Oxford, 1829), book I, chap. 2, sect. 5. 'From thence . . . are' is Ralegh's translation of the Ovid.

8 *limits*: also in the financial sense of being at the end of one's line of credit.

slough: swampy ground, but probably a reference also to the Slough of Despond in John Bunyan, *The Pilgrim's Progress* (1678) (New York, 1811), chap. 1, where the burdened Christian becomes bogged down in despair of his sins.

nutshell: cf. Shakespeare, *Hamlet*, II. ii. 258–60: 'O God, I could be bounded in a nutshell and count myself a king of infinite space, were it not that I have bad dreams.'

9 *southern overseer*: the comparison is between black slaves and white industrial workers. H points out that T was one of the first Americans to protest against the factory system.

divinity in man: refers to Ralph Waldo Emerson's doctrine of the perpetual revelation, which denied the special revelation of God's purpose in Jesus Christ and affirmed that men could become 'acquaint[ed] . . . at first hand with the Deity' (Address to the Harvard Divinity School, 1838).

divinity stir: cf. Joseph Addison, *Cato* (1713), v. i. 7–8: ' 'Tis the divinity that stirs within us | 'Tis heaven itself that points out our hereafter' (Chalmers, vol. ix); Cato is about to fall on his sword.

Squire Make-a-stir: a character suggestive of, rather than a specific allusion to, *Pilgrim's Progress*, or even a Restoration comedy. The title of 'Squire' was still used locally for magistrates in T's time.

West Indian . . . Wilberforce: William Wilberforce (1759–1833), English philanthropist and member of Parliament, fought a long parliamentary battle (and the first English campaign for the public conscience on a secular issue) against the slave trade (not slavery itself, as T implies in his reference to emancipation) in the British West Indies. The bill finally became law in 1807.

toilet cushions: as in toilet articles or toilet tables: i.e. cushions in a lady's dressing-room (*DAE*)

catechism: the Shorter Catechism, authorized by the General Assembly of the Commonwealth of Massachusetts, was included in *The New England Primer* (c.1683), the popular and much reprinted school book. The first question of the Catechism is 'What is the chief end of man?' The answer is 'Man's chief end is to glorify God, and to enjoy him forever.'

10 *dry wood . . . whirled*: refers to the railroad, powered then by woodburning steam locomotives.

11 *prætors*: magistrates appointed to supervise the administration of the Roman civil law.

Evelyn ". . . neighbor": John Evelyn, *Silva; or, a Discourse of Forest-Trees*, 3rd edn. (London: 1679). Evelyn refers not to Solomon but Solon, the fifth-century BC Athenian lawgiver. Only the first ten words (except for the misreading 'Solomon') are taken verbatim from Evelyn; the rest is T's paraphrase.

Hippocrates . . . cut our nails: no one has yet been able to trace this regime of the finger nails to Hippocrates.

"be not afflicted . . . undone?": cf. Horace Haymen Wilson (trans.), *The Vishnu Purana* (London, 1840), 87 (H).

12 *Confucius ". . . true knowledge"*: cf. *The Confucian Analects*, book II, chap. xvii, in Pauthier.

13 *Darwin ". . . such a roasting"*: cf. Charles Darwin, *Journal of researches . . . during the voyage of HMS Beagle round the world* (2 vols.; New York, 1846), i. 284.

New Hollander: probably recalling his reading of *The three voyages of Capt. James Cook round the world* (7 vols.; London: 1821), particularly vol. iii.

14 *Liebig*: Justus von Liebig (1803–73), Professor of Chemistry, University of Giessen; cf. e.g. his *Animal chemistry . . .* (New York, 1842), part I, chap. ii.

Elysian life: the Elysian Fields: here not the realm of the heroic dead but in the earlier (Homeric) sense of the happy land west of the earth, near Ocean, where there is no cold or wet weather, and which is always fanned by a delightful breeze.

Yet some: the clipper-ship trade between New England and the Far East was at its height when T was writing (H).

15 *à la mode*: literally, after the fashion (Fr). The point of T's joke is a reference to central heating, then being introduced into the homes of the better off.

degenerate ever: possibly a reference to the hypothesis of Georges-Louis Leclerc, Count Buffon (1707–88), Cornelius DePauw (1739–99), and other European scientists that men and animals degenerated in size, number, and variety when transplanted from the Old World to the New.

16 *nobler plants*: a reference to the theory of the transcendentalist and educational theorist Bronson Alcott (1799–1888) that plants which aspire to the sun are 'nobler' than those that put most of their energy into their roots (H).

esculents: simply edible plants, hence vegetables; from *edere* (L). The onion is a good example of the humble biennial esculent here referred to.

17 *notch*: as Robinson Crusoe kept track of time in the novel of that name (1719) by Daniel Defoe.

the meeting of two eternities: i.e. his life; cf. Thomas Carlyle, *On heroes, hero worship, and the heroic in history: Six lectures* (London, 1841), Lecture V of which defines the individual life as 'a little gleam of time between two eternities'.

I long ago lost . . . themselves: a much discussed passage. H (270–2) summarizes the various interpretations and sources that have been offered for T's remark about the hound, bay horse, and turtle-dove. What is clear is that the three lost items are figurative and that T had no intention of assigning specific equivalents to any of the three symbolic animals.

18 *capital*: probably in the sense of his time and accumulated knowledge, which, for a surveyor, were his capital, literally.

mannawise: manna, the bread with which the Lord provided the children of Israel every morning of their journey through the wilderness, melted 'when the sun waxed hot'. Cf. Exodus 16: 13–25.

reporter to a journal: could be a joke about his own journal, which of course had no 'circulation' at all, or a reference to his work for *The Dial*, edited by Margaret Fuller, a quarterly magazine of literature, philosophy, and religion that ran for four years only (1840–4) and never sold more than a few hundred copies. Though sympathetic to T's writing, *The Dial* rejected a number of T's submissions.

For many years . . . utility: in part a joke, though T really did make detailed notes of natural phenomena, as his Journal makes clear. In the last fifteen years or so of his life he earned more as a surveyor than from the family pencil business.

red huckleberry . . . yellow violet: there has been some confusion as to what T's point is here. Not all these species were rare, or in especial need of care, in the Concord and W area (see Appendix). The point seems to be that T looked after everything, common as well as exceptional, more or less as he came across it.

town officers: some officers in Concord were appointed by the Town Manager, who was head of the Selectmen; others, like the Selectmen themselves, were elected. Not long before he died, T was appointed Town Surveyor.

19 *room in the court house*: i.e. as a town officer. Before the Concord Town House and the new Court House were built in 1850 on the opposite side of the Green, the old Court House served for town offices.

private business: i.e. to write *Week*, an account of a journey made by small boat with his brother John to the White Mountains in New Hampshire. The 'business' was 'private' because John had died in 1842, and *Week* was going to be his memorial.

20 *business habits*: not wholly facetious. T had managed his father's pencil and graphite business quite sucessfully.

Celestial Empire: i.e. China.

Salem harbor: once a chief port of the China trade, Salem had already passed its peak by the time T was writing, due in part to the decline in America's trade with the world following the panic of 1839, and in part to the extension of railheads to ports with deeper harbours.

native bottoms: ships originating from the same place from which the exports are sent; hence, American vessels. Ice, pine timber, and granite were typical New England exports.

Jersey shore: the coast of New Jersey was noted for shipwrecks (H).

telegraph: presumably T has the old form of telegraph in mind here, a system of semaphores or other visual signals repeated by stages over however great a distance the message needed to be sent, but S. F. B. Morse's electromagnetic telegraph (as it was called then so as to distinguish it from the visual kind) was already in use by the time T was writing.

speaking . . . vessels: hailing ships to speak with the captain.

logarithmic tables to be corrected: logarithms are 'artificial' numbers corresponding to 'real' numbers, enabling the latter to be multiplied by addition, divided by subtraction, raised to any power by simple multiplication, and their root found by simple division. Before the Harrison chronometer came into general use in merchant shipping, allowing navigators to determine the ship's exact time—hence distance—east or west of the Greenwich meridian, logarithms were used to facilitate calculations to correct (they were not themselves corrected) estimates of longitude. Nineteenth-century textbooks on navigation contained elaborate logarithmic tables of such data as the moon's 'apparent altitudes' and 'horizontal parallax'.

La Perouse: Jean François Gallup (1741–88), French explorer whose ship was wrecked in the New Hebrides (H).

Hanno: the Carthaginian navigator of the fifth century BC, author of the *Periplus* (literally: 'sailing around'), an account of a voyage out beyond the Straits of Gibraltar.

tare and tret: the two standard deductions made to arrive at the net weight of a commodity shipped (Web).

21 *post*: T's copy: *port*

Neva marshes: founded in 1703, St Petersburg (after 1924, Leningrad, now returned to its original name) was built on the marshes of the River Neva where it empties into the Gulf of Finland. Canals for drainage were cut through the marshy left bank of the Admiralty side, south of the final bend in the river.

Clothing: H says that Thoreau was obviously indebted to Carlyle's *Sartor Resartus* here. Emerson produced an edition in Boston in 1836.

22 *Madam Pfeiffer*: Ida Pfeiffer (1795–1857), the Viennese author of the popular *Eine Frauenfahrt um die Welt* (Vienna, 1850), whose five English editions included *A Woman's Journey Round the World* (New York, 1852).

woman's dress . . . done: cf. the American proverb, 'A woman's work is never done.'

hero: another hint of Carlyle's influence.

23 *new wine in old bottles*: cf. Matthew 9: 17: 'Neither do men put new wine into old bottles, else the bottles break.' Jesus is saying that a new religion requires new practices.

mortal coil: i.e. body; cf. *Hamlet*, III. i. 66–7: 'For in that sleep of death what dreams may come | When we have shuffled off this mortal coil?'

like exogenous plants: i.e. like trees or shrubs, which grow by successive additions to the outside of the wood (Web).

integument: outer covering, as skin, or the shells of crustaceous animals (Web).

cortex: bark, as of a tree (Web); outer or superficial part of an organ.

liber: the inner, newly formed bark of a plant (Web). It is here, in the cambium layers, that the tree's growth takes place, forming the record of yearly growth in the annual rings seen in the cross-section of the trunk.

24 *old philosopher*: identified in the Journal as Bias (fl. 550 BC), one of the so-called Seven Sages of Greece. Cf. J 1.157–8 (NS).

thick coat . . . winter cap: compared with the earnings of a contemporary unskilled labourer of $1 per day and with those of a skilled labourer of between $1.50 and $2 per day, these prices are not especially cheap.

Fates: the Moirae (Gr) or Parcae (L), the three women in classical mythology who determine the length and course of the individual life; in most forms of the myth one is represented as spinning the thread of life, another as weaving it, and the third as cutting it off.

Graces: the Carites (Gr) or Gratiae (L), three goddesses who enhanced the enjoyment of life by refinement and gentleness.

head monkey: identified in the Journal as Count D'Orsay, after whom a sort-of winter garment for men was named.

25 *Egyptian wheat*: refers to contemporary efforts to get wheat found in Egyptian burials to germinate (H).

was: T's copy: *is said to have been*

Harlequin: a stock character in parti-coloured clothes in the Italian popular burlesque form, the *commedia dell'arte*, descended from the comic servant and paralleled in a character of the slapstick pantomime known in England as harliquinade.

26 *Samuel Laing*: (1780–1808), a much read author and traveller. His *Journal of a residence in Norway during the years 1834, 1835, & 1836*, 2nd edn. (London, 1837), which Thoreau cites here, treats the moral and political economy of Norway, and the condition of its inhabitants.

27 *labyrinth*: generally, any 'edifice or place full of intricacies, or formed with winding passages' (Web); specifically the building in which the Cretans kept the Minotaur, half man and half bull. After killing the monster, Theseus was able to escape through the Labyrinth following the thread (the 'clew') which Ariadne had provided him to lay down on his way in.

almshouse . . . prison: see Introduction.

Penobscot Indians: from northern Maine, frequently visited Concord, camping outside the town (H).

28 *freedom in his love*: cf. 'To Althea, From Prison' (1649), by Richard Lovelace, the royalist courtier and poet, who was imprisoned for seven weeks in 1642 for petitioning Parliament in favour of the Bishops and the Book of Common Prayer. T found the poem in Robert Anderson, *The Works of the British Poets . . .* (14 vols.; London, 1795–).

28 *Gookin* "... *English houses*": Daniel Gookin, 'Historical Collections of the Indians in New England ...', *Collections of the Massachusetts Historical Society*, 1st series (1792, rpt. 1806), i. 141–227.

29 *nests ... holes*: cf. Matthew 8: 20: 'And Jesus saith unto him, "The foxes have holes, and the birds of the air have nests; but the Son of man hath not where to lay his head."'

Rumford fireplace: Benjamin Thompson, Count Rumford (1753–1814) invented a chimney with a smoke shelf that prevents downdrafts from blowing smoke into the room; the principle is still used in fireplaces today (VDS).

back plastering: i.e. plastering the back, or inside, of the outer wall for insulation and to form a smooth surface for the walls of the rooms inside.

eight hundred dollars: revised downwards; MS 98 has $1,000.

encumbered: had also a specifically financial sense; an estate was said to be 'encumbered' with a mortgage or with a widow's dower, tax liens, or other debts (Web).

one dollar a day ... more ... less: $1 per day was the average wage for unskilled labour; a painter or carpenter might receive from $1.50 to $2; an immigrant labourer, such as an Irishman working on the railroad, $.50.

30 *poor ... with you*: cf. Matthew 26: 11: 'For ye have the poor always with you; but me ye have not always.'

What mean ye "... *shall die*": Ezekial 18: 2–4.

encumbrances: see 'encumbered', above.

assessors: i.e. tax assessors; still a town office today in Concord.

31 *repudiation*: also the technical term for reneging on debts. The economy of the early American Republic was mobile, speculative, and innovative; so repudiation of debt was as common as overnight fortunes, and should almost constitute, according to Emerson, a feature of American local colour. Cf. 'The Poet' (1844).

Middlesex Cattle Show: see Introduction.

suent: or suant, a dialect word for proceeding regularly. Thoreau's usage was rare enough to be cited in 'most dictionaries' (H). Cf. J 3. 272 (OS).

hair springe: i.e. with a very delicate tripping mechanism.

[*verse*]: George Chapman, *The Tragedy of Caesar and Pompey* (printed 1631), v. ii (H).

32 *Momus*: the classical god of mockery and censure, H identifies T's citation as from John Lempriere, *Bibliotheca Classica* (New York, 1842).

Minerva: the Rom equivalent of Athene, Gk goddess of wisdom.

"silent poor": i.e. the genteel poor, who received relief in their own homes; cf. (below) the 'degraded poor', or paupers, kept in the almshouse; see Introduction.

shanties: the cheap wooden dwellings of Irish labourers brought in to build the railroads.

33 *workhouse of the world*: as the pioneer of the Industrial Revolution, England described itself as the workshop of the world, following Prime Minister Benjamin Disraeli's coinage of the phrase in a speech in the House of Commons on 15 March 1838. T's wordplay comments on the miseries of the factory system.

Irish: when the potato crop failed in 1845 the Irish population of over 8 million people faced widespread famine.

34 *glowshoes*: folk etymology for galoshes, or rubber over-shoes for wet weather, from the Fr for clog, or wooden shoe, *galoche*.

Aurora: the Rom equivalent of Gk Eos, goddess of the dawn.

Memnon: son of Eos and the mortal Tithonus, but in Egypt the name was connected with the colossal stone statue of Amenhotep III which after being damaged by an earthquake emitted a musical sound like the twang of a harp string when touched by the warming rays of the morning sun. The sound was taken to be the voice of Memnon responding to the greeting of his mother, Eos.

Sardanapalus: the last king of Assyria, noted for his effeminacy and sybaritic way of life.

oriental: reflects on the contemporary vogue for oriental decor and furnishings inspired by the clipper-ship trade with China (H).

Jonathan: the American equivalent of John Bull: a popular and journalistic name for the ordinary citizen.

35 *malaria*: literally, bad air (It), by which the often fatal disease was then thought to be caused.

agri-culture: the L root of agriculture signifies the tilling of fields; here T deconstructs the term to mean a culture that valorizes the ownership of land. The idea is that 'we' moderns have modified Christianity into a creed of property.

mansion: cf. John 14: 2: 'In my father's house are many mansions.' Jesus means rooms, but T uses the word in its modern sense of palatial houses, so as to accentuate the modern world's inversion of the original Christian disdain for worldly possessions.

36 *or the three*: p. proof: *or of the three*

Old Johnson '. . . long season': Edward Johnson's *Wonder-working providence of Sions saviour in New England (A History of New-England . . .* , London, 1654) chronicled the first two decades of New England settlement as a beacon of reformation to other nations wishing to follow the English Revolution into the millennial kingdom. T probably took this quote from the extensive citation from Johnson in Barber's chapter on Concord.

37 *"those . . . thousands"*: E. B. O'Callaghan, *The documentary history of the state of New York . . .* (4 vols.; Albany, NY, 1850–1), vol. iv; a set of this work was in T's own library.

degenerate: J 2.180 (NS) puts this in more ample context, with a separate retrospect on wigwams. The wigwams were not the skins stretched over poles meeting at the top so familiar to western moviegoers, but domed or arched structures of saplings bound together and covered with bark.

38 *I borrowed an axe*: Amos Bronson Alcott (the educational theorist and father of Louisa May) claimed to have lent T the axe, but another account identifies Emerson as the owner. See H 276 for a fuller note.

winter . . . discontent: cf. the first lines of Shakespeare's *Richard III*: 'Now is the winter of our discontent | Made glorious summer by this sun of York.'

39 *studs and rafters*: T's Walden house was a miniature nineteenth-century New England braced-frame house; studs were wooden uprights running from floor to horizontal tie beam, and rafters framed the pitched roof, running from gutter to the ridge board at the peak.

[verse]: the poem is T's own.

bark on: the point is that, lacking a sawmill, T could not saw squared timbers lengthwise from a log; so he used an axe or adze to strip away one or two of the curved sides of the log to form a flattened face where the timber had to present a smooth surface, so as to receive other boards laid against it.

mortised or tenoned: a mortise is a square notch cut in a board to receive the tenon, a square peg in the end of another board to be set at right angles to it. The 'other tools' T needed for this work would include a chisel and a saw.

40 *James Collins . . . Fitchburg Railroad*: Collins was one of the Irish labourers brought in to build the railroad running from Boston through Concord to Fitchburg, Massachusetts. T was writing this in the autumn and winter of 1845/6 (see J 2.134–5 (NS)), or just over a year after the railroad had been completed to Concord; so Collins was either out of a job, or moving on to work further west. T needed his shanty for his roof, walls, window and door frames—the milled timber which he could not produce with his axe and other hand tools.

four dollars and twenty-five cents: despite T's apparent pleasure at his bargain, he had to disburse roughly the same again to bring his stock of boards up to the quantity required to clad the cabin. In all he spent a little less than half of what he would have paid for new boards covering the same area. See note to p. 45.

41 *Patrick*: like Paddy, slang generic name for an Irishman.

removal of the gods of Troy: cf. Virgil, *Aeneid*, ii. 715; when Troy is overrun by the Greeks and Aeneas escapes with his father on his back, they take with them the household idols representing the country's gods.

acquaintances: identified in H 276, following George Willis Cooke, as Emerson, Alcott, W. E. Channing (qqv), Burrill and George Curtis (former residents of Brook Farm now living in Concord), and T's farmer friend Edmund Hosmer and his three sons.

42 *4th of July:* the nation's birthday, anniversary of the proclamation of the Declaration of Independence.

feather-edged and lapped: planed so that one long edge is thinner than the other to allow the boards to be overlapped like shingles. This is standard New England clapboard siding.

43 *tailor . . . man*: from the seventeenth-century English and American proverb, 'nine tailors make a man'.

Trinity Church: the Anglican church in the perpendicular Gothic style, dating from 1698, on Broadway, just north of Wall St., New York City. Then at the heart of the fashionable residential district, it has long since been over-topped by skyscrapers.

architecture: apparently used throughout this passage in the sense, not of the design of the building, but following John Ruskin's definition at the beginning of his *The Seven Lamps of Architecture* (New York, 1849), 'the art which . . . adorns the edifices raised by men'.

44 *belles-lettres . . . beaux-arts*: eighteenth- and early nineteenth-century terms for literature and art history.

44 *"carpenter"* . . . *"coffin-maker"*: this was often literally the case in small New England towns.

45 [*price list*]: the hair was for reinforcing the plaster. A 'mantle-tree' was a mantelpiece over a fireplace, and the iron the two pieces let into the wall or chimney for its support. T fails to cost his labour and the 'rent' of his land. His bargains in second-hand materials are not what they seem. Contemporary prices for $5\frac{1}{2}$-in. boards were $.03 per foot; for shingles, $7.45 per one-and-a-quarter thousand. Given the size of T's cabin, 10 ft. by 15 ft., and the proportion of sides and end taken up by windows and door, his need for lumber could hardly have exceeded 688 board feet of siding; so with no discount for quantity, T could have bought enough new wood to have built the hut from scratch for $20.64 at most. At contemporary prices, $3.90 would have got T 130 pounds of nails. $2.40 was not 'high' for lime; George L. Prescott sold two casks for the same amount in 1853, and in the same year 2,000 shingles for $6.75. T's $4 for 1,000 used bricks does not compare very favourably with Prescott's bill of $4.50 for 500 'Hard Brick'—that is, new brick of engineering standard. Nor is T's $1.25 for laths (for back plastering inside the hut) much of a bargain compared to Prescott's $.75 for 300 laths.

46 *Cambridge College*: Harvard University in Cambridge, Massachusetts, from which T was graduated in 1837. The Harvard student's rent was not $30 per year, but rose from $10 in 1827 to still only just $19 by 1859—over twenty years after Thoreau was graduated—for the 30-week term. Tuition may have been 'an important item in the term bill', but it was only around a quarter of the student's annual expenditure, compared with the cost of 'commons' (or food), which came in at around half.

47 *motes*: cf. Luke 6: 41: 'And why beholdest thou the mote that is in thy brother's eye, but perceivest not the beam that is thine own eye?'

monsters . . . vinegar: i.e. bacteria seen through a microscope.

Rogers' penknife: Joseph Rodgers (not Rogers) & Sons, Sheffield, England, were among 'the most noted cutlers' (H).

48 *navigation!*: 'nautical astronomy' was part of the second-year mathematics curriculum at Harvard in the 1830s (H).

political economy: the administration of the revenues of a nation (Web); i.e. economics.

Smith, Ricardo, and Say: Adam Smith, David Ricardo, and Jean Baptiste Say were influential economists of the eighteenth and early nineteenth centuries.

Princes Adelaide: wife of the Duke of Clarence, later (from 1830), William IV of England.

locusts and wild honey: cf. Matthew 3: 4, of John the Baptist, whose 'meat was locusts and wild honey'.

Childers: English racehorse in the eighteenth century, reputed to be able to run a mile in a minute.

take the cars: contemporary American usage for 'to travel on the railway'.

49 *Fitchburg*: a manufacturing town in northern central Massachusetts, then end of the Boston-to-Fitchburg railway that ran past the southern end of Walden Pond.

"melancholy accident": typical headline in the popular press.

Englishman: possibly Robert Clive, Baron Clive of Plassey (H).

50 *no manure whatever on*: p. proofs: *no manure on*

lot . . . sold . . . squatter: the lot was sold to Emerson in 1844. T was not a squatter, but allowed to use the land provided he worked to clear the undergrowth, plant new trees, and keep the townspeople from raiding it for timber.

51 *$8.71½*: this was just a little over the average *weekly* wage of a farm labourer.

Arthur Young: author of many volumes on agriculture, including *Rural Œconomy* (London, 1733) (H).

52 *halls for . . . free speech*: in 1844, when Emerson wanted to address a gathering of abolitionists in Concord, none of the churches would allow the meeting to be held on their premises. T finally secured the courthouse for the convention (H).

Bhagvat-Geeta: the classic of Hindu religious literature, T's favourite oriental work (H).

53 *Arcadia*: figuratively, as the land of the golden age, when nature provided for human needs without the intervention of agriculture.

Thebes: not the Greek, but the Egyptian city, said to be the oldest in the world.

rod: American for pole or perch, a unit of measurement equal to 5 yards (Web).

United States Bank: designed by Samuel Blodgett (1754–1814), the First Bank of the United States in Philadelphia was the first American public building to be given a classic Greek portico. The Second US Bank, designed by William Strickland (1787–1854) was in the full-blown Greek-revival style, its portico based on that of the

Parthenon. By the time T was writing Andrew Jackson had suppressed the Bank of the US as a conspiracy of moneyed eastern and foreign interests against the common people.

53 *garlic . . . butter*: the pyramid-builders were thought to have been fed on garlic; bread and butter was a common expression for wages.

Vitruvius: the influential Rom architect of the 1st century BC.

Mr. Balcom . . . Dobson & Sons: if these refer to real people, they have yet to show up in the standard histories of, or reference books on, architecture and building.

54 [*food costs itemized*]: here again T seems occasionally to have paid over the odds. A local grocer sold six gallons of molasses for $1.50; and against Thoreau's 3 cents for salt, the grocer sold one peck— that is, eight *quarts*—of the commodity for only 13 cents. Charles Anderson, in *The Magic Circle of Walden* (New York: Holt, Rinehart, and Winston, 1968), 27, points out that the total of $8.74 is around one-fifth the usual cost of feeding one person for a year, and that it comes to exactly the same as the annual estimate for a vegetable and cereal diet advocated by a leading food reform group in *The Young Housekeeper* (Boston, 1838). T's accounting omits the numerous meals he had in return for odd jobs and out of simple hospitality at the house of his mother and sister, and at the Emersons', while he stayed at W.

transmigration, as a Tartar: refers to the belief that at death the soul of the animal departs the body to cross over to another just being born. It is not clear why T connects the idea with 'a Tartar'.

55 *oil*: coal oil, or kerosene, for lighting and heating.

washing and mending: done at home by his mother and sister.

56 *certain completeness*: not true; see Introduction.

drinking water: cf. J 2.142 (NS).

57 *hoe-cakes*: small cakes made of 'Indian meal' or corn-meal, so-called because originally baked on the blade of a large hoe.

hatching eggs: H cites Diodorus (Siculus, the first-century BC historian), I.74.4 on the skill with which Egyptians hatched eggs 'artificially by their own wit'.

souring of the dough: the idea is that unleavened dough left over from a baking session began to ferment, so that when it was used in the next batch, it made the new dough rise, thus providing an accidental lesson in how to make leavened bread.

spiritus: soul, inner life, animating principle (L).

vestal fire: Vesta, the Rom goddess of the hearth and household, had a temple in the Forum in which an eternal flame burning on the altar was tended by virgin priestesses.

cerealian: T's own coinage (the *OED* credits him with the word), but not just an adjective for 'cereal'; there is also a sense of ritual celebration implied in the word (as used here and elsewhere in T), based on the *Cerealia*, the ancient Rom games in honour of the goddess of fertility, Ceres.

58 *sal soda . . . alkali*: rising agents, other than yeast, used in bread making.

Cato ". . . sub testu": (Cato Censorius, or Cato Major), *De agri cultura*, lxxiv, in *Scriptores*.

hominy: corn-meal ground coarser than usual, then mixed with water and boiled.

molasses . . . beets: pumpkin, beetroot, and the sap of the sugar maple contain sugar (the last in much higher concentrations), which can be isolated by painstakingly boiling down the raw material.

59 *[verse]*: from the railing New England folksong about the hardships of frontier life, 'New England's Annoyances' (1643); T probably got it from the chapter on Lynn, Mass., in Barber.

operative: contemporary English usage for labourer, artisan, or factory worker.

fuel . . . encumbrance: presumably because in a new country fuel was wood, not coal; hence the need for a woodlot, and the labour of cutting, splitting, and stacking the wood for burning.

thirds in mills: in contemporary practice a wife would inherit one-third of her husband's property (many owned mills), and the oldest son the rest.

60 *My furniture*: now in the Concord museum. It consists of a school desk with a lifting lid, a caned pallet frame, possibly of Chinese origin, with short legs added to it, and a chair with turned legs.

exuviæ: castoffs, discards (L).

muskrat . . . free: unlike Aesop's fox, not a fable or a figure, but hard fact. See J 2.127 (NS).

dead set: a fixed state or condition which precludes further progress (Web).

61 *trig*: trim, neat; not in common use, says Web.

bed and walk: Matthew 9: 6: Christ tells the man he has cured of the palsy, 'Arise, take up thy bed, and go unto thine house.'

61 *moon . . . milk . . . meat of mine*: an allusion to the old superstition.

fade my carpet: economical New England housewives used to keep the curtains drawn in their parlours to keep their carpets from fading in the sunlight (H).

62 *"The evil . . . them"*: Shakespeare, *Julius Caesar*, III. ii. 76.

bonfire: originally a purifying fire making good (Fr *bon*) a source of corruption; hence a fire for burning books or heretics; Web gives it the American association with public celebration—say, after an election or on the Fourth of July.

auction: the root sense (*auctio*) is increase (L); hence to augment a sale price by bidding.

dust holes: figurative for cellar pits; this is a kind of prolepsis, since often the only sign remaining of an abandoned farmhouse would be the cellar pit in the overgrown woods.

kicks the dust: contemporary slang from accounts of Indian warfare (VDS).

Bartram: William Bartram (1729–1823), like his father, John, a distinguished American botanist, *Travels through North and South Carolina* . . . (Philadelphia, 1791), part IV, chap. 3.

63 *sacrament . . . defines it*: this is the definition in Web.

working: T is being self-depreciating about his skill as a teacher; his other work included land surveying, odd jobs for the Emersons, laying out foundations, building fences and hen houses. When living at home, he fed the pig and cut wood for his mother and sister in return for room and board.

picking huckleberries: though selling for as little as 2 cents per quart in a glut, they grew wild and had to be gathered by hand; so T could have made something of a living picking them, for two weeks to a month in August.

64 *Admetus*: king of Pheres in Thessaly. The Gk god Apollo, banished from Parnassus, was made to tend his flocks.

Grecian or the Gothic: roughly speaking, the American equivalent to Georgian and Victorian styles of building and decoration. The former was giving way to the latter as T wrote.

free papers: refers to immigrants who arrived as indentured servants, contracted to work for a specified period before being allowed to remain as free agents.

In short . . . than I do: taken almost verbatim from a letter T wrote to the New York editor, Horace Greeley on 19 May 1848 (H).

65 *keeps . . . eye*: as the fugitive slave did, following the north star to the free states; cf. the old slave song 'Follow the Drinking Gourd'; the polestar is at the tip of the tail of *Ursa Minor*, or the handle of the Little Dipper, as it is called in America.

not proportionally more expensive than a small one: p. proofs: *not more expensive than a small one, in proportion to its size.*

operate at all: the pun depends on the old usage of 'operative' for worker; see above, p. 327.

66 *devil . . . idle*: cf. the old English and American proverb dating from the seventeenth century at the latest: 'The Devil finds work for idle hands.'

67 *Robin Goodfellow*: or Puck, the imp of English folklore who causes domestic chaos; see Shakespeare, *A Midsummer Night's Dream*, II. i. for an account of his mischief.

darkness visible: cf. Milton, *Paradise Lost*, i. 63: 'No light, but rather darkness visible.'

truer philosophy: 'philosophy' here is as in 'natural philosophy': hence, physics. The Copernican model of the solar system, in which the planets circle the sun, supplanted the Ptolemaic, in which the earth was supposed to lie at the centre.

Phaeton: or Phaethon, son of the sun in Gk mythology, who borrowed his father's chariot, grew fearful at his altitude, and veered so near the earth that he scorched it, leaving the Sahara Desert in his wake.

68 *Howard*: John H. (1726–90), English philanthropist and prison reformer (H).

cut ice: to remove and store for refrigeration in the summer; see below, 'The Pond in Winter'.

extra . . . intra: puns on L for outside and inside.

69 *tenth part*: as members of the Concord (First Parish) Church did in T's time. This was the tithe generally recommended by American protestant churches at mid-century.

Penn . . . Fry: for Howard, see above; William Penn (1644–1718) and Elizabeth Fry (1780–1845) were Quaker reformers; the former also founder of Pennsylvania.

70 *charity . . . sins*: cf. I Peter 4: 8: 'charity shall cover a multitude of sins.'

71 *left hand*: cf. Matthew 6: 3: 'let not thy left hand know what thy right hand doeth'; Jesus is advising almsgivers to keep their charity secret.

71 *manners . . . saints*: cf. I Corinthians 15: 33: 'evil communications corrupt good manners' (H).

72 *Gulistan . . . cypress*: the *Rose Garden* (*sic*) of Saadi, a Persian poet of the thirteenth century, was particularly popular among the Transcendentalists (H).

74 [*verse*]: Thomas Carew, from his masque *Coelum Britannicum*, in Chalmers, vol. v. The words are those of the god Mercury; the title is T's own. Chalmers has 'servile minds' for T's 'servile mind'.

76 *Hollowell place*: an old farm on the west side of the Sudbury River, just below Hubbard's Bridge, Concord (H).

[*verse*]: William Cowper (1731–1800), 'Verses Supposed to be Written by Alexander Selkirk During his Solitary Abode in the Island of Juan Fernandes', C. xvii, in *Poems of William Cowper . . .* (New York, [1850]). Selkirk was the real-life original of Robinson Crusoe. T italicizes 'survey' to pun on his own profession.

77 *Atlas*: the giant of Gk mythology who held the sky upon his head and upraised arms; by the nineteenth century converted to an emblem showing a giant holding the round earth itself.

Cato . . . "Cultivator": *De re rustica* (or more accurately *De rerum rusticarum*) (*On Country Matters*) is by Marcus Terentius Varro, not Cato, but in *Scriptores* it was bound up with Cato's *De agri cultura* (*On agriculture*); 'Cultivator' alludes to various contemporary farming journals, like *The Boston Cultivator* and *The New England Cultivator*.

78 *two years . . . one*: T lived at W (interruptedly) for two years, from 4 July 1845 to September 1847, but wrote up the experience as falling within the seasonal cycle of one year.

dejection: see note to p. 5, above.

Independence day: unmarked on p. proofs, so survives in 1st edn.; Prince and LA correct to 'Independence Day'.

certain house: cf. J 1.361 (OS).

Olympus: Mount Olympus, home of the gods in Gk mythology.

boat: T built it himself and used it for the excursion he made with his brother on the Concord and Merrimack Rivers in 1839. In 1842 he sold it to Nathaniel Hawthorne, then living in the Old Manse, near the Old North Bridge, for $7.

79 *Harivansa*: an appendix to the Hindu epic, the *Mahabharata*. The English is T's own translation from a French edition of the epic, *Mahabharata. Harivansa, ou Histoire de la famille de Hari . . .* (2 vols.; Paris, 1834–5).

mile and a half: this is the distance from Walden to the centre of Concord. To the Emerson's house, which T visited frequently, the walking distance was closer to a mile.

Battle Ground: W lies just over two miles to the south-east of the Old North Bridge, where on 19 April 1775 the skirmish took place between American Minutemen and British regulars that began the American Revolutionary War.

tarn: here in the English sense of a mountain lake (*OED*) rather than the American of a lake, a bog, or marsh (Web).

conventicle: an assembly of dissenters; hence Web (defensively): 'In the *United States* this word has no appropriate application, and is little used, or not at all.' Here T seems to hint at a forest meeting of witches.

80 *distant . . . ranges*: the Peterborough Range in southern New Hampshire (H).

keeps butter cool: butter was lowered into wells to keep it from melting and spoiling in the summer (H).

Sudbury meadows: the Sudbury River joins the Concord River just to the north-west of Concord, less than half a mile upstream of the battleground. Both rivers flood their meadows in the spring.

81 *"There are . . ."—said Damodara*: from the *Harivansa* (i. 283), T's translation from the French. Damodara was another name for the Hindu god Krishna (H).

Cassiopeia's Chair . . . Altair: common names of constellations, open clusters, and individual stars.

[verse]: from 'The Shepherd's Love for Philliday', in Thomas Evans (ed.), *Old Ballads, Historical and Narrative*, rev. edn. (4 vols.; London, 1810), i. 248.

Aurora: the Rom counterpart of Gk *Eos*, goddess of the dawn.

"Renew thyself": from Confucius, *The Great Learning*, 'Commentary of the Philosopher Tsang', chap I, part 1 (H).

Morning . . . heroic ages: a Gk pun on ἥέριος (epic Gk for early in the morning) and ἥρως (hero), as J 2.234 (NS) makes clear.

82 *trumpet . . . fame*: cf. 'And the trumpet that sings of fame', from 'The Landing of the Pilgrims', by Felicia Hemans, a contemporary British poet very popular in the US, in *The poetical works of Mrs. F. Hemans* (2 vols.; Philadelphia, 1832).

Iliad . . . wanderings: *The Iliad* and *The Odyssey*, the Homeric epics, begin by announcing their topics as, respectively, the wrath of Achilles and the wanderings of Odysseus.

82 *till forbidden*: (or t.f. for short), a printer's sign to keep an advertisement 'standing', or set up in type, until further notice.

Vedas: the four oldest sacred books of the Hindus.

Memnon: see above, p. 321.

83 *Spartan-like*: the ancient Spartans were noted for their simplicity, frugality, and self-discipline.

84 *it is "... forever"*: as the opening of the Shorter Catechism of *The New England Primer* (*c.*1683) put it.

ants ... men: Aeacus, son of Zeus and Aegina, was given the island named after his mother; wanting a population to rule, he asked Zeus to turn all the ants on the island into people.

cranes: at the beginning of *Iliad*, book III, the Trojan battalions are compared to cranes; cf. J 2.234 (NS).

clout upon clout: cloth upon cloth, i.e. endlessly repaired. Cf. 'New England's Annoyances' (see above, p. 327): 'If we can get a garment to cover without, | Our other in-garments are clout upon clout.'

dead reckoning: i.e. navigating by time elapsed, speed, and direction sailed, unchecked by bearings taken from landmarks or heavenly bodies.

internal improvements: American nineteenth-century usage for large capital projects, like canals and turnpikes.

85 *sleepers*: the English word for American 'ties' was not unknown in the States, where so many railroads were financed and designed by the British, but T prefers English usage here to facilitate his pun.

heaven in season: alludes to Nathaniel Hawthorne's story 'The Celestial Railroad' (1843), an ironic reworking of *Pilgrim's Progress* in which the devout narrator travels to the Celestial City on a train.

riding on a rail: with the additional meaning of riding someone out of town on a rail—usually tarred and feathered—a form of public humiliation in American frontier settlements.

Saint Vitus' dance: a disease of the nervous system 'affecting with irregular movements the muscles of voluntary motion' (Web).

setting the bell: in New England towns the church bell was rung in coded groups of rings to direct volunteer fire-fighters to the location of a blaze. When it called worshippers to meeting, the bell was 'set'—i.e. rung conventionally over a longer period, allowing the momentum of the swing to maintain the ringing.

we ... did not set it on fire: a defensive reaction to T's reputation in Concord as the man whose camping fire was allowed to get out of control and set a large part of W Woods on fire. T tells the story in J 3.75–8 (NS).

86 *Wachito River*: or Ouachita, in southern Arkansas; eye gouging was popularly supposed to be a common tactic of unarmed combat on the frontier.

mammoth cave: refers to the blind fish in Mammoth Cave, Kentucky (H).

penny-post: of British origin; it began in 1680 as a private operation serving greater London, then was reformed in 1840 as a service to send letters to all parts of the UK and the Empire for a penny.

87 *Don Carlos . . . Granada*: people and places involved in the Spanish power struggle of the 1830s and early 1840s.

1649: when the Commonwealth under Cromwell abolished the monarchy.

Kieou-he-yu: corrected in Prince and AL to Kieou-pe-yu; Confucian Analects, XIV. xxvi. 2, in Pauthier.

come to the end of them: T's copy: *accomplish it*

88 *Hindoo book*: as yet unidentified.

"Mill-dam": at the centre of Concord, site of an Indian fishing weir, then a dam for a water-powered mill, along which Main St., Concord, was laid out.

89 *nerves*: here in the older English sense of sinews or tendons.

tied to the mast like Ulysses: Ulysses, or Odysseus, had himself tied to the mast so that he could hear the song of the Sirens without being free to steer towards the rocks towards which they had lured other ships. His crew had to put wax in their ears and were told to respond to no orders from him.

whistles . . . bell rings: i.e. on a railroad locomotive.

point d'appui: bedrock, basis (Fr).

Nilometer: a device for measuring the rise and fall of the River Nile.

90 *cimeter*: scimitar.

91 *Mîr Camar Uddîn Mast*: a Hindu poet of the eighteenth century; T's translation, from Joseph Héliodore Garcin de Tassy, *Histoire de la littérature hindoui et hindoustani* (2 vols.; Paris, 1839–47).

92 *Delphi and Dodona*: ancient Gk oracles.

93 *waste paper*: refers to the fact that some of the ancient classics survived only because churchmen of the Middle Ages used the manuscripts as scrap paper for their own notes (H).

94 *Alexander . . . casket*: according to Plutarch in his life of Alexander (H).

95 *Homer . . . Virgil even*: not literally true; only that the results so far had failed to transmit the force of the original (H).

Vaticans: i.e. the great libraries, like the Vatican Library.

Vedas and Zendavestas: the sacred scriptures of the Hindus and Zoroastrians, respectively.

pile . . . heaven: as the people of Babel intended, with their brick tower (Genesis 1: 9), for which the Lord punished them by mixing up their languages.

foremost form: a form is a bench without a back; students in the first, or most elementary class, in a country school would sit on the foremost form; English secondary schools still use 'form' for class.

96 *Circulating Library . . . Little Reading*: the catalogue of the Concord Social Library for 1836 lists a *Much Instruction from Little Reading* (H). The pun is on Reading as a place-name, as in Reading and Little Reading in Berkshire, England, and Pennsylvania. Concord had a form of circulating library from as early as 1672. See Introduction.

Zebulon and Sephronia: typical characters in the sentimental fiction of T's day (H).

true love run smooth: cf. Shakespeare, *A Midsummer Night's Dream*, I. i. 132–3: 'Ay me! For ought that I could read I . . . the course of true love never did run smooth.'

I will not stir: able-bodied members of the community, summoned by the church bell, were expected to turn out to fight fires.

meeting-house: in New England towns the most important public building, since it was both the principle church and the venue for town meetings and other business. The Concord Meeting House was the First Church of Christ, which changed its name after the Revolution to the First Parish Church. In 1835 town business was moved away from the church meeting-house.

'The Skip . . . Tan': the first of these facetious titles may be a dig at James Fenimore Cooper's romance, *The Wept of Wish-ton-Wish* (1829); the second from a children's rhyme (H and others).

gizzard . . . sharpening: it is not clear why the corrugations lining a bird's gizzard should be in need of sharpening; they act to secure the pebbles and grit which act as teeth to macerate seeds and other food.

bencher: see note on 'form', above.

97 *deliquium*: a failure of vitality, given as archaic in Web; VDS says T is mocking contemporary medical language here.

English classics: as collected, for example, in Chalmers.

98 *Easy Reading*: T may have been thinking of *Easy Reading for Little Folks* (Boston, n.d.) (H).

"*Little Reading*": see note above, p. 324.

tit-men: i.e. diminutive men, as in titmouse or tomtit.

99 *Zoroaster . . . men*: Zoroaster, the Iranian religious reformer (c.628–556 BC) was once thought to have introduced monotheism, by placing the god Ahura Mazda at the centre of his system promising immortality in return for a good life.

common schools: primary schools supported by town funds. From 1647 New England towns of over fifty families were required to have a common school; when T was writing Concord had benefited recently from a state-wide campaign for common-school reform in Massachusetts.

Lyceum: a forum for debates and lectures, a common institution of adult and further education in New England towns. Concord's was founded in 1828. T lectured there often and served on its committee.

library . . . state: the Concord Town Library, which the Commonwealth of Massachusetts offered to fund provided it met certain specifications.

Abelard: Peter Abelard (1079–1144), the controversial logician, moral philosopher, poet, and theologian; what is evoked here is not his tragic affair with Héloïse, or even his learning as such, but the peripatetic nature of his teaching; as was common then, he lectured and tutored at Paris, Melun, Corbeil, and elsewhere.

100 *town-house*: in New England where the town's public business is transacted in legal meeting (Web). Begun in 1850, T himself surveying and laying out the foundations, Concord's new town-house was (still is) a handsome Greek-revival structure at the centre of town.

annually subscribed: T acted as treasurer of the Concord Lyceum for the year 1842/3.

"*neutral family*" *papers*: presumably fit for all the family to read and supported no political party.

"*Olive-Branches*": a Methodist weekly paper.

Harper . . . Redding & Co.: the New York publisher and the Boston bookseller.

100 *philosophical instruments*: 'philosophical', as in the context of the older usage of natural philosophy, hence scientific instruments.

102 *corn in the night*: in the short, hot and humid New England summers Indian corn grows so rapidly that it almost seems to have gained in height overnight.

103 *heathen deity*: apart from Sunday and Monday the English days of the week are named after Teutonic or Roman gods; hence e.g. Tiw's day (Tuesday, or the day of the god of war), Saturn's day (Saturday, or the day of the old Italic god of the Golden Age), etc.

Puri Indians ". . . day": natives of eastern Brazil; T is citing Ida Pfeiffer, *A Lady's Voyage Round the World* (see above, p. 22).

104 *half a dozen rods*: equal to 99 ft., therefore an underestimate; T's cabin was actually over 200 feet from the shore.

105 *tantivy*: rapid gallop, as at a fox-hunt, so called because of the tune blown on the huntsman's horn at that point.

pigeons: passenger pigeons, already scarce in T's day (H); now extinct, they were shot in great numbers because of their appetite for newly sown crops.

beat of a partridge: the male partridge, or ruffled grouse in British usage, attracts the female by beating the air in rapid oscillations to make a hollow, rumbling sound.

[*verse*]: from William Ellery Channing (nephew of the Unitarian minister), 'Walden Spring', in *The Woodman, and Other Poems* (1849). T called his style 'sublimoslipshod'.

106 *laden*: cf. Matthew 11: 28: 'Come unto me, all ye that labour and are heavy laden, and I will give you rest.'

raked: cranberry meadows were flooded at time of harvest so that the berries would float to the surface, from which they would be skimmed off by large wooden rakes.

cloud-compeller: T is trying for epic periphrasis here, or Anglo-Saxon kenning, as in 'Earth-Shaker' for Poseidon in *The Odyssey*, or *hwael-weg* or *wael-weg* (whale-road or death-road), for 'sea' in the *The Seafarer* (?ninth century).

107 *celestial train*: see note on 'heaven in season', p. 332, above.

drill-barrow: English usage for American drill-box: the reservoir holding the seed sown by the drill.

108 *Dismal Swamp*: although reminiscent of a station on an allegorical journey like *Pilgrim's Progress* or *The Celestial Railroad*, also an actual place in south-east Virginia and north-east Carolina; frequently mentioned in literature as the refuge for escaped slaves.

set their clocks: as in Great Britain, railroad schedules imposed strict uniformity of time (within broad zones) on regions until then content to set their clocks by the sun at the zenith.

depot: American for railroad station.

are on hand when the bell rings: T's copy: *were on hand when the bell rang*

when the bell rings: i.e. on the engine when the train was about to depart.

read the riot act: an English expression; the Riot Act of 1714 commanded 'all persons being assembled immediately to disperse' in the name of 'our Sovereign Lady, the Queen'; anyone present an hour after the act was read could be charged with a felony.

Atropos: of the three Fates (see above, p. 319), it was Atropos who cut off the thread of life.

bolts: the missiles fired by crossbows.

Tell: William Tell, the legendary Swiss hero (?thirteenth century), a peasant from the Guri district who resisted the Austrian governor and was made to shoot an apple off his son's head with a crossbow. Later he ambushed the governor, providing the impetus for the Swiss to rise up against Austrian rule.

Buena Vista: battlefield in northern Mexico where in 1847 American forces withstood a severe attack in the Mexican War (H). T opposed the Mexican War as a war to extend the boundaries of slavery. Cf. 'On the Duty of Civil Disobedience'.

109 *Bonaparte . . . rarest*: cf. H 290 n. 23.

Great Snow: the great snow of 20 February 1717 was described at length in a letter by Cotton Mather, from which T quotes extensively in J 8.163–5 (os). See below, pp. 236–7.

veto: literally, 'I forbid' (L), but the means by which the President, as a last resort, could kill a Congressional bill he opposed. Andrew Jackson, President from 1829 to 1837, was the first to use the power repeatedly to block legislation.

mould-board: the wooden or metal plate on a plough which turns over the furrow's slice.

Long Wharf . . . Champlain: i.e. from Boston Harbour to the Canadian border.

palm-leaf: woven to make summer hats (H).

cocoa-nut husks: the contemporary American spelling. Coir, or the rough fibre on the outside of the coconut shell, was used for doormats, ships' fenders, etc.

109 *gunny*: American for hessian or jute.

torn sails: used in paper making.

freshet: spring flood, which either breaks up the logs being floated downstream to the mill, or washes them out to sea.

Thomaston: in Maine, a major source of lime in T's day (H).

110 *slacked*: slaked, i.e. mixed with water so a chemical reaction can take place (Web).

no longer . . . Milwaukie: i.e. no longer praised as the height of fashion, unless in outlying districts.

Grand Banks: the shoals off Newfoundland, a great source of cod for European and American fishermen, who salted the fish, then dried it as hard, shingle-like planks.

dun: (of fish), cured (Web).

"*A cur's tail . . . form*": Charles Wilkins (trans.), *Fables and Proverbs from the Sanskrit . . .* (London, 1787, 1830, or 1844), chap. II, fable 9, 'The Lion and the Rabbit'. The original has '. . . will return to its original form' rather than T's '. . . will retain its natural form'.

111 *Green Mountains*: in Vermont.

the Connecticut: i.e. the Connecticut River, which divides Vermont from New Hampshire.

[*verse*]: Milton, *Paradise Lost*, i. 293–4; in this usage, 'Ammiral' (admiral) is a capital ship, not an officer.

cattle . . . hills: cf. Psalm 50: 10: 'the cattle upon a thousand hills.'

mountains . . . lambs: cf. Psalm 114: 4: 'The mountains skipped like rams, and the little hills like lambs.'

Peterboro' Hills: range of hills in south-west New Hampshire, visible from the hills around Concord (H).

112 [*verse*]: T's own; J 3.107 (NS) has 'a flowing' for 'a-blowing'.

Lincoln . . . Concord: towns around W. As the crow flies (and sound carries) they are respectively two, six, five, and one and a half miles distant from the pond.

113 *ancient u-lu-lu*: 'ancient' possibly because L *ululo* is 'howl', 'shriek', or 'wail', giving English 'ululate' or 'ululation'.

Ben Jonsonian: H conjectures that T might have been thinking of 'We give thee a shout: Hoo!' (Ben Jonson, *Masque of Queens*, ll. 317–18), but this seems a long shot.

tu-whit tu-who: cf. Shakespeare's song about winter in *Love's Labour's Lost*, V. ii. 928–9: 'Then nightly sings the staring owl, Tu-whit, | Tu-who! a merry note, | While greasy Joan doth keel the pot.'

115 *single spruce*: T's copy: *double spruce*.

wassailers: singers of drunken songs; from the AS drinking salutation 'wassail', or *waes hael* (good health).

catch: a song which one singer picks up as another stops, sometimes with ironic effect.

Stygian: like the River Styx, which the dead crossed on their way to the Underworld in Gr mythology.

116 *Indian pheasant*: chickens were not once 'wild Indian pheasant[s]', but red jungle fowls in SE Asia.

early to rise: cf. Benjamin Franklin, Maxims prefixed to *Poor Richard's Almanack* for 1735: 'Early to bed and early to rise, makes a man healthy, wealthy and wise.'

sailor: seamen took coops of chickens on long voyages to supplement their diet with eggs and fresh meat.

117 *Great Snow*: see above, p. 337.

118 *[title]*: Anderson (see note to food costs, above, p. 236), 62, suggests that the title may have come from Johann Zimmerman's *Solitude* (Albany, NY, 1796), a copy of which T owned, but in any case T's solitude owes more to the romantic trope than to actuality.

119 *abandoned . . . by men*: not true; it had recently been bought by Emerson, for whom T was caring for it.

"the world . . . me": cf. Thomas Gray, 'Elegy Written in a Country Churchyard' (1751), in *Poems of Mr. Gray . . .* (4 vols.; New York, 1778), vol. iii: 'The ploughman homeward plods his weary way, | And leaves the world to darkness and to me.'

black melancholy: one of the four humours, according to ancient and medieval physiology; *melancholy* is Gr for black bile.

Æolian music: i.e. the sound of the Æolian harp, produced when wind (Aeolus was god of the winds in Gr mythology) played across the horizontal strings tuned so as to produce a complex mixture of primary notes and harmonics. The instrument was a favourite of the English romantics and American transcendentalists. T had one.

121 *[verse]*: from Patrick MacGregor's translation of Ossian (James Macpherson), *The genuine remains of Ossian* (London, 1841), 193.

north-east rains: a 'north-easter' brings the strongest winds and longest rains and snows to New England.

Beacon Hill: not the conservation area of today, but the rise above the Common on which the Old State House stands, hence the centre of Boston business and government.

112 *Five Points*: a busy intersection, now obliterated, north of the City
     Hall, New York, notorious for its slums and crime (H).

     Walden road: now Route 126; runs from Concord eastwards past W.

     Brighton: a suburb of Boston, then the site of numerous slaughter-
     houses (H).

     Bright-town: 'bright' was a common name for a prize or favoured ox
     (H).

     "*How . . . sides*": Confucius, *The Doctrine of the Mean*, XIV, in
     Pauthier.

123 "*Virtue . . . neighbors*": Confucius, *Analects*, book IV, xxv (H).

     Indra: in Hindu mythology, the Vedic god of the middle realm, the
     air (H).

124 *Society*: here in sense of company, fellowship (Web).

     girls . . . factory: probably refers to mills in Lowell, Massachusetts,
     where worked young, unmarried women, chaperoned and living in
     dormitories and given ample opportunities to develop their talents
     and knowledge. In the 1840s and 1850s Lowell was on the itinerary
     of reformers from all over the world, and generally approved, even
     by critics of other aspects of American life. T was sceptical, though,
     and so was Herman Melville in his short story, 'The Tartarus of
     Maids' (1855), which explores the establishment in images of
     thwarted sexuality.

125 *blue devils*: the blues, lowness of spirit, hypochondria (Web).

     mock sun: parhelion, a bright spot where two solar halos intersect,
     giving the illusion of another sun alongside the actual one.

     devil . . . legion: cf. Mark 5: 9, where Jesus asks the name of the
     unclean spirit he has just cast out, and the spirit answers, 'My name
     is Legion; for we are many.'

     Mill Brook: see note on Mill Dam, p. 333, above.

     old settler: H conjectures Pan, the Gk god of nature.

     Goffe or Whalley: Edward Whalley and his son-in-law William
     Goffe, both implicated in the indictment and execution of Charles I,
     escaped to New England after the Restoration.

     elderly dame: Mother Nature (H).

126 *Parrs*: Thomas Parr, supposedly born in Shropshire in 1483, died in
     1653, aged 152 (H).

     Acheron: a name given to several rivers which in Gk mythology led
     to the Underworld; also in a general sense, the Underworld itself.

stopples: stoppers; the reference is to secondary fermentation working so as to pop the corks from bottles of beer or wine kept in too warm an environment.

Aurora: see above, p. 331.

Hygeia: Gk personification of health.

Æsculapius: Gk god of medicine.

Hebe: Gk personification of youth.

127 *Tremont . . . House*: three luxury hotels in Boston, New York, and Concord, respectively.

128 *enough room*: p. proofs: *room enough*

withdrawing room: or drawing-room, originally a room behind another room, for retirement (Web).

sun rarely fell: see note above on 'fade . . . carpet', p. 328.

hasty-pudding: traditional American dish made of corn-meal mixed with water and boiled, or (Web) flour and milk boiled.

129 *Cerberus*: in Gk mythology the three-headed dog that guarded the entrance to the Underworld.

[verse]: Edmund Spenser, *The Faerie Queene*, book I, canto i, stanza 35, in Chalmers, vol. iii. T modernizes spelling.

Winslow: Edward Winslow (1595–1655) served three times as Governor of Plymouth Plantation, in 1633, 1636, and 1644.

Massassoit: chief of the Wampanoag Indians, with whom Plymouth Colony had made a peace treaty in 1621.

130 *"He laid . . . asleep"*: 'A Journey to Pokanoket, the habitation of the great King Massasoit . . .', in *Mourt's Relation* (London, 1622), reprinted in *Collections of the Massachusetts Historical Society*, 1st series, 8 (1802), 203–39. The original has 'planks' for T's 'plank'.

Paphlagonian man: identified in J 2.160 as T's French Canadian friend Alex Therien. The Paphlagonians were one of the most ancient peoples of Anatolia.

131 *[verse]*: from *Iliad*, beginning of book XVI, in *The Iliad of Homer, from the text of Wolf with English notes and Flaxman's designs*, 2nd edn. (Boston, 1834); Patroclus comes to Achilles' tent in tears to beg to be allowed to go out wearing Achilles' armour to fight for the Greeks.

white-oak bark: used as an astringent.

133 *threescore*: sixty; cf. Psalm 90: 10: 'The days of our years are threescore years and ten.'

134 *Vermont gray*: homespun cloth made in Vermont.

135 *pecunia*: money (L), derived from *pecus*, cattle.

Plato's . . . man: it was Diogenes the Cynic (d. *c*.320 BC), the Gk philosopher of self-sufficiency in poverty, who confronted Plato with the plucked chicken.

136 *He*: i.e. his example.

dipper: American for ladle.

almshouse: at the time the almshouse was on the old Stratton Farm, on the Walden Road between the Pond and Concord, about a mile from T's hut.

overseers: officials who looked after the almshouse and assigned the poor to various houses to work.

selectmen: elected officials of the town.

137 *exalted*: cf. Matthew 23: 12: 'And whosoever shall exalt himself shall be abased; and he that shall humble himself shall be exalted.'

hospitalality: i.e. charity of the sort offered by a hospital. In T's time a hospital was a building for the reception of sick, infirm, and helpless paupers supported by charity (Web).

fox in the fable: possibly a reference to Aesop's fable of 'The Fox and the Woodman', in which a fox begs a woodchopper to hide him from a pursuing hunter.

138 *"O Christian . . . back?"*: if this is a citation, say of a poem or song, it has yet to be identified.

runaway slave: T once hid a runaway slave in his family home in Concord, before sending him on his way north to Canada.

White Mountains: a popular resort area in New Hampshire.

the rub: as Hamlet says (III. i. 66) of the unknowableness of death before being sampled.

139 *Dr. B.*: Dr Josiah Bartlett, a Concord doctor for over half a century (H).

com-munity: a pun on the L syllable *com* for 'together' and L *munitio*, 'fortification'.

[verse]: T's own variant on the cumulative nursery rhyme, 'This is the house that Jack built.'

"Welcome, Englishmen!": on 16 March 1621, a Native American walked out of the forest and spoke these words to the startled settlers at Plymouth. He was Samoset (d. *c*.1653), a sachem of the Pemaquid tribe, who according to Barber, 'had learned something

of the English tongue from the captains of the fishing vessels who resorted [to the coast of Maine]'.

140 *Herculean labor*: see above, p. 313.

Antæus: en route to his final labour to fetch the golden apples of the Hesperides, Hercules encountered this giant, who, being the son of Earth, revived every time he was struck down. Hercules finally choked him to death by hoisting him on his shoulders.

why should I raise them?: see Introduction.

141 *cords*: measurement of wood by volume; a cord forms a pile 8 ft. long by 4 ft. high and 4 ft. wide (Web).

throw dust . . . heads: the Old Testament sign of mourning and humiliation; cf. Job 2: 12 '. . . they . . . knew him not . . . and sprinkled dust upon their heads toward heaven.'

plastic artist: sculptor.

142 *agricola laboriosus*: hard-working farmer (L).

Lincoln and Wayland: Wayland is five miles due south of Concord, Lincoln past W about three miles to the SW.

gigs: light carriages pulled by one horse (Web).

when others . . . hoe: i.e. when others had begun to chop weed.

ministerial husbandman: the Revd. Henry Coleman (1785–1849), a leading proponent of agricultural reform.

Mr. Coleman's report: Coleman published four surveys of Massachusetts agriculture from 1838 to 1841 (H).

English hay: grass crops of imported varieties grown for fodder, as distinct from the less nutritious native grasses used for animal bedding (H).

143 *Rans des Vaches*: the lyric by Friedrich Schiller (1759–1805), opening his drama, *Wilhelm Tell* (1804), in *Songs and Ballads: translated from Uhland, Korner and other German poets* (Boston and London, 1842). A *ranz des vaches* was a traditional song sung by Swiss herdsmen to call their cattle home.

Paganini: Niccolò Paganini (1784–1840), the violinist of legendary virtuosity.

144 *guns*: the Concord Artillery was a unit of the State Militia (H).

"trainers": artillerymen.

144 *tintinnabulum*: bell (L). H points out that the word does not appear in Virgil.

    Middlesex: the county containing Concord and W.

145 *the trumpet . . . fame*: see above, p. 331.

    spit a Mexican: T's opposition to the Mexican War (1846–8) was well known, and in July 1846, he spent a night in prison for refusing to pay his taxes to support it, producing 'Civil Disobedience' as his apology.

    know beans: cf. the New England expression, 's/he doesn't know beans', meaning s/he is ignorant or inexperienced.

146 *cranes*: see above, p. 332.

    Hector: the Trojan hero in *The Iliad*, whose helmet was crested with horsehair; when Achilles finally strikes him down with a lance through the neck in book XXII, he falls and rolls in the dust.

    Pythagorean: Pythagoras, the sixth-century Gk philosopher and mathematician, forced his followers into a strict diet that included a prohibition of fish and beans.

    porridge or voting: i.e. whether they be used for pottage or as markers for votes.

    Evelyn: John Evelyn, *Terra: a philosophical discourse of earth* (York, 1778). Evelyn's 'Vicars succedanous' is a figure for a substitute twice removed.

    Sir Kenelm Digby: (1603–65), courtier, philosopher, diplomat, and scientist, was the first to recognize the importance of oxygen to plant life.

    Mr. Coleman: see above, p. 343.

147 *outgoes . . . income*: the salient points are that T, even though paying only half the going rate for labour—and not accounting for either his own labour or the cost of his land—produced far fewer, poorer beans per acre than the average contemporary farmer would have done. See Introduction.

    patrem . . . oportet: the head of the family should have the habit of selling, not buying (L). Cato, *De agri cultura*, ii. 7, in *Scriptores*.

148 *Congress . . . land*: it was a popular custom in T's time for Congressmen to distribute free seeds to their constituents (H).

149 [*verse*]: Francis Quarles, *The shepheards oracles: delivered in certain eglogues . . .* (London, 1646), eclogue v.

    Cattle-shows: the Middlesex County Fair, held in Concord each year, included a cattle show.

Ceres: Rom equivalent of Gk Demeter, goddess of fertility and the harvest.

Jove: Jupiter, the Rom equivalent of Gk Zeus, father of the gods, god of the heavens.

infernal Plutus: a pun turning Plutus, Gk personification of wealth, into Pluto, the ritual title of Hades, god of the Underworld, or infernal regions.

Varro: M. Terenti Varronis (116 BC–28 BC), said to be the most learned of the Romans, *Rerum Rusticarum* ('On Agriculture'), III. i, 5, in *Scriptores*.

Cato . . . Saturn: *De agri cultura* [Introduction]. 4, in *Scriptores*.

Saturn: mythical king of ancient Italy, thought to have introduced agriculture; also the father of Jove or Jupiter, but much more closely associated with fertility than the Gk equivalent father of the gods, Chronos.

150 *The ear (. . . bearing)*: a paraphrase of Varro, *Rerum Rusticarum*, I. xlviii. 1–3, in *Scriptores*.

151 *stint*: in American usage, most commonly a certain, limited task, after which the worker is free (Web).

homœopathic doses: i.e. very small, as medicines are prescribed in homeopathy.

Redding & Company's: see above, p. 335.

Etesian winds: in ancient Gk reference, the winter north wind that blew down the Aegean.

152 *caryatides*: columns or pilasters carved in the shape of draped female forms, such as hold up the roof of the Erechtheum in Athens.

Orpheus . . . Sirens: the legendary singer of Gk mythology, personification of the enchanting musician, was a member of the crew of the Argo on the pursuit of the Golden Fleece. His song calmed the seas and was even more melodious than that of the sirens, whose singing lured mariners to destruction on the rocks. The translation, which H conjectures is T's own, is from the *Argonautica* by Apollonius of Rhodes (4.903).

153 *avenues*: the shortest route from T's mother's house to W (about one and a quarter miles) was via back streets to the south of Concord, then along the railroad embankment; even closer (at one mile) was the Emersons' house, from which a path led up Brister's Hill through W Woods to T's hut.

153 *"as I sailed."*: refrain of the old American 'Ballad of Robert Kidd' (H).

elsewhere related: in 'Resistance to Civil Government' (1849), now better known by the title of its first reprint, 'Civil Disobedience' (1866).

155 *odd-fellow society*: a reference to the International Order of Odd Fellows, one of several secret and benevolent societies formed for social union and mutual help. Founded in London in the eighteenth century, the IOOF opened its first American lodge 1821.

Fair-Haven Hill: about one-half mile south-west of W on the east bank of the Sudbury River.

volume of Homer . . . improperly gilded: according to H, it was vol. i of Alexander Pope's translation of Homer (5 vols.; Baltimore, 1814) that went missing; 'improperly' here probably in the sense of unsuitably or unbecomingly (Web), rather than imperfectly.

soldier of our camp: a story from the 'Sayings of Confucius', which T edited for *The Dial*, 3 (1843), 494, in which a soldier who had lost his shield comforted himself with the thought that another soldier would find and make use of it. See longer note in H.

156 *[verse]*: H notes that Evelyn's *Silva* (see above, p. 315) cites these same two lines from the *Elegies of Tibullus* (III. xi. 7–8) and supplies nearly the same translation.

"You . . . bends": Confucius, *Analects*, book XII, xix, in Pauthier.

157 *"to fresh . . . new"*: John Milton, 'Lycidas', l. 193, in *The Poetical Works of John Milton . . .* (6 vols.; London, 1801), vol v.

huckleberries . . . cow-boy: plentiful but hard to gather, hence uncommercial; the 'cow-boy' here is literally a boy who looks after cows, not the more seasoned worker of western cattle drives.

three hills: Copp's, Fort, and Beacon, where Boston was first built (H).

bloom: not the flower, but the yeasty covering on the skin of the fruit.

Cænobites: a religious order; and (H thinks) one of T's best puns: 'See? No bites.'

158 *rods*: see above, p. 325.

159 *perennial spring*: true, in a sense; although water does not flow into W up a hole in the bottom of the pond. Eugene H. Walker ('Walden's Way Revealed', *Man and Nature* (Dec. 1971), 12–20) has shown that W is 'fed by seepage from sand and gravel'.

160 *"to be . . . solid"*: source so far unidentified.

watered . . . silks: silks that catch the light in wavy patterns and colours, made that way by watering, then passing them through a hot press (Web).

161 *fit studies for a Michael Angelo*: Michelangelo (1475–1564), the Italian renaissance sculptor, painter, and architect, made numerous sketches of the male nude showing the musculature in anatomical relief.

neighorhood: uncorrected in p. proofs, so survived into 1st edn. Prince and LA correct to 'neighborhood'.

162 *bottomless*: H writes, 'Seemingly every community in New England has its so-called "bottomless ponds".' T mentions others in J 3.115 (NS).

Castalian Fountain?: a spring sacred to Apollo, named after the young woman who jumped into the spring while being pursued by the god.

163 *Golden Age?*: in Gk and Rom legend, the first great age of man, when people lived without pain or old age, the earth yielded up its fruits without cultivation, and private property, travel, trade, and warfare were unknown.

path: still used by people visiting W to walk or swim.

alto-relievo: high-relief (It), the projection of a figure in sculpture, by half or more of its depth, without being entirely detached (Web), or the illusion of same in painting; cf. *bas-relief* (Fr), or low-relief, in which the figure is projected by less than half.

The pond rises and falls: not necessarily responding immediately to the rate of precipitation, since its level depends more on the water table than on rainfall run-off.

chowder: in New England a dish in which biscuit, salt pork, and usually fish are boiled in milk (Web). Walker provides geological evidence to suggest that the chowder party probably took place in 1823.

164 *sympathize with Walden*: this is untrue, since Flint's (or Sandy) Pond depends proportionately much more on the immediate run-off of precipitation than does W.

White Pond: about two and a half miles SW of W as the crow flies.

165 *the tradition*: see H (299) for possible sources of this story.

ancient settler: see above, p. 340.

stones . . . paver: the stones are glacial till and the paver glacial action.

166 *Boiling Spring*: not a hot spring but one that bubbles up from the bottom (H).

few breams, and: T's copy: *few breams, Pomotis obesus [Nov. 26–58] one trout weighing a little over 5 lbs (Nov. 14–57), and*

167 *reticulatus*: with a net-like pattern (L).

guttatus: speckled (L).

skim over it, and the peetweets: p. proofs: *skim over it, kingfishers sound their alarm around it, and the peetweets*; T's copy: *skim over it, kingfishers dart away from its coves, and the peetweets*

Fair Haven: Fairhaven Bay, a widening in the Sudbury River, less than one mile SW of W.

168 *nests of the chivin*: T was right; see note in H.

fluviatile: belong to, or existing in rivers (Web), from L *fluvius*, river.

169 *skater insects, at equal intervals*: T's copy: *skater insects (Hydrometer), at equal intervals*

piscine: relating to fish (Web), from L *piscis*, fish.

murder will out: cf. 'Mordre wol out', Chaucer, 'The Prioress's Tale', l. 1766, in Chalmers, vol. i.

170 *quicksilver*: mercury. From the sixteenth century, mirror-makers 'silvered' the backs of mirrors with an amalgam of tin and mercury, but this process was superseded in 1835 when Von Liebig (see above, p. 315) discovered how to coat mirrors with metallic silver by a chemical reduction of silver-ammonia compounds.

173 *workshop . . . desk*: T had worked in his father's small pencil factory and as a schoolteacher.

instead . . . drink: the prediction was too pessimistic. In 1867 the Railroad, who owned the land, built a little pleasure park at Hawthorne's Cove at the south-west end of the Pond, where people could swim, dance, and go out in canoes. W is still a popular swimming and boating centre for the people of the surrounding towns.

Ganges: the great river of the plains in north India, sacred to the Hindus.

pipe: at one time the citizens of Concord planned to run water from W to Concord, but decided instead to use Sandy Pond as the reservoir, because it was higher up and the water from it would run under greater pressure (H).

174 *Iron Horse . . . shore*: refers to the railroad steam engines' enormous appetite for water to boil and wood to burn.

Trojan horse: as the story is told in *Aeneid*, book II, the hollow horse in which the Gk soldiers hid, and from which they descended to open the gates of Troy for their fellow-invaders, after the Trojans had pulled the horse into the city thinking it was an offering to Athene.

Moore of Moore Hall: cf. 'But More of More-Hall, with nothing at all, | He slew the dragon of Wantley', from 'The Dragon of Wantley', in Bishop Percy's *Reliques of Ancient Poetry* (3 vols.; London, 1765).

Deep Cut: on the stretch of railroad between W and Concord.

thrust . . . lance: the Trojan Laocoön, trying to warn his fellow-countrymen of the treacherous gift born by the Greeks, threw his lance into the side of the Trojan Horse, but the gods kept it from penetrating the wooden wall.

175 *[verse]*: T's own composition.

engineers . . . brakemen: American railroad terms for drivers, firemen, guards.

State-street: financial district in Boston.

"God's Drop": as Emerson refers to W in his Journal for 9 April 1840 (H); 'drop' probably as in the diamond in an earring (Web).

waste its sweetness: cf. Gray, 'Elegy in a Country Churchyard', ll. 154–5: 'Full many a Flower is born to blush unseen, | And waste its Sweetness on the desert Air', in *Poems of Mr. Gray . . .* (4 vols.; New York, 1778).

176 *pad*: the metaphor is of a lily-pad, or the leaf of any floating water plant (Web).

Indian file: i.e. in a single line, as parties of Indians were said to walk along paths in the forest.

Some skin-flint: citing another source, H (301 n. 44) points out that T had first hoped to build his hut on the shores of Flint's Pond, but the plan was thwarted by Mr Flint, the Pond's owner.

177 *English hay*: see above, p. 343.

privilege: landowners in New England had to obtain a written 'privilege' from the community before damming up a stream for water power (H).

177 *landscape*: cf. Emerson, 'Nature' (1836): 'Miller owns this field, Locke that, and Manning the woodland beyond. But none of them owns the landscape.'

A *model farm*: for T's contempt for modern methods of intensive cultivation, see above, pp. 142–3, and Introduction.

178 *Icarian Sea*: surrounding the island of Samos, where according to Gk legend, Icarus, son of Daedalus fell to his death when trying to escape from Crete on wings crafted by his father.

"*brave attempt resounds*": cf. William Drummond of Hawthornden (1585–1649), in 'Icarus': 'For still the shore my brave attempt resounds', in Chalmers, vol. v.

on my way: i.e. east of W.

Fair-Haven . . . Concord River: now Sudbury River; it does not become the Concord River until it joins the Assabet at Concord itself.

lake country: the Lake District in Cumberland and Westmorland, England: i.e. T's equivalent to Wordsworth's actual and symbolic landscape.

Since . . . profaned Walden: in J 3.24 (NS) T originally wrote: 'Now, since the railroad & and the Irish have prophaned Walden . . .'

dog-day: when Sirius, the dog-star, rises and sets with the sun; hence the hot, humid summer from the end of July to the beginning of September.

sand-paper: T's family business manufactured sandpaper as well as pencils.

Virid: poetic for green.

"*Topographical Description . . . Concord*": by William Jones in vol. i of the 1st series (1792).

180 *diamond of Kohinoor*: the legendary diamond of over 100 carets, now one of the Crown Jewels.

181 [*title*]: the residence of James Baker of Lincoln Township, Baker Farm lay about a mile east of W.

Druids: learned priests among the ancient Celts who held their rites in oak groves.

Valhalla: in Scandinavian mythology, a hall reserved for those who have died in battle; by extension, any post-mortem reward proposed for the deserving.

white-spruce trees: T's copy: *black-spruce trees*

182 *halo ... shadow*: cf. J 3.85 (NS), where it is related matter-of-factly.
'Cellini's halo' is a ring of bright light above or around the shadow of
the observer's head on wet grass when the sun is low in the sky.

elect ... natives: Protestant, especially Puritan term for the chosen
of God; by 'natives' T means, not the Indians, but the original
Puritan settlers of Massachusetts; cf. the Irish, later arrivals who
were predominantly Roman Catholic.

Benvenuto Cellini: (1500–71), Florentine sculptor and goldsmith.
The halo effect named after him is recounted in chap. 26 of his
Memoirs ... written by himself, trans. Thomas Roscoe (2 vols.; New
York, 1845).

castle of St. Angelo: the Castel Sant'Angelo, originally the Tomb of
the Emperor Hadrian, the drum-shaped fortress on the west bank
of the Tiber in Rome, was converted into a fortress in the fifth
century.

Pleasant Meadow ... Baker Farm: just under a mile south of W, on
the east shore of Fair Haven Bay.

183 *[verse]*: 'Baker Farm' by William Ellery Channing, the younger
(1818–1901), *The Woodman, and other poems* (Boston, 1849). *musquash* is the Indian word for (and origin of) muskrat.

[verse]: Channing, 'Baker Farm', with slight variations.

sibyl-like: derived from Sibyl, a class of ancient Gk priestesses who
interpreted the oracles of Apollo, of whom the best known was the
one at Cumae, near Naples. Apollo granted her wish for a long life,
but she neglected to ask for prolonged youth to accompany it.

184 *"bogging"*: not in Web or *OED*; presumably Field's own word for
turning over the soil, by extension from the practice of digging peat
from bogs in Ireland. 'Bog-Irish' was a common pejorative for the
poorest Irish immigrants to Massachusetts.

ten dollars an acre: cf. the contemporary farm labourer's wages of
$1 per day for native workers, 50 cents for immigrants.

185 *dead reckoning*: navigation by estimates drawn from distance and
direction sailed, unverified by bearings taken from landmarks or
celestial bodies.

rout: probably in the sense of to scoop, gouge, or hollow out
(W7NWD), hence to fashion the grooves in a column.

186 *Remember ... youth*: cf. Ecclesiastes 12: 1.

English hay: see above, p. 343.

187 *[verse]*: Channing, 'Baker Farm', with slight variations. 'Guy Faux'
(Guy Fawkes) was the conspirator caught red-handed in the

Catholic Gunpowder Plot (5 November 1605) to blow up the Houses of Parliament, whose effigy is still burned in England on the anniversary of the plot's discovery.

188 *talaria*: winged sandals (L), of the sort worn by the Rom messenger of the gods and patron of commerce, Mercury.

189 [*title*]: the Transcendentalists used this phrase to refer to laws of conscience, informed by God or the Oversoul, transcending the laws of the state.

Falls of St. Mary: it is not clear what specific geographical feature T is referring to here.

190 *public holidays*: the New England Puritan heritage did restrict the number of holidays by comparison, say, to a Catholic European country, but Concord allowed a number of mainly secular holidays, like Thanksgiving, election days, the annual 'muster day' (when citizens joined their militia company to drill), and some court days.

fowling piece: i.e. a (smoothbore) shotgun.

common: meant to evoke an English, rather than American, element in town geography. Though Boston still has a common, it is not a regular feature of more modern American towns.

191 *fishers of men*: as Jesus said he would make his apostles; see Mark 1: 17.

[*verse*]: Chaucer, *The Canterbury Tales*, Prologue, ll. 177–8, in Chalmers, vol. i. In fact, it refers not to the nun, but to the monk 'that loved venerye' (hunting, with a pun on venery, or the pursuit of sexual pleasure).

Algonquins: one of the New England Indian tribes.

philanthropic: the Gk root sense of 'philanthropy' is the love of mankind, so by breaking the word and italicizing 'anthropy' T is making the point that his love for creation is not confined to men.

Good Shepherd: i.e. Christ.

192 *council*: in the Commonwealth of Massachusetts an executive council advises the governor.

legislature: in Massachusetts, as in other states, the legislature is a branch of government distinct from the executive (governor and council).

scullion: a menial kitchen servant (Web).

193 *contemporaries*: H points out that '[t]he 1840's and 1850's were a period of food reform, with Dr. Sylvester Graham of "graham cracker" fame at the helm'.

Kirby and Spence: William Kirby and William Spence, *Introduction to Entymology* . . . (4 vols.; London, 1815–26).

194 *carniverous*: thus spelled in p. proofs and 1st edn. Prince and LA correct to 'carnivorous'.

genius: here used in the sense of 'nature' or 'structure of mind' (Web).

195 *ebriosity*: habitual drunkenness; the word is so rare that the *OED* gives only two uses of it: this one and Sir Thomas Browne's *Pseudodoxia Epidemica* (1646), 270.

"*he . . . distress*": Raja Rammohun Roy (trans.), *Translation of Several . . . of the Veds* (London, 1832), 21 (H).

196 "*The soul . . . of food*": Confucius, *The Great Learning*, 'Commentary of the Philosopher Tsang', chap. vii, in Pauthier.

Not that . . . man: cf. Matthew 15: 11–19, where the comment is a critique of the Pharisees' strict insistence on dietary laws.

197 *reptile*: as an adjective, grovelling, low, vulgar (Web).

"*That in . . . carefully*": in Pauthier.

"*A command . . . to God*": in Rammohun Roy's translation, see above.

fauns and satyrs: in Gk myth, satyrs were beings associated with the worship of Dionysus, representing the vital powers of nature, usually described as of human form with the tails and feet of goats, pointed ears, and small horns on their heads. Roman writers often confused them with the fauns of L mythology who were demigods of similar appearance, associated with fields and shepherds.

198 [*verse*]: John Donne (1572–1631), 'To Sir Edward Herbert at Julyers', ll. 9–10, 13–17, in Chalmers, vol. v, with minor variations.

199 *Hindoo lawgiver*: refers to the Hindu *Laws of Menu* or the *Vishnu Purana*, which he had read in the H. H. Wilson translation (London, 1840) (H).

201 *companion*: William Ellery Channing, the younger; see note above, p. 351. In the dialogue that follows, T is the hermit and Channing the poet.

Bose: a popular dog's name in the mid-nineteenth century (H).

202 *Con-fut-see*: a way of transcribing the name Kung-fu-tseu, known better by his L name Confucius.

Mem.: memo or memorandum.

203 *Pilpay*: or Bidpai, reputed author of a collection of east-Indian fables; see H 304 n. 7.

203 *a wild native kind not found*: T's copy: *a wild native kind (mus leucopus) not found*

distinguished naturalist: Louis Agassiz (1807–73), the Swiss-born scientist and collector, Professor of Natural History at Harvard from 1848. T corresponded with him and sent him specimens.

up my clothes: J 2.162 (NS) is even more alarming: 'up my pantaloons inside.'

bo-peep: peek-a-boo.

204 *dishabille*: a loose, negligent dress for the morning (Web).

callow: used here in its primary sense of 'bald', 'unfledged'.

205 *Brister's Hill*: just north of W.

206 *duellum . . . bellum*: duel . . . war (L).

Myrmidons: warlike Achaean race in ancient Thessaly who followed their leader Achilles to the Trojan War.

republicans: in the nineteenth-century sense of members of a commonwealth in which power is lodged in representatives elected by the people.

imperialists: in the nineteenth-century sense of subjects or soldiers of an emperor (Web), hence (cf. the republicans) soldiers ordered or paid to fight, rather than taking up (their own) arms in defence of their commonwealth.

mother . . . upon it: as, according to the Gk biographer and historian Plutarch (in 'Sayings of Spartan Women', *Moralia*, iii. 241.16), a Spartan matron commanded her son to do. The Gk city-state was legendary for its military culture.

207 *Achilles . . . Patroclus*: 'The Wrath of Achilles', the theme of the *Iliad*, was Achilles' quarrel with Agamemnon, leader of the Achaeans' expedition against Troy, and his consequent refusal to fight. When his lover Patroclus was killed in battle, Achilles turned his wrath against the Trojans and re-entered the campaign.

Austerlitz or Dresden: sites of two of Napolean's bloodiest battles (H).

Concord . . . Hosmer: for the skirmish at Concord Bridge on 19 April 1775, see Introduction. The two killed on the American side were Abner Hosmer from Acton and his Captain, Isaac Davis, both brought down by the second volley fired by the British. Luther Blanchard was slightly wounded in the first volley. Jonathan Buttrick commanded the American Minutemen at the engagement at Concord Bridge.

tax on their tea: the hated Townshend Acts imposed by the British Exchequer included a tariff on glass, lead, paper, paint, and tea exported to the colonies, the revenues to be used for the salaries of colonial judges and governors.

Bunker Hill: actually fought on the adjacent Breed's Hill near Boston in July 1775. The British advanced up the hill in ordered ranks of redcoats, to be potted at by the Americans at the top, firing only when they could see 'the whites of their eyes'. The British took the position but lost over a thousand killed or wounded doing so— nearly half of their attacking force.

208 *Hotel des Invalides*: the home for 7,000 disabled soldiers in Paris, founded by Louis XIV. Later the tomb of Napoleon, his son and brothers.

Kirby and Spence: (see note above, p. 353), pp. 361–2.

Huber: François Huber (1750–1831), Swiss entymologist.

209 *Æneas Sylvius ... Christiern the Second*: Aeneas Sylvius (1405–64) was Pope Pius II; Eugenius IV was Pope from 1431 to 1447; Olaus Magnus (1490–1558) was Archbishop of Uppsala; Christiern II ruled Denmark and Norway in the sixteenth century (H).

Polk: James K. Polk was President for one term, 1845–9.

Webster's ... Bill: the Fugitive Slave Act 1850, which outraged many northern progressives, including T, because it called for slaves who had escaped to the non-slaveholding states to be shipped back to their owners without a trial, prompted Harriet Beecher Stowe to write *Uncle Tom's Cabin* (1852). Daniel Webster (1782–1852) was not the author of the FSA but a prominent New England Whig who supported the bill in order to dampen down sectional disputes threatening the Union.

Bose: see note above, p. 353.

farm-houses ... Gilian Baker's: probably the house of Jacob Baker, just over half a mile south of W along the old Walden Road; not to be confused with the Baker Farm of Chap. 10.

210 *winged ... horse*: a reference to Pegasus, the winged horse of Gk myth.

Mill-dam: in the centre of Concord; see note above, p. 333.

211 *checker*: 'checkers' is American for draughts.

213 *bayous*: in Louisiana a bayou is the marshy outlet of a lake, or a channel for water; also one of the channels through which the Mississippi crosses its delta to the Gulf of Mexico.

214 *rake*: for the method of harvesting cranberries, see above, p. 336.

jammed: i.e. turned into jam.

butchers . . . prairie grass: in the first decades of the nineteenth century the North American bison, by this time already extinct east of the Mississippi, were sought increasingly by hide hunters, meat packers, and tongue picklers.

sleep: punning on 'sleepers', English for American railroad ties.

215 *Ceres or Minerva*: Ceres was the Rom name for Gk Demeter, protectress of agriculture and all the fruits of the earth; Minerva the Rom equivalent of Gk Athena, protectress of the arts and trades, and personification of the power to think and invent.

many a tale . . . told!: cf. Thomas Moore (1779–1852), 'Those Evening Bells', '. . . those evening bells! | How many a tale their music tells'.

216 *avoiding . . . cold*: H points out that *Iliad*, iii. 4 'is often translated as "avoiding winter and unspeakable cold" '.

winter quarters: T lets it drop almost as an aside that he did not live at W through the harshest part of winter.

Nebuchadnezzar: King of Babylon, who looted the Temple at Jerusalem; when his son Belshazzar held a great feast, drinking and eating out of the sacred vessels of the Temple, God's finger wrote on the wall of his palace to warn him of the collapse of his dynasty. See Daniel 5.

217 *poet . . . room*: William Ellery Channing, the younger; see above, p. 351. 'room and board' is American for food and lodging, with (here) a pun on room as space.

218 *keeping-room*: New England term for common parlor or sitting-room (Web).

Cato ". . . and glory": Cato, *De agri cultura*, iii. 2, in *Scriptores*. The original has 'Patrem familiae', and 'so that . . . hard times' would be better translated: 'so that he may hold his produce for good prices.'

firkin: in England a quarter of a barrel, 9 gallons of beer or 8 of ale; seldom used in America except as a small vessel of indeterminate size (Web).

ginger-bread work: wooden scrollwork following the line of the eaves, fashionable on mid- and late-Victorian American houses.

rafters . . . queen posts: elements in the roof structure in a New England wooden-frame house.

Saturn: the ancient king of Italy in Rom myth, associated with agriculture and the introduction of cultivation in all its senses; he is of an 'older dynasty' because he is father of the Rom gods Jupiter, Neptune, Pluto, etc., just as Chronos is father of the Gk equivalents.

settles: benches with high wooden backs (Web).

219 *keeping*: with a pun on the sense of 'to keep' as in to keep lodgers (Web).

backing out: as one does when departing from a king or queen, to show respect by never showing the back to the monarch.

parlors . . . parlaver: from Fr *parler*, 'to speak'; Sp *palabra*, Port *palavra*, a word; 'palaver' is slang for idle talk, time-wasting chat (Web); also pigeon for conference or deliberation.

220 *dumb-waiters*: small hand-pulled elevators, to lift food from kitchen to dining-room.

North West Territory: i.e. the north central part of Canada, that extends beyond the Arctic Circle.

parliamentary: depends on the root sense of 'parliament' as the act of speaking.

hasty-puddings: see above, p. 341.

freezing weather: when T returned to Concord for the better part of a month. See J 1.387 (os), where he says he left 'on account of plastering'.

221 *limestone . . . burned*: limestone is converted to lime for plastering by heating in a furnace so as to drive out the carbonic acid.

223 *Fair Haven*: see above, p. 348.

Vulcan . . . Terminus: Rom gods, respectively, of fire, and of boundaries and frontiers.

bread . . . sweet: cf. Proverbs 9: 17: 'Stolen waters are sweet, and bread eaten in secret is pleasant.'

224 *withe*: twig, usually applied to willow twigs (Web).

Gilpin . . . forest: William Gilpin, *Remarks on forest scenery . . . illustrated by the scenes of New Forest, in Hampshire*, 3rd edn. (London, 1808).

vert: literally, green (Fr), but a technical term in English forest law: everything that grows and bears a green leaf within the forest. It is the verderer's responsibility to preserve vert and venison (Web).

Lord Warden: i.e. of the forest.

224 *though . . . accident*: a sidelong reference to T's having accidentally set the W woods on fire. See above, pp. 232-3.

lucum conlucare: to let light into (L) a grove, by thinning it; Cato, *De agri cultura*, cxxxix, in *Scriptores*: 'To thin a grove, the Roman formula to be observed is to sacrifice a pig and utter the following prayer . . .'

225 *Michaux ". . . plains"*: F. Andrew Michaux, *The North American silva* (3 vols.; trans. from the Fr by the author; Paris, 1819).

Goody . . . Gill: cf. William Wordsworth 'Goody Blake and Harry Gill: A True Story' (1798), in Wordsworth's *Complete poetical works* (Philadelphia, 1837). Blake, a poor old woman desperate to keep warm in the winter, steals sticks from the hedgerow of Gill, a prosperous farmer; when he apprehends her, she prays that he may never be warm again, and from then on his teeth chatter incessantly, no matter how thick his clothing.

226 *to "jump" it*: a process by which an old axe can be renovated by heating, then beating to narrow the aperture around the handle, or to thin out the edge for further resharpening.

sapwood: the alburnum or exterior part of the trunk, next to the bark (Web).

[*verse*]: T's own, one of his best known; for *Icarian*, see above, p. 350.

227 *hair*: used to reinforce plaster.

cut their threads: see note on the three Fates, above, p. 319.

228 *Cold Fridays*: 19 January 1810; cf. J 9.230 (os).

Great Snows: see above, p. 337.

[*verse*]: Ellen Hooper, 'The Wood Fire', ll. 18-35, in *The Dial*, 1 (October 1840).

229 *this way to Lincoln*: the old road from Concord to Lincoln, now a footpath branching off Route 126, ran past T's hut. VDS points out that this, the shortest way between the two towns, was often taken by people who needed to keep out of sight, like escaped slaves.

foundation of logs: old New England roads were often first 'paved' with a corduroy of cedar logs.

Stratten: T's copy: *Stratton*

Stratten . . . Hill: in 1800 the old Stratton Farm, on the road between Concord and W just opposite the present-day town police and fire station at the foot of Brister's Hill, was bequeathed to the town for an almshouse.

230 *Duncan Ingraham*: Concord's wealthiest citizen in the late nineteenth century (H); cf. J 2.212 (NS).

Cato, not Uticensis: the personification of Rom republican virtue, so-called because he committed suicide at Utica in North Africa.

whiter speculator: a man called Richardson, according to J 2.221 (NS).

cellar hole: New England country houses had cellars beneath them to keep things cool; then as now, the cellar hole was often the only remnant of an abandoned farm overgrown by woods.

English . . . parole: English soldiers captured during the War of 1812 were kept 'on parole' in Concord, awaiting exchange (H).

Squire Cummings: (d. 1788), Concord physician of Scottish parentage (H). For more on Cummings and Brister Freeman, see J 2.212, 221 (NS).

his epitaph: on his grave still standing, the epitaph reads 'In memory of Sippio Brister a man of Color who died Nov. 1, 1820 Et. 61' (H); T is confusing two black neighbours here: Scipio Brister and Brister Freeman, the freed slave of Squire Cummings, from whom Brister's Hill took its name. He also mis-transcribed Scipio's death as 'AEt 64' in J 3.74 (NS).

231 *Stratten*: see above, p. 358.

Breed's location: John C. Breed was a Concord barber, according to J 2.20 (OS). Breed was a drunkard addicted to rum; see H 308 n. 13.

Election night: election days were holidays and the nights often an occasion for mischief (H).

Davenant's Gondibert: Sir William Davenant (1606–68); *Gondilbert: An Heroick Poem* is his unfinished verse-epic, a tale of chivalry.

uncle: Charles Dunbar, T's maternal uncle; cf. J 8.229–46 (OS) for T's extended appreciation on the occasion of his death.

232 *Chalmers' collection*: vol. vi of Chalmers contains all 1,700 quatrains of *Gondilbert*.

Nervii: pun on nerves, sinews, and the Germanic tribe occupying what is now part of Belgium, conquered by Caesar after great slaughter in 58 BC.

leaped the brook: i.e. Mill Brook, which runs just behind Emerson's House, where T was staying when the fire happened.

"Baker's barn . . . Codman Place": both over two miles south of Concord.

232 *agent . . . Company*: the Middlesex Mutual Fire Insurance Co., established in Concord in 1826 (H); houses and farms covered with paid-up premiums were guaranteed the services of the firemen, no matter how distant from the station.

"tub": a hand-drawn fire-engine.

soul's powder: from 'The Author's Preface' to Davenant's *Gondilbert*; see above, p. 359.

233 *well-sweep*: piece of timber mounted on a fulcrum over a well to lever the bucket down and up (Web).

"rider": interior rib to strengthen the frame of a structure, usually of a ship (Web).

"attach a chip": when no possessions remained on which to place a lien, the Sheriff would 'attach a chip' to the property. 'Chip' is not in W, but it may be a distant relative of AS *cypan*, to buy or sell.

234 *potter's clay and wheel*: T's copy: *potter's wheel*

potter's clay . . . Scripture: cf. the many references in the Bible to God working man as the potter works clay: e.g. Job 10: 9; 23: 16; Isaiah 41: 25; 64: 8; and Jeremiah 18: 1–6, where the faulty clay vessel is scrapped for another as the Lord threatens to do to Israel.

Quoil: T writes at length on him, e.g. in J 2.207–10 (NS); the pun is on 'coil' as noise, tumult, confusion (Web).

St. Helena: the island in the South Atlantic to which Napoleon was exiled after his forces were defeated in the Battle of Waterloo (1815).

pipe . . . fountain: cf. Ecclesiastes 12: 6.

Brister's Spring: just to the east of the Walden Road, at the foot of Brister's Hill.

administrator: official put in charge of the estate of one dying intestate (Web).

Reynard: stock literary name for the fox.

235 *"fate . . . absolute"*: one of the Devils' punishments in *Paradise Lost* (ii. 558–61) were endless seminars on these topics.

"pulled wool": that is, in the local idiom, they stripped sheepskins of their wool to prepare them for tanning.

stirp: stock, race, family (Web).

236 *Concord . . . ground*: a significant late alteration of MS 217, which reads: 'Concord grows apace'; see Introduction.

privileges: pun on the common sense of the word and its specialized use in connection with water rights; see above, p. 349.

corn-parching: the Indians parched (scorched) corn so as to be able to carry it dry and eat it without seething or grinding; elsewhere T writes of parching as bursting the seed (*Week*, 182.16), so here corn-parching is probably corn-popping.

wilderness to blossom: cf. Isaiah 35: 1.

Sutton: about ten miles south of Worcester, in south central Massachusetts.

great snow: see above, p. 337. T's account of the family being rescued by the Indian comes from the chapter on Sutton in Barber.

238 *causeway*: the railway connecting W and Concord runs along a raised embankment over low meadows as it nears town.

cheek . . . turned: cf. Matthew 5: 39.

meadow mouse: T's copy: *meadow deer mouse*; Prince accepts 'deer' as a substitution for 'meadow'.

woodchopper: Alek Therien; see above, pp. 130–6.

239 *long-headed farmer*: Edmund Hosmer, according to H, quoting VDS.

"crack": dialect term for chatty talk, gossip (W7NWD).

poet: see above, p. 351.

Walden vale: presumably T's coinage; not an actual place-name.

Broadway: i.e. the busiest street in New York, then the largest city in the US.

"bran new": more common American usage is 'brand new', or 'bright as the brand of a fire' (Web).

another welcome visitor: Amos Bronson Alcott (1799–1888), father of Louisa May, who had been a Yankee pedlar in the South before becoming a 'philosopher'. Cf. J 2.223–34 (NS) and J 5.130–1 (OS) for contexts. For T's distinction between 'philosophers' and 'professors of philosophy', see *Week*, 334.

240 *"How . . . serenity!"*: 'Wolseius Triumphans', in Thomas Park, *Halconia, Comprising a selection of English poetry of the Elizabethan age . . .* (3 vols.; London, 1815).

Old Mortality: the wandering inscription carver, supposedly the source of Walter Scott's novel of the same title (1816) about the Scottish Covenanters in the late seventeenth century.

241 *rounding a fable*: pun on 'to round' (as of a headland or cape) and 'to
     sing a round'.

     New . . . Entertainment: cf. *The Arabian Nights' Entertainment*.

     hermit . . . settler: see note above on the 'old settler', p. 340.

     oakum: sailors off watch would yarn while unravelling old hemp
     ropes into strands of 'oakum' to stuff between the wooden planks to
     keep the ship watertight.

     one other: Emerson, who lived just over a mile away from T's hut at
     W.

     Vishnu Purana: *The Vishnu purana, a system of Hindu mythology
     and tradition . . .* (London, 1840).

     man . . . town: 'But never more could see the man approaching
     from the town', from the English ballad, 'The Children in the Wood'
     (H).

242 *Flints' Pond*: see above, pp. 175–8.

     Baffin's Bay: or Baffin Bay, in the Arctic Ocean, west of Greenland.

     Lincoln hills: Cedar Hill, Three Friends' Hill, and Bear Hill, all in
     the township of Lincoln, ring Flint's Pond to the east, south, and
     south-west.

     Goose . . . way: for Goose Pond (to the north-east of W) to be in T's
     way, he would have to be taking the northerly circuitous path from
     W to the north shore of Flint's Pond.

     moose-yard: not remembered as a local expression and not in Web,
     but T's copy of Chester Dewey and Ebenezer Emmons, *Reports on
     the herbaceous plants and on the quadrupeds of Massachusetts*
     (Boston, 1840) says that in winter moose 'herd together . . . [to]
     occupy a space or enclosure . . . [in which] the snow is trodden
     down, and the space gradually extended as the food diminishes
     within' (p. 76).

     lingua vernacula: native, or indigenous language (L).

243 *Hudson's Bay*: the Canadian goose migrates south from this large
     inland lobe of the Atlantic, north of Ontario.

     citadel: refers to the incident narrated in Livy, *History of Rome*,
     5.47, when geese alarmed the citadel when the Gauls took Rome in
     390 BC (H).

     whooping: caused by contraction and expansion of the ice on the
     pond.

244 *somerset*: a sudden leap (Web), later somersault.

brisk: p. proofs: set as: *pick*; corrected to: *frisk*; correction misread as *brisk*. T's copy: *frisk*

trigonometrical: concerned with the measurement of triangles, hence moving zig-zag, in a triangular trajectory.

246 "*sometimes . . . two*": J 3.54–5 (NS) attributes this to John James Audubon (1785–1851), the American naturalist and wildlife artist, but the reference is still untraced.

diet-drink: medicated liquor; drink prepared with medicinal ingredients (Web).

247 *Actæon*: the hunter of Gk legend. When he happened upon the maiden divinity Artemis bathing naked with her nymphs in the woods, she changed Acteon into a stag, and his hunting dogs tore him apart.

Lexington: about six miles east of Concord.

Wayland: a town about ten miles south of Concord.

248 *Well-Meadow*: on the shore of Fair Haven Bay.

Weston: a town about seven miles south of Concord.

249 *Nutting*: for a fuller reference to 'Fox Nutting, Old Fox, he was called', see J 5.16 (OS).

Burgoyne: doubtless named after the British general in the War of Independence, John ('Gentleman Johnny') Burgoyne (1722–92).

"*Wast Book*": an account book T found 'in Deacon R. Brown's garret since his death'. T quotes from the book extensively in J 6.77–80 (OS). The denominations are in pounds, shillings, and pence.

Nimrod: literary periphrasis for hunter; from Genesis 10: 9.

250 *dropsical*: dropsy: the unnatural collection of water in any part of the body (Web).

venison: the flesh of any edible beast of the chase (Web); by T's time this was English usage; in America the word was already confined to deer meat.

light-foot, some think: 'Lucius Aelius thought that the hare received its name *lepus* because of its swiftness, being *levipes*, nimblefoot' (Varro, *Rerum Rusticarum*, III. xii. 6, in *Scriptores*).

251 *twiggy fences*: used to divert the wild animal into a snare (H).

cow-boy: see above, p. 346.

252 "*O Prince . . . ether*": from the *Harivansa*, an appendix to the *Mahabharata*, the classical Sanskrit epic (*c*.200 BC–AD 200), in

M. A. Langlois (trans.), *Harivansa, ou Histoire do la famille de Hari* . . . (2 vols.; Paris, 1834–5).

253 *fear-naughts*: woollen cloths of great thickness (Web).

scale of being: an allusion to the renaissance figure for creation of being, a hierarchical chain of being running from the angelic to the inanimate.

254 *trumpeted*: fishmongers sounded a trumpet to advertise their wares as they walked through the streets (H).

Waldenses: pun on Walden as it might specify a fanciful local pickerel genus (*Esox waldeniensis*) and the medieval French protestant sect, the Waldenses.

compass and chain: the surveyor's instruments for establishing direction and distance; T was a professional surveyor from around 1845, and shortly before his death in 1862 was made Town Surveyor of Concord.

256 *Bottomless Ponds*: cf. J 2.68 (OS) and J 3.115 (NS) for more detail.

Styx: in Gk myth, the river that flows seven times around the Underworld, or land of the dead.

"fifty-six": half a hundred-weight (112 lb.), a standard used for weighing on balance scales, and for calibrating scales.

before . . . underneath: i.e. after the weight had broken free from the muddy bottom and, being wholly surrounded by the water it displaced, felt lighter.

five feet . . . risen since: for the source of water in W and its fluctuations in surface level, see above, pp. 162–5.

[*measurements*]: a survey in 1939 confirmed the accuracy of T's measurement of the depth of W. Cf. H 312 n. 8, and VDS 408.

257 *Gilpin . . . Loch Fyne*: William Gilpin, *Observations on . . . the High-lands of Scotland*, 3rd edn. (2 vols.; London, 1808), vol. ii.

[*verse*]: Milton, *Paradise Lost* (Philadelphia, 1808), vii. 288–90. The Angel Raphael is telling Adam how God created the earth.

plain have been: the grammar requires 'has', but 'have' was set and remained uncorrected in p. proofs, so came through to 1st edn. Prince and LA correct to 'has'.

258 *ten rods*: i.e. 50 yards. T's 'Reduced Plan' of W (1846), showing its size and shape and profile of its depth through two sections of length and width, was drawn to a scale of 40 rods (200 yards) to an inch and published, as here, in the first edition of W.

259 *Every harbor . . . bar*: this may be generally true of harbours in the American East, where the land is slowly emerging from the sea; it is not true of submergent coasts like the northern coast of California, or the west coast of Norway.

260 *trend*: to run, stretch, or tend, as of the shore of the sea (Web); T uses the word six times in this sense.

Achillean shore: in the sense of the legendary Achaean hero's terrible extremes of temperament: his ups and downs.

261 *ice-men . . . '46–7*: Frederic Tudor and Nathaniel Wyeth were partners in the largest Massachusetts ice-harvesting operation until the two fell out in 1840. Since Wyeth had already secured the rights to the ice of Fresh Pond, Cambridge, Tudor took advantage of the extension of the Fitchburg Railroad to Concord, shipping ice direct to the Boston docks. But he soon found other, cheaper sources, so left the W ice harvest to melt on the shore. See Richard Cummings, *The American Ice Harvests* . . . (Berkeley: University of California Press, 1949), *passim*, but especially pp. 45–6.

"leach hole": the root sense of 'to leach' is 'to leak' (Web).

level cannot . . . on ice: because, as T shows and measures, it moves up and down. Before the introduction of electronic instruments surveyors used a small telescope mounted on a tripod to sight a graduated staff held vertical at a distance by an assistant. The telescope was levelled by reference to a plumb bob hanging down between the tripod's legs; so the mark on the distant staff cut by the hairline in the surveyor's telescope gave the exact height of the telescope with reference to the ground on which the staff was held. Using his surveying skills, therefore, T was able to determine the amount of fluctuation of the ice on the pond.

262 *lays up no treasures*: cf. Matthew 6: 19–20: 'Lay not up for yourselves treasures upon earth . . . But lay up for yourselves treasures in heaven.'

pit-fashion: i.e. where a log is laid over a pit, in which one sawyer stands; the other, taking the other end of the saw, stands on the surface.

Hyperborean: far northern; in Gk fable, a people living beyond the north wind, and therefore not exposed to its cold.

drill-barrows: not in Web, but a drill-box contained seed for sowing; a drill-harrow was used for smoothing the ground between rows or drills in which seed had been planted; a barrow was a small, light carriage, or a box carried or on wheels (Web).

262 *turf-knives*: not in Web but turf-spade (for cutting and digging turf) and other variants are.

263 *New-England . . . Cultivator*: *The New England Farmer*, *The New England Cultivator* and *The Boston Cultivator* were agricultural journals published in Boston at the time (H).

gentleman farmer: Frederic Tudor; see above, p. 365.

ploughing . . . rolling: three stages in preparing the earth: breaking up sod; breaking up clods; smoothing out the soil preparatory to making furrows into which to sow the seed.

model farm: see above, pp. 142–3, and Introduction.

terra firma: fixed earth (L).

cutting . . . bog: because the ice cutters were Irish immigrants.

Tartarus: in Gk myth, that part of the Underworld to which the souls of the wicked were consigned, as far below Hades as Hades was below the earth's surface.

ninth part of a man: see above, p. 323.

Cambridge: around fifteen miles to the east, via the railroad.

too well known: the 'method', which gave T the basis of his agricultural fantasy, consisted of horse-drawn, square-framed, harrow-like contraptions, lined on the underside with chisels in rows about 20 in. apart. Pulled one way, these scored parallel grooves in the ice; pulled again at a right angle to the first direction, they formed squares in the ice which were then separated with long-handled chisels or saws.

264 *Valhalla*: see above, p. 350.

old man: in the 1850s, issues of the *Old Farmers' Almanac* showed Winter as an old man on the page for January (H).

estivate: spend the summer (Web).

cars: freight or goods wagons.

265 *Fresh Pond*: in Cambridge, where the workers came from and the local ice-harvesting industry was centred.

fable of the lark: H conjectures La Fontaine's fable (iv. 22) on 'The Lark and her Young Ones with the Owner of the Field', in which the mother bird demonstrates to her children that they are safe in the wheat so long as the owner relies on others to cut it for him.

parable of the sower: cf. Matthew 13: 3–8, on which only some seeds fall upon good ground for growing.

Charleston . . . Calcutta: this is not exaggerated; New England ice really was shipped to all these ports, and more.

266 *Bhagvat Geeta*: one of the major Hindu writings.

Brahma . . . Indra: the three major gods of the Hindu religion.

Ganges: sacred river of the Hindus.

Vedas: sacred scriptures of the Hindus.

Atlantis: according to ancient legend, a great island west of the Pillars of Hercules (the entrance to the Mediterranean), a prosperous state whose inhabitants grew so wicked and impious that the gods caused the sea to swallow it up in a day and a night.

Hesperides: in Gk myth, the nymphs guarding the golden apples that grew in the Garden of the Blest, at the western end of the earth.

periplus of Hanno: the celebrated Carthaginian navigator wrote an account of his voyage out beyond the Pillars of Hercules; written in Punic, it was translated into Gk as the *Periplus*, which is Gk for 'sailing around'.

Ternate and Tidore: the Moluccas, or Spice Islands, south of the Philippines. Cf. Milton, *Paradise Lost*, ii. 638–40: 'the isles | Of Ternate and Tildore, whence merchants bring | Their spicy drugs'. Satan is exploring the limits of Hell.

Alexander: thought to be the most widely travelled man of his time.

268 *reflected heat*: the reason why pond ice melts, or is thinner around a protruding rock or log is not because the object reflects heat but the exact opposite: being dark it retains more, and reflects less, of the heat from the sun.

experiment at Cambridge: at Harvard University, according to one source; see H 314 n. 4.

269 *full of cracks*: i.e. with plenty of expansion joints, the ice no longer resonates as a single, drum-head-like surface.

air . . . elastic: the oldest sense of 'elastic' is pertaining to or causing the spontaneous expansion of air (*OED*).

papillæ: literally, 'little paps' or nipples; minute projecting filaments, being the terminations of nerves (Web).

270 *stocks*: here the framework supporting a ship while under construction.

Methuselah: cf. Genesis 5: 27: 'And all the days of Methuselah were nine hundred sixty and nine years.'

271 *Sudbury . . . Fair-Haven Pond*: via the Sudbury River a distance of about five miles.

phenomena: for the observational basis of T's fantasy on sand vegetation, see J 2.382–3 (NS). T had trouble with the phrasing of this section of *W*; MS 906–11 shows it to be one of the most heavily revised parts of the MS (VDS).

deep cut . . . village: T's most direct route from W to the house in Concord where his mother and sister lived was along the track of the Fitchburg Railroad, a distance of a little over a mile and a half.

272 *laciniated*: adorned with fringes; in botany, jagged (Web).

imbricated: bent or hollowed, like a roof or gutter tile; in botany, laying over each other in regular order, like tiles, as the scales on the cup of some acorns (Web).

thalluses: not in Web, but Gk for green twigs; this is T's only use of the word.

excrements: anything excreted or ejected.

grotesque: here in the sense of wildly formed, whimsical, extravagant (Web).

typical: not just representative, but predicative of fulfilments to come later in time, as in the exegetical practice of finding links between the Old and New Testaments.

acanthus: the leaf figured in capitals of the Corinthian order; H says that the other plants mentioned here have also formed classical architectural motifs.

273 *laboratory*: here in the root sense of workshop; for the most part this is how T uses the word elsewhere, e.g. in *Week*.

foliaceous: both leafy and consisting of thin leaves or laminae (Web).

labors: both works and gives birth to.

lobe: the word was indeed first used of the divisions in lungs and livers, as of the lobe of the ear, and it is related to Gk *lobos* and L *lobus*. English globe is descended from L *globus*. The Gk *leibo* is connected with flowing, at least to the extent of 'to pour forth' (as of a libation). The L verb *labor* means 'to glide, slide or slip' and its past participle *lapsus* means 'fallen' in the moral sense. Beyond that T's etymology here is more fanciful than scholarly; the words are related more by the accidents of their English sounds than by their language family. 'Labo[u]r' comes from OF *labour* for work; 'lap' is from OE *lappa*, not the L past participle of *labor*; 'flap' originates in onomatopoeia and can be compared to Du *flappen*, to strike, and *flap*, a blow.

ova . . . axils: the eggs of insects left in the upper angle between branch and stem.

274 *You here see . . . are formed*: again, this is figurative or fanciful, not scientifically accurate.

silicious: of or pertaining to silicic acid, mostly impure, as found in e.g. flint, quartz, and most sands; silica, the pure form of silicic acid, was originally thought to be 'one of the primitive earths' (Web).

umbilicaria: lichens.

lip: from OE *lippa*; it is cognate to, but not derived from L *labium*, which in turn is unrelated to *labor*.

275 *Champollion*: Jean François Champollion (1790–1832) deciphered the Rosetta Stone found in Egypt which bore the same message inscribed in Gk and hieroglyphics, thus providing a key to the hieroglyphics in ancient Egyptian tombs.

hieroglyphic: emblematic, expressive of some meaning by characters, pictures, or figures (Web); its root meaning of sacred writings, apparently lost by mid-century.

bowels: in the old physiology, thought to be the seat of pity or kindness (Web).

slag . . . furnace . . . "in full blast": the allusion is to the process of iron and steel making.

exuviæ: here the cast skin, shells, or other coverings of animals (Web).

clay . . . potter: cf. Jeremiah 18: 6: 'Behold, as the clay is in the potter's hand, so are ye in mine hand, O house of Israel.'

Thor: Scandinavian god of thunder.

276 *weeds*: garments (survives mainly in 'widow's weeds' as plain, seemly clothing), with a pun on the 'volunteer' plants mentioned above.

types: see note above on *typical*, p. 268.

277 *"et primitus . . . evocata"*: Varro, *Rerum Rusticarum*, II. ii. 14, in *Scriptores*: 'And the grass springs up, called forth by the early rains.'

all watered . . . palace floor: possibly alluding to the marble patterns in palace floors.

278 *Walden was dead . . . again*: cf. Luke 15: 24, of the Prodigal Son: 'For this my son was dead, and is alive again.'

279 *"plump"*: sixteenth- and seventeenth-century English usage for a group of animals that go in flocks; sufficiently rare for this instance to be recorded in the *OED*.

279 *martins . . . afford me any*: presumably because martins normally nest in or near inhabited areas, in barns or under the eaves of houses.

Cosmos . . . Golden Age [and verse following]: the process is outlined in Ovid, *Metamorphoses*, i. 61–2, 78–81, in *Publii Ovidii Nasonis metamorphosen . . .* (Philadelphia, 1823).

Nabathacaque: p. proofs: *Nabathæaque*

Aurora: L for Gk Eos, the personification of the dawn; hence (here), the East.

Nabathæan kingdom: ancient Arab country, in classical times situated on the north-east coast of the Red Sea and the Idumaean Mountains.

And . . . the ridges: a more idiomatic translation would be: 'And the Persian hills . . .'

280 *While such . . . may return*: cf. Isaac Watts, *Hymns and Spiritual Songs*, i. 88: 'And while the lamp holds out to burn | The vilest sinner may return' (H).

Even he . . . his Lord: cf. Matthew 25: 23: 'His lord said unto him, "Well done, good and faithful servant . . . enter thou into the joy of thy lord."'

281 *"A return . . . sentiments of man"*: Works of Mencius, book VI, 'Kaou Tsze', part I, chap. viii, in Pauthier.

[*verse*]: Ovid, *Metamorphoses*, i. 89–96, 107–8, in *Publii Ovidii Nasonis metamorphosen . . .* (Philadelphia, 1823). 'Enforcer' or 'defender' might be closer than T's 'avenger' and his 'flowers born without seed' would be better rendered as 'flowers born unseeded'.

Nine-Acre-Corner bridge: crosses the Sudbury River about a mile and a half south-west of Concord centre.

sticks . . . fingers: 'bones' or 'clappers' which children use as a rhythmical instrument (H).

282 *cupreous*: copper-coloured.

O Death . . . victory, then?: cf. I Corinthians 15: 55: 'O death, where is thy sting? O grave, where is thy victory?'

284 *Calidas' . . . Sacontala*: Calidas, *Sacontala; or, The Fatal Ring*, trans. William Jones (13 vols.; London, 1807), vol. ix.

285 *rail-fences . . . stone-walls*: i.e. if we substitute more permanent boundaries for more temporary, less substantial ones.

town-clerk: the officer who keeps the records of the town (Web).

Tierra del Fuego: literally, 'land of fire', a group of islands at the southern tip of South America.

tafferel: or taff rail, the rail at the stern of the ship; the idea is that passengers, having nothing to do, would lean on the taffrail contemplating the ship's wake.

sailors picking oakum: see above, p. 362.

great-circle sailing: a great circle is the shortest distance between any two points on the globe.

[verse]: William Habington (1605–54), 'To My Honoured Friend, Sir Ed. P. Knight', ll. 37–40, in Chalmers, vol. vi; orthography modernized by T; 'eye right inward' should read 'eye sight inward', but because the 'r' was indistinct in p. proof, it was not caught; so it emerged into 1st edn. Prince and LA correct 'right' to 'sight'.

white on the chart: unexplored regions were left blank on maps.

286 *North-West Passage*: the search for a sea passage connecting the Atlantic with the Pacific Oceans over the top of North America was a long-standing project for explorers from Frobisher (1576) to Wilkes (1838–42) and Franklin (1845). It was finally accomplished by the Norwegian explorer Roald Amundsen (1872–1928) in 1906.

Franklin: Sir John Franklin (1786–1847), the English naval officer and explorer credited with establishing the possibility of a Northwest Passage; who set out to cross the Arctic Ocean from east to west in 1845, but his ships were caught and crushed by the ice midway between the Atlantic and Pacific. In 1847 his second wife, Lady Jane, financed a rescue mission which found the crews' skeletons and an account of the voyage to April 1848.

Grinnell: another mission to rescue Franklin was promoted by the American shipping magnate, Henry Grinnell (1799–1874).

Mungo Park: (1771–1806), Scottish explorer in Africa.

Lewis and Clarke: Meriwether Lewis (1774–1809) and William Clark (not Clarke; 1717–1838), the two Americans who explored and surveyed the Louisiana Purchase of the greater part of the American West in 1803–4.

Frobisher: Sir Martin Frobisher (1535–94), the English navigator and explorer, who set out in 1576 to find the Northwest Passage, and discovered Hudson's Bay.

shiploads . . . meats: the Grinnell expedition to the Arctic found 600 cans of preserved meat left by Franklin (H).

286 *hummock*: or hommock, a small eminence of a conical form, sometimes covered with trees (Web, who supposed it to be 'an Indian
word'). By 'left by the ice' T may mean deposited by a glacier,
though the sense suggests something smaller in scale.

South-Sea Exploring Expedition: an American venture led by the
naval officer Charles Wilkes (1798–1877) to explore the Antarctic
and the Pacific Northwest. From 1838 to 1842 they charted the
archipelagoes and mapped over 1,600 miles of the coastline of
Antarctica.

[*verse*]: from the short poem by Claudian (fourth century AD), '*De
sene Veronensi qui Suburbium numquam egressus est*' ('Of the Old
Man of Verona who never left his Home'), ll. 21–2. T's translation
substitutes 'Australians' for Claudian's 'Hiberos' ('Spaniards') to
update the 'outlandishness' of the distant country; see J 1.310 (NS).

cats in Zanzibar: an allusion to Charles Pickering's *The Races of
Man* (London, 1851), an account of a world tour which T read in
1853, and which does indeed report on domestic cats in Zanzibar
(H).

"Symmes' Hole": in 1818 John Cleves Symmes (1780–1829) of St
Louis published the theory that the earth was hollow and that the
interior could be entered through holes in the poles. Energetically
promoted, the idea caught the public imagination and helped to
finance the American Antarctic Exploring Expedition of 1829 to test
it. Edgar Allen Poe uses the notion to facilitate the plot of *The
Narrative of Arthur Gordon Pym of Nantucket* (1837).

287 *bark*: or barque (generally) a small sailing craft; (specifically) a
sailing ship with three masts, the first two square rigged and the
third fore-and-aft.

Sphinx . . . stone: in Gk legend the Sphinx was a monster with the
head of a woman, the body of a lion and the wings of a bird, which
would ask passers-by to solve riddles, then kill them when they
could not. When Oedipus answered the Sphinx's riddle successfully,
the monster threw herself off the top of a rock.

Explore thyself: 'Know thyself' (*gnõthe seauton*), a maxim ascribed
to various Gk philosophers, was inscribed on the Temple of Apollo
at Delphi.

western way: T was writing just after the first flush of mass transcontinental western migration, which followed the opening of the
Oregon Trail and the discovery of gold in California late in 1848.

Mirabeau: Honoré-Gabriel Riqueti, comte de Mirabeau (1749–91),
historian, constitutional theorist, moderate democratic politician, a

leading figure in the National Assembly in the early phases of the French Revolution. J 3.325–6 (NS) for 21 July 1851 quotes a longer extract of this passage, citing as its source *Harper's Monthly*, 1 (1851), 648.

foot-pad: highwayman, or robber on foot (Web).

I left . . . reason: H says (316 n. 20) because 'Emerson planned to go abroad on a lecture tour and wanted Thoreau to take over the care of his house and family'; but J 3.216 (OS) is more enigmatic: 'I must say that I do not know what made me leave the pond. I left it as unaccountably as I went to it. To speak sincerely, I went there because I had got ready to go; I left it for the same reason.'

288 *moonlight . . . mountains*: refers to a trip on the Hudson River that T made with his close friend William Ellery Channing (the younger), when the moon was so bright they preferred to spend the night on the ship's bow rather than return to their cabin (H).

cabin passage . . . below now: on contemporary sailing ships 'cabin passage' and 'below' were for passengers; 'before the mast' and 'on deck' for working sailors.

hush and who, which Bright: probably sounds to make a draft ox ('Bright' was a common name for an ox) to turn to the right and left; cf. 'gee' and 'haw'.

289 *extra-vagant*: a pun on excess wandering and wandering outside (L).

yarded: confined, as cattle in a yard (Web).

translated: as the italics hint, another pun: on 'translate' as to carry from one place to another; to change; to remove or convey to heaven; to render into another language (Web). T uses the word in all these senses in his published work.

"They pretend . . . Vedas": T's translation of M. Garcin de Tassy, *Histoire de la Littérature hindui et hindoustani* (2 vols.; Paris, 1839–47).

potato-rot: the potato rot or 'late blight' struck Ireland in 1845 and 1846, causing widespread damage to the crop on which the economy of Ireland depended, followed by famine and large-scale emigration to the US.

290 *Walden . . . Cambridge ice*: refers to the competition between Tudor and Wyeth; see the note above to 'the ice-men', p. 365.

dog . . . lion: Ecclesiastes 9: 4: 'For to him that is joined to all the living there is hope: for a living dog is better than a dead lion.'

artist . . . of Kouroo: a great *locus classicus* of W interpretation, the issues of which are summarized in H 317 n. 29. No source for the story has been found, and it is probably safest to assume that it is T's own self-consciously 'mythical' allegory, along the lines of Herman Melville's *Mardi* (1849).

291 *Kalpa*: in Hindu literature, not a star but a great period of time, though the Hindus knew that over time the pole-star changed (H).

Brahma had awoke: a day of Brahma lasted two billion, one hundred and sixty million years, after which he slept (H).

292 *Tom Hyde*: possibly a Massachusetts version of Tom Tiddler (Tom t'idler) or Tom Tipple, the highwayman. 'Tom' is a common sobriquet for vagrants in English and American folklore.

philosopher ". . . thought.": Confucius, *Analects*, ix. 25, in Pauthier.

"and lo! . . . view": 'And lo! Creation widened to man's view' (Joseph Blanco White, 'Sonnet on Night and Death', in *The life of . . . Joseph Blanco White, written by himself* (3 vols.; London, 1845)).

Croesus: ancient king of Lydia, reputed to be the richest man in the world.

293 *tintinnabulum*: little bell (L); hence, the sound of same.

the Hon. Mr. . . . Massachusetts: i.e. a US Senator from one or other of these states.

Mameluke bey: refers to the order that went out from Pasha of Egypt to massacre all the Mamelukes 'They were trapped in a citadel, but one escaped by leaping on his horse from the ramparts and fleeing to Syria' (H). A bey is a governor of a province. The Mamelukes were the warrior class (originally slave soldiers) who ruled several Muslim states during the Middle Ages; their descendants remained politically powerful even under Ottoman occupation.

Webster: Daniel Webster, the Whig Senator for Massachusetts (see above, p. 355), was renowned for his oratory.

294 *kittlybenders*: or kidleybenders, ice which undulates under the feet of skaters; or the children's game of running on thin ice that bends underfoot.

[traveller . . . boy . . . swamp]: 'This joke about the swamp had appeared in the *Concord Yeoman's Gazette* on Nov., 22, 1828' (VDS).

furrowing: thin strips of board nailed onto a surface to bring it level for lathing, boarding, etc. (Web).

invoke the Muse: as the author of an epic might do at the beginning of his work.

man . . . tree: still not identified, though he seems to have existed; H (318 n. 43) quotes another reference to him, by Emerson.

295 *seven-years' itch*: American colloquialism for anything that irritates or goes on far too long (*DAE*).

pellicle: thin skin or film (Web).

296 *drown . . . muskrats*: Muskrats build their dwellings just above the water level, so may be drowned if the water table rises.

bug . . . old table: apparently an actual event, from the chapter on Williamstown in Barber, telling of 'a strong and beautiful *bug* eat out of a table made from an apple-tree', and goes on to provide details of provenance and dating that T follows closely.

alburnum: the white and softer part of the wood of the living tree, between the inner bark and the hard wood, or duramen (Web).

handselled: of the many meanings of this word, which seems originally to have denoted an auspicious gift, the use that best suits the sense here is something too cheap to sell; hence given away.

John or Jonathan: journalistic representative names for the British and the Americans.

noon . . . Prescott and Identical, though he seems to have existed. H (1318 n. 3), quotes another reference to him, by Emerson.

895 since years: Indo-American colloquialism for anything that carries or goes on far too long. DAE.

polled: thin slab or film (Web).

200 chosen . . . muskrats: Muskrats build their dwellings just above the water level, so may be drowned if the water table rises.

bug . . . old table: apparently an actual event, from the chapter on Withamstown in Thacher telling of a strong and beautiful bug eat out of a table made from an apple tree; and goes on to provide details of provenance and dating that T follows closely.

otherwise: the white and yellow part of the wood of the living tree between the inner bark and the hard wood, or duramen (Web).

townkeld: of the many meanings of this word, which seems essentially to have alluded to an auspicious gift, the one that best suits the sense here. Is something too clever to sell. held up in a show.

John, or Jonathan: journalistic representative names for the British and the Americans.

THE WORLD'S CLASSICS

A Select List

HANS ANDERSEN: Fairy Tales
Translated by L. W. Kingsland
Introduction by Naomi Lewis
Illustrated by Vilhelm Pedersen and Lorenz Frølich

ARTHUR J. ARBERRY (Transl.): The Koran

LUDOVICO ARIOSTO: Orlando Furioso
Translated by Guido Waldman

ARISTOTLE: The Nicomachean Ethics
Translated by David Ross

JANE AUSTEN: Emma
Edited by James Kinsley and David Lodge

Mansfield Park
Edited by James Kinsley and John Lucas

Northanger Abbey, Lady Susan, The Watsons,
and **Sanditon**
Edited by John Davie

HONORÉ DE BALZAC: Père Goriot
Translated and Edited by A. J. Krailsheimer

CHARLES BAUDELAIRE: The Flowers of Evil
Translated by James McGowan
Introduction by Jonathan Culler

WILLIAM BECKFORD: Vathek
Edited by Roger Lonsdale

R. D. BLACKMORE: Lorna Doone
Edited by Sally Shuttleworth

KEITH BOSLEY (Transl.): The Kalevala

DANIEL DEFOE: Robinson Crusoe
Edited by J. Donald Crowley

THOMAS DE QUINCEY:
The Confessions of an English Opium-Eater
Edited by Grevel Lindop

CHARLES DICKENS: Christmas Books
Edited by Ruth Glancy

David Copperfield
Edited by Nina Burgis

The Pickwick Papers
Edited by James Kinsley

FEDOR DOSTOEVSKY: Crime and Punishment
Translated by Jessie Coulson
Introduction by John Jones

The Idiot
Translated by Alan Myers
Introduction by W. J. Leatherbarrow

Memoirs from the House of the Dead
Translated by Jessie Coulson
Introduction by Ronald Hingley

ARTHUR CONAN DOYLE:
Sherlock Holmes: Selected Stories
Introduction by S. C. Roberts

ALEXANDRE DUMAS *père*:
The Three Musketeers
Edited by David Coward

ALEXANDRE DUMAS *fils*:
La Dame aux Camélias
Translated by David Coward

MARIA EDGEWORTH: Castle Rackrent
Edited by George Watson

A London Life *and* The Reverberator
Edited by Philip Horne

The Spoils of Poynton
Edited by Bernard Richards

RUDYARD KIPLING: The Jungle Books
Edited by W. W. Robson

Stalky & Co.
Edited by Isobel Quigly

MADAME DE LAFAYETTE: The Princesse de Clèves
Translated and Edited by Terence Cave

WILLIAM LANGLAND: Piers Plowman
Translated and Edited by A. V. C. Schmidt

J. SHERIDAN LE FANU: Uncle Silas
Edited by W. J. McCormack

CHARLOTTE LENNOX: The Female Quixote
Edited by Margaret Dalziel
Introduction by Margaret Anne Doody

LEONARDO DA VINCI: Notebooks
Edited by Irma A. Richter

MIKHAIL LERMONTOV: A Hero of our Time
Translated by Vladimir Nabokov with Dmitri Nabokov

MATTHEW LEWIS: The Monk
Edited by Howard Anderson

JACK LONDON:
The Call of the Wild, White Fang, and Other Stories
Edited by Earle Labor and Robert C. Leitz III

NICCOLÒ MACHIAVELLI: The Prince
Edited by Peter Bondanella and Mark Musa
Introduction by Peter Bondanella

ORIENTAL TALES
Edited by Robert L. Mack

OVID: Metamorphoses
Translated by A. D. Melville
Introduction and Notes by E. J. Kenney

FRANCESCO PETRARCH:
Selections from the Canzoniere and Other Works
Translated by Mark Musa

EDGAR ALLAN POE: Selected Tales
Edited by Julian Symons

JEAN RACINE: Britannicus, Phaedra, Athaliah
Translated by C. H. Sisson

ANN RADCLIFFE: The Italian
Edited by Frederick Garber

The Mysteries of Udolpho
Edited by Bonamy Dobrée

The Romance of the Forest
Edited by Chloe Chard

THE MARQUIS DE SADE:
The Misfortune of Virtue and Other Early Tales
Translated and Edited by David Coward

PAUL SALZMAN (Ed.):
An Anthology of Elizabethan Prose Fiction

OLIVE SCHREINER: The Story of an African Farm
Edited by Joseph Bristow

SIR WALTER SCOTT: The Heart of Midlothian
Edited by Claire Lamont

Waverley
Edited by Claire Lamont

HORACE WALPOLE : The Castle of Otranto
Edited by W. S. Lewis

IZAAK WALTON and CHARLES COTTON:
The Compleat Angler
Edited by John Buxton
Introduction by John Buchan

OSCAR WILDE: Complete Shorter Fiction
Edited by Isobel Murray

The Picture of Dorian Gray
Edited by Isobel Murray

MARY WOLLSTONECRAFT:
Mary *and* The Wrongs of Woman
Edited by Gary Kelly

VIRGINIA WOOLF: Mrs Dalloway
Edited by Claire Tomalin

Orlando
Edited by Rachel Bowlby

ÉMILE ZOLA:
The Attack on the Mill and Other Stories
Translated by Douglas Parmée

Nana
Translated and Edited by Douglas Parmée